IET CONTROL, ROBOTICS AND SENSORS SERIES 83

Robust and Adaptive Model Predictive Control of Nonlinear Systems

Other volumes in this series:

Volume 8	**A History of Control Engineering, 1800–1930** S. Bennett
Volume 18	**Applied Control Theory, 2nd Edition** J.R. Leigh
Volume 20	**Design of Modern Control Systems** D.J. Bell, P.A. Cook and N. Munro (Editors)
Volume 28	**Robots and Automated Manufacture** J. Billingsley (Editor)
Volume 33	**Temperature Measurement and Control** J.R. Leigh
Volume 34	**Singular Perturbation Methodology in Control Systems** D.S. Naidu
Volume 35	**Implementation of Self-Tuning Controllers** K. Warwick (Editor)
Volume 37	**Industrial Digital Control Systems, 2nd Edition** K. Warwick and D. Rees (Editors)
Volume 39	**Continuous Time Controller Design** R. Balasubramanian
Volume 40	**Deterministic Control of Uncertain Systems** A.S.I. Zinober (Editor)
Volume 41	**Computer Control of Real-Time Processes** S. Bennett and G.S. Virk (Editors)
Volume 42	**Digital Signal Processing: Principles, Devices and Applications** N.B. Jones and J.D.McK. Watson (Editors)
Volume 44	**Knowledge-Based Systems for Industrial Control** J. McGhee, M.J. Grimble and A. Mowforth (Editors)
Volume 47	**A History of Control Engineering, 1930–1956** S. Bennett
Volume 49	**Polynomial Methods in Optimal Control and Filtering** K.J. Hunt (Editor)
Volume 50	**Programming Industrial Control Systems Using IEC 1131-3** R.W. Lewis
Volume 51	**Advanced Robotics and Intelligent Machines** J.O. Gray and D.G. Caldwell (Editors)
Volume 52	**Adaptive Prediction and Predictive Control** P.P. Kanjilal
Volume 53	**Neural Network Applications in Control** G.W. Irwin, K. Warwick and K.J. Hunt (Editors)
Volume 54	**Control Engineering Solutions: A Practical Approach** P. Albertos, R. Strietzel and N. Mort (Editors)
Volume 55	**Genetic Algorithms in Engineering Systems** A.M.S. Zalzala and P.J. Fleming (Editors)
Volume 56	**Symbolic Methods in Control System Analysis and Design** N. Munro (Editor)
Volume 57	**Flight Control Systems** R.W. Pratt (Editor)
Volume 58	**Power-Plant Control and Instrumentation: The Control of Boilers and HRSG Systems** D. Lindsley
Volume 59	**Modelling Control Systems Using IEC 61499** R. Lewis
Volume 60	**People in Control: Human Factors in Control Room Design** J. Noyes and M. Bransby (Editors)
Volume 61	**Nonlinear Predictive Control: Theory and Practice** B. Kouvaritakis and M. Cannon (Editors)
Volume 62	**Active Sound and Vibration Control** M.O. Tokhi and S.M. Veres
Volume 63	**Stepping Motors, 4th Edition** P.P. Acarnley
Volume 64	**Control Theory, 2nd Edition** J.R. Leigh
Volume 65	**Modelling and Parameter Estimation of Dynamic Systems** J.R. Raol, G. Girija and J. Singh
Volume 66	**Variable Structure Systems: From Principles To Implementation** A. Sabanovic, L. Fridman and S. Spurgeon (Editors)
Volume 67	**Motion Vision: Design of Compact Motion Sensing Solution for Autonomous Systems** J. Kolodko and L. Vlacic
Volume 68	**Flexible Robot Manipulators: Modelling, Simulation and Control** M.O. Tokhi and A.K.M. Azad (Editors)
Volume 69	**Advances in Unmanned Marine Vehicles** G. Roberts and R. Sutton (Editors)
Volume 70	**Intelligent Control Systems Using Computational Intelligence Techniques** A. Ruano (Editor)
Volume 71	**Advances in Cognitive Systems** S. Nefti and J. Gray (Editors)
Volume 72	**Control Theory: A Guided Tour, 3rd Edition** J.R. Leigh
Volume 73	**Adaptive Sampling with Mobile WSN** K. Sreenath, M.F. Mysorewala, D.O. Popa and F.L. Lewis
Volume 74	**Eigenstructure Control Algorithms: Applications to Aircraft/Rotorcraft Handling Qualities Design** S. Srinathkumar
Volume 75	**Advanced Control for Constrained Processes and Systems** F. Garelli, R.J. Mantz and H. De Battista
Volume 76	**Developments in Control Theory towards Glocal Control** L. Qiu, J. Chen, T. Iwasaki and H. Fujioka (Editors)
Volume 77	**Further Advances in Unmanned Marine Vehicles** G.N. Roberts and R. Sutton (Editors)
Volume 78	**Frequency-Domain Control Design for High-Performance Systems** J. O'Brien
Volume 80	**Control-Oriented Modelling and Identification: Theory and Practice** M. Lovera (Editor)
Volume 81	**Optimal Adaptive Control and Differential Games by Reinforcement Learning Principles** D. Vrabie, K. Vamvoudakis and F. Lewis
Volume 83	**Robust and Adaptive Model Predictive Control of Nonlinear Systems** M. Guay, V. Adetola and D. DeHaan
Volume 84	**Nonlinear and Adaptive Control Systems** Z. Ding
Volume 88	**Distributed Control and Filtering for Industrial Systems** M. Mahmoud
Volume 89	**Control-based Operating System Design** A. Leva *et al.*
Volume 90	**Application of Dimensional Analysis in Systems Modelling and Control Design** P. Balaguer
Volume 91	**An Introduction to Fractional Control** D. Valério and J. Costa
Volume 92	**Handbook of Vehicle Suspension Control Systems** H. Liu, H. Gao and P. Li
Volume 93	**Design and Development of Multi-Lane Smart Electromechanical Actuators** F.Y. Annaz
Volume 94	**Analysis and Design of Reset Control Systems** Y.Guo, L. Xie and Y. Wang
Volume 95	**Modelling Control Systems Using IEC 61499, 2nd Edition** R. Lewis and A. Zoitl

Robust and Adaptive Model Predictive Control of Nonlinear Systems

Martin Guay, Veronica Adetola
and Darryl DeHaan

The Institution of Engineering and Technology

Published by The Institution of Engineering and Technology, London, United Kingdom

The Institution of Engineering and Technology is registered as a Charity in England & Wales (no. 211014) and Scotland (no. SC038698).

First published 2015

The Institution of Engineering and Technology
Michael Faraday House
Six Hills Way, Stevenage
Herts, SG1 2AY, United Kingdom

www.theiet.org

British Library Cataloguing in Publication Data
A catalogue record for this product is available from the British Library

ISBN 978-1-84919-552-2 (hardback)
ISBN 978-1-84919-553-9 (PDF)

Typeset in India by MPS Limited

Contents

List of figures

List of tables

Acknowledgments

The authors gratefully acknowledge the National Science and Engineering Research Council of Canada and Queen's University for financial support received, as well as other colleagues at Queen's University for the many years of stimulating conversations. In particular, we would like to thank Professors James McLellan, Kim McAuley, Denis Dochain, Michel Perrier and Drs. Nicolas Hudon and Kai Hoeffner for their valuable feedbacks on the contents of this book and the rewarding discussions I had with them.

On a personal note, the authors would like to make the following acknowledgements.

Veronica Adetola:

I thank my family for their unreserved support, patience and understanding.

Darryl DeHaan:

For Evelyn and the munchkins, who constantly prove that the idea of control is all just an illusion.

Martin Guay:

For Bonnie, Samuel and Mélanie. Thank you for your love and support.

Chapter 1

Introduction

Most physical systems possess parametric uncertainties or unmeasurable disturbances. Examples in chemical engineering include reaction rates, activation energies, fouling factors, and microbial growth rates. Since parametric uncertainty may degrade the performance of model predictive control (MPC), mechanisms to update the unknown or uncertain parameters are desirable in application. One possibility would be to use state measurements to update the model parameters offline. A more attractive possibility is to apply adaptive extensions of MPC in which parameter estimation and control are performed online.

The literature contains very few results on the design of adaptive nonlinear MPC (NMPC) [1, 125]. Existing design techniques are restricted to systems that are linear in the unknown (constant) parameters and do not involve state constraints. Although MPC exhibits some degree of robustness to uncertainties, in reality, the degree of robustness provided by nominal models or certainty equivalent models may not be sufficient in practical applications. Parameter estimation error must be accounted for in the computation of the control law.

This book attempts to bridge the gap in adaptive robust NMPC. It proposes a design methodology for adaptive robust NMPC systems in the presence of disturbances and parametric uncertainties. One of the key concepts pursued is set-based adaptive parameter estimation. Set-based techniques provide a mechanism to estimate the unknown parameters as well as an estimate of the parameter uncertainty set. The main difference with established set-based techniques that are commonly used in optimization is that the proposed approach focusses on real-time uncertainty set estimation. In this work, the knowledge of uncertain set estimates are exploited in the design of robust adaptive NMPC algorithms that guarantee robustness of the NMPC system to parameter uncertainty. Moreover, the adaptive NMPC system is shown to recover nominal NMPC performance when parameters have been shown to converge to their true values.

The book provides a comprehensive introduction to NMPC and nonlinear adaptive control. In the first part of the book, a framework for the study, design, and analysis of NMPC systems is presented. The framework highlights various mechanisms that can be used to improve computational requirements of standard NMPC systems. The robustness of NMPC is presented in the context of this framework.

The second part of the book presents an introduction to adaptive NMPC. Starting with a basic introduction to the problems associated with adaptive MPC, a robust set-based approach is developed. The key element of this approach is an internal model identifier that allows the MPC to compensate for future moves in the parameter estimates and more importantly, the uncertainty associated with the unknown model parameters. It is shown that the proposed adaptive NMPC can recover the performance of the nominal MPC problem once the parameters have converged to their true value.

The third part of the book is dedicated to the practical realization of the adaptive NMPC methodology. An alternative approach to adaptive parameter estimation is first developed that yields a systematic set-based parameter estimation approach. This approach is integrated to the design of adaptive NMPC and robust adaptive NMPC control systems. An application for the design of adaptive economic NMPC systems is presented.

The last part of the book presents a treatment of the discrete-time generalization of the continuous-time algorithms proposed in the third part of the book. It is shown how the set-based estimation can be extended to the discrete-time case. The discrete-time techniques can be integrated easily using the concept of internal model identifiers to provide designs of adaptive robust NMPC systems.

Chapter 2
Optimal control

When faced with making a decision, it is only natural that one would aim to select the course of action which results in the "best" possible outcome. However, the ability to arrive at a decision necessarily depends upon two things: a well-defined notion of what qualities make an outcome desirable and a *previous* decision defining to what extent it is necessary to characterize the quality of individual candidates before making a selection (i.e., a notion of when a decision is "good enough"). Whereas the first property is required for the problem to be well defined, the latter is necessary for it to be tractable.

The process of searching for the "best" outcome has been mathematically formalized in the framework of optimization. The typical approach is to define a scalar-valued *cost function*, that accepts a decision candidate as its argument, and returns a quantified measure of its quality. The decision-making process then reduces to selecting a candidate with the lowest (or highest) such measure.

2.1 Emergence of optimal control

The field of "control" addresses the question of how to manipulate an *input* u in order to drive the *state* x of a dynamical system

$$\dot{x} = f(x, u) \tag{2.1}$$

to some desired target. Ultimately this task can be viewed as decision-making, so it is not surprising that it lends itself toward an optimization-based characterization. Assuming that one can provide the necessary metric for assessing quality of the trajectories generated by (2.1), there exists a rich body of "optimal control" theory to guide this process of decision-making. Much of this theory came about in the 1950s and 1960s, with Pontryagin's introduction of the Minimum (a.k.a. Maximum) Principle [135] and Bellman's development of Dynamic Programming [19, 20]. (This development also coincided with landmark results for linear systems, pioneered by Kalman [83, 84], that are closely related.) However, the roots of both approaches actually extend back to the mid-1600s, with the inception of the calculus of variations.

The tools of optimal control theory provide useful benchmarks for characterizing the notion of "best" decision-making, as it applies to control. However applied directly, the tractability of this decision-making is problematic. For example, Dynamic Programming involves the construction of a n-dimensional surface that satisfies a challenging nonlinear partial differential equation, which is inherently plagued by the so-called *curse of dimensionality*. This methodology, although elegant, remains generally intractable for problems beyond modest size. In contrast, the Minimum Principle has been relatively successful for use in offline trajectory planning, when the initial condition of (2.1) is known. Although it was suggested as early as 1967 in Reference 103 that a stabilizing feedback $u = k(x)$ could be constructed by continuously re-solving the calculations online, a tractable means of doing this was not immediately forthcoming.

2.2 MPC as receding-horizon optimization

Early development [43, 141, 142] of the control approach known today as MPC originated in the process control community, and was driven much more by industrial application than by theoretical understanding. Modern theoretical understanding of MPC, much of which developed throughout the 1990s, has clarified its very natural ties to existing optimal control theory. Key steps toward this development included such results as in References 33, 34, 44, 85, 124, and 128, with an excellent unifying survey in Reference 126.

At its core, MPC is simply a framework for implementing existing tools of optimal control. Taking the current value $x(t)$ as the initial condition for (2.1), the Minimum Principle is used as the primary basis for identifying the "best" candidate trajectory by predicting the future behavior of the system using model (2.1). However, the actual quality measure of interest in the decision-making is generally the total future accumulation (i.e., over an infinite future) of a given instantaneous metric, a quantity rarely computable in a satisfactorily short time. As such, MPC only generates predictions for (2.1) over a finite time horizon, and approximates the remaining infinite tail of the cost accumulation using a penalty surface derived from either a *local* solution of the Dynamic Programming surface, or an appropriate approximation of that surface. As such, the key benefit of MPC over other optimal control methods is simply that its finite horizon allows for a convenient trade-off between the online computational burden of solving the Minimum Principle, and the offline burden of generating the penalty surface.

In contrast to other approaches for constructive nonlinear controller design, optimal control frameworks facilitate the inclusion of constraints, by imposing feasibility of the candidates as a condition in the decision-making process. While these approaches can be numerically burdensome, optimal control (and by extension, MPC) provides the only real framework for addressing the control of systems in the presence of constraints—in particular those involving the state x. In practice, the predictive aspect of MPC is unparalleled in its ability to account for the risk of future constraint violation during the current control decision.

2.3 Current limitations in MPC

While the underlying theoretical basis for MPC is approaching a state of relative maturity, application of this approach to date has been predominantly limited to "slow" industrial processes that allow adequate time to complete the controller calculations.

There is great incentive to extend this approach to applications in many other sectors, motivated in large part by its constraint-handling abilities. Future applications of significant interest include many in the aerospace or automotive sectors, in particular constraint-dominated problems such as obstacle avoidance. At present, the significant computational burden of MPC remains the most critical limitation toward its application in these areas.

The second key weakness of the model predictive approach remains its susceptibility to uncertainties in the model (2.1). While a fairly well-developed body of theory has been developed within the framework of robust-MPC, reaching an acceptable balance between computational complexity and conservativeness of the control remains a serious problem. In the more general control literature, adaptive control has evolved as an alternative to a robust-control paradigm. However, the incorporation of adaptive techniques into the MPC framework has remained a relatively open problem.

2.4 Notational and mathematical preliminaries

Throughout the remainder of this book, the following is assumed by default (where $s \in \mathbb{R}^s$ and $\mathbb{S}$ represent arbitrary vectors and sets, respectively):

- All vector norms are Euclidean, defining balls $B(s,\varepsilon) \triangleq \{s' \mid \|s - s'\| \leq \varepsilon\}$, $\varepsilon \geq 0$.
- Norms of matrices $\mathcal{S} \in \mathbb{R}^{m \times s}$ are assumed induced as $\|\mathcal{S}\| \triangleq \max_{\|s\|=1} \|\mathcal{S}s\|$.
- The notation $s_{[a,b]}$ denotes the entire continuous-time trajectory $s(\tau)$, $\tau \in [a,b]$, and likewise $\dot{s}_{[a,b]}$ the trajectory of its forward derivative $\dot{s}(\tau)$.
- For any set $\mathbb{S} \subseteq \mathbb{R}^s$, define
  i. its closure $\mathrm{cl}\{\mathbb{S}\}$, interior $\mathring{\mathbb{S}}$, and boundary $\partial\mathbb{S} = \mathrm{cl}\{\mathbb{S}\} \setminus \mathring{\mathbb{S}}$
  ii. its orthogonal distance norm $\|s\|_{\mathbb{S}} \triangleq \inf_{s' \in \mathbb{S}} \|s - s'\|$
  iii. a closed ε-neighborhood $B(\mathbb{S},\varepsilon) \triangleq \{s \in \mathbb{R}^s \mid \|s\|_{\mathbb{S}} \leq \varepsilon\}$
  iv. an interior approximation $\overleftarrow{B}(\mathbb{S},\varepsilon) \triangleq \{s \in \mathbb{S} \mid \inf_{s' \in \partial\mathbb{S}} \|s - s'\| \geq \varepsilon\}$
  v. a (finite, closed, open) cover of $\mathbb{S}$ as any (finite) collection $\{\mathbb{S}^i\}$ of (open, closed) sets $\mathbb{S}^i \subseteq \mathbb{R}^s$ such that $\mathbb{S} \subseteq \cup_i \mathbb{S}^i$
  vi. the maximal closed subcover $\overline{\mathrm{cov}}\{\mathbb{S}\}$ as the infinite collection $\{\mathbb{S}^i\}$ containing *all possible* closed subsets $\mathbb{S}^i \subseteq \mathbb{S}$; that is, $\overline{\mathrm{cov}}\{\mathbb{S}\}$ is a maximal "set of subsets."

Furthermore, for any arbitrary function $\alpha : \mathbb{S} \to \mathbb{R}$ we assume the following definitions:

- $\alpha(\cdot)$ is C^{m+} if it is at least m-times differentiable, with all derivatives of order m yielding locally Lipschitz functions.
- A function $\alpha : \mathbb{S} \to (-\infty, \infty]$ is *lower semicontinuous* ($\mathcal{LS}$-continuous) at s if it satisfies (see Reference 40):

$$\liminf_{s' \to s} \dot{\alpha}(s') \geq \alpha(s). \tag{2.2}$$

- A continuous function $\alpha : \mathbb{R}_{\geq 0} \to \mathbb{R}_{\geq 0}$ belongs to *class* $\mathcal{K}$ if $\alpha(0) = 0$ and $\alpha(\cdot)$ is strictly increasing on $\mathbb{R}_{>0}$. It belongs to *class* $\mathcal{K}_\infty$ if it is furthermore radially unbounded.

- A continuous function $\beta : \mathbb{R}_{\geq 0} \times \mathbb{R}_{\geq 0} \rightarrow \mathbb{R}_{\geq 0}$ belongs to *class* $\mathcal{KL}$ if (i) for every fixed value of τ, it satisfies $\beta(\cdot, \tau) \in \mathcal{K}$, and (ii) for each fixed value of s, then $\beta(s, \cdot)$ is strictly decreasing and satisfies $\lim_{\tau\rightarrow\infty} \beta(s, \tau) = 0$.
- The scalar operator $\text{sat}_a^b(\cdot)$ denotes saturation of its arguments onto the interval $[a, b]$, $a < b$. For vector- or matrix-valued arguments, the saturation is presumed by default to be evaluated element-wise.

2.5 Brief review of optimal control

The underlying assumption of optimal control is that at any time, the pointwise cost of x and u being away from their desired targets is quantified by a known, physically meaningful function $L(x, u)$. Loosely, the goal is to then reach some target in a manner that accumulates the least cost. It is not generally necessary for the "target" to be explicitly described, since its knowledge is built into the function $L(x, u)$ (i.e., it is assumed that convergence of x to any invariant subset of $\{x \mid \exists u \ s.t. \ L(x, u) = 0\}$ is as acceptable). The following result, while superficially simple in appearance, is in fact the key foundation underlying the optimal control results of this section, and by extension all of MPC as well. Proof can be found in many references, such as Reference 143.

Definition 2.5.1 (Principle of Optimality). *If $u^*_{[t_1,t_2]}$ is an optimal trajectory for the interval $t \in [t_1, t_2]$, with corresponding solution $x^*_{[t_1,t_2]}$ to (2.1), then for any $\tau \in (t_1, t_2)$ the sub-arc $u^*_{[\tau,t_2]}$ is necessarily optimal for the interval $t \in [\tau, t_2]$ if (2.1) starts from $x^*(\tau)$.*

2.5.1 Variational approach: Euler, Lagrange, and Pontryagin

Pontryagin's Minimum Principle (also known as the Maximum Principle [135]) represented a landmark extension of classical ideas of variational calculus to the problem of control. Technically, the Minimum Principle is an application of the classical Euler–Lagrange and Weierstrass conditions[1] [78], which provide first-order *necessary* conditions to characterize extremal time-trajectories of a cost functional.[2] The Minimum Principle therefore characterizes minimizing trajectories $(x_{[0,T]}, u_{[0,T]})$ corresponding to a constrained finite-horizon problem of the form

$$V_T(x_0, u_{[0,T]}) = \int_0^T L(x, u)\, d\tau + W(x(T)) \tag{2.3a}$$

$$s.t. \ \forall \tau \in [0, T]:$$

$$\dot{x} = f(x, u), \quad x(0) = x_0 \tag{2.3b}$$

$$g(x(\tau)) \leq 0, \quad h(x(\tau), u(\tau)) \leq 0, \quad w(x(T)) \leq 0 \tag{2.3c}$$

[1] Phrased as a fixed initial point, free endpoint problem.

[2] That is, generalizing the nonlinear program (NLP) necessary condition $\frac{\partial p}{\partial x} = 0$ for the extrema of a function $p(x)$.

where the vector field $f(\cdot,\cdot)$ and constraint functions $g(\cdot)$, $h(\cdot,\cdot)$, and $w(\cdot)$ are assumed sufficiently differentiable.

Assume that $g(x_0) < 0$, and, for a given $(x_0, u_{[0,T]})$, let the interval $[0, T)$ be partitioned into (maximal) subintervals as $\tau \in \bigcup_{i=1}^{p} [t_i, t_{i+1})$, $t_0 = 0$, $t_{p+1} = T$, where the interior t_i represent intersections $g < 0 \Leftrightarrow g = 0$ (i.e., the $\{t_i\}$ represent changes in the active set of g). Assuming that $g(x)$ has constant relative degree r over some appropriate neighborhood, define the following vector of (Lie) derivatives: $N(x) \triangleq [g(x), g^{(1)}(x), \ldots, g^{(r-1)}(x)]^T$, which characterizes additional tangency constraints $N(x(t_i)) = 0$ at the corners $\{t_i\}$. Rewriting (2.3) in multiplier form

$$V_T = \int_0^T \mathcal{H}(x,u) - \lambda^T \dot{x}\, d\tau + W(x(T)) + \mu_w w(x(T)) + \sum_i \mu_N^T(t_i) N(x(t_i)) \quad (2.4a)$$

$$\mathcal{H} \triangleq L(x,u) + \lambda^T f(x,u) + \mu_h h(x,u) + \mu_g g^{(r)}(x,u) \quad (2.4b)$$

Taking the first variation of the right-hand sides of (2.4a) and (2.4b) with respect to perturbations in $x_{[0,T]}$ and $u_{[0,T]}$ yields the following set of conditions (adapted from statements in References 24, 28, and 78) which necessarily must hold for V_T to be minimized.

Proposition 2.5.2 (Minimum Principle). *Suppose that the pair ($u^*_{[0,T]}$ and $x^*_{[0,T]}$) is a minimizing solution of* (2.3). *Then for all* $\tau \in [0, T]$, *there exists multipliers* $\lambda(\tau) \geq 0$, $\mu_h(\tau) \geq 0$, $\mu_g(\tau) \geq 0$, *and constants* $\mu_w \geq 0$, $\mu_N^i \geq 0$, $i \in \mathcal{I}$, *such that*

i. *Over each interval* $\tau \in [t_i, t_{i+1}]$, *the multipliers* $\mu_h(\tau)$, $\mu_g(\tau)$ *are piecewise continuous,* $\mu_N(\tau)$ *is constant,* $\lambda(\tau)$ *is continuous, and with* $(u^*_{[t_i,t_{i+1}]}, x^*_{[t_i,t_{i+1}]})$ *satisfies*

$$\dot{x}^* = f(x^*, u^*), \quad x^*(0) = x_0 \quad (2.5a)$$

$$\dot{\lambda}^T = \nabla_x \mathcal{H} \quad \text{a.e., with} \quad \lambda^T(T) = \nabla_x W(x^*(T)) + \mu_w \nabla_x w(x^*(T)) \quad (2.5b)$$

where the solution $\lambda_{[0,T]}$ *is discontinuous at* $\tau \in \{t_i\}$, $i \in \{1, 3, 5 \ldots p\}$, *satisfying*

$$\lambda^T(t_i^-) = \lambda^T(t_i^+) + \mu_N^T(t_i^+) \nabla_x N(x(t_i)) \quad (2.5c)$$

ii. $\mathcal{H}(x^*, u^*, \lambda, \mu_h, \mu_g)$ *is constant over intervals* $\tau \in [t_i, t_{i+1}]$, *and for all* $\tau \in [0, T]$ *it satisfies (where* $\mathcal{U}(x) \triangleq \{u \mid h(x,u) \leq 0$ *and* $(g^{(r)}(x,u) \leq 0$ *if* $g(x) = 0)\}$)

$$\mathcal{H}(x^*, u^*, \lambda, \mu_h, \mu_g) \leq \min_{u \in \mathcal{U}(x)} \mathcal{H}(x^*, u, \lambda, \mu_h, \mu_g) \quad (2.5d)$$

$$\nabla_u \mathcal{H}(x^*(\tau), u^*(\tau), \lambda(\tau), \mu_h(\tau), \mu_g(\tau)) = 0 \quad (2.5e)$$

iii. *For all* $\tau \in [0, T]$, *the following constraint conditions hold*

$$g(x^*) \leq 0 \qquad h(x^*, u^*) \leq 0 \quad w(x^*(T)) \leq 0 \quad (2.5f)$$

$$\mu_g(\tau) g^{(r)}(x^*, u^*) = 0 \quad \mu_h(\tau) h(x^*, u^*) = 0 \quad \mu_w w(x^*(T)) = 0 \quad (2.5g)$$

$$\mu_N^T(\tau) N(x^*) = 0 \qquad (\text{and } N(x^*) = 0,\ \forall \tau \in [t_i, t_{i+1}],\ i \in \{1, 3, 5 \ldots p\}) \quad (2.5h)$$

The multiplier $\lambda(t)$ is called the *co-state*, and it requires solving a two-point boundary-value problem for (2.5a) and (2.5b). One of the most challenging aspects to locating (and confirming) a minimizing solution to (2.5) lies in dealing with (2.5c) and (2.5h), since the number and times of constraint intersections are not known *a priori*.

2.5.2 *Dynamic programming: Hamilton, Jacobi, and Bellman*

The Minimum Principle is fundamentally based upon establishing the optimality of a *particular* input trajectory $u_{[0,T]}$. While the applicability to offline, open-loop trajectory planning is clear, the inherent assumption that x_0 is known can be limiting if one's goal is to develop a *feedback* policy $u = k(x)$. Development of such a policy requires the consideration of *all possible* initial conditions, which results in an optimal cost *surface* $J^* : \mathbb{R}^n \rightarrow \mathbb{R}$, with an associated control policy $k : \mathbb{R}^n \rightarrow \mathbb{R}^m$. A constructive approach for calculating such a surface, referred to as *Dynamic Programming*, was developed by Bellman [19]. Just as the Minimum Principle was extended out of the classical trajectory-based Euler–Lagrange equations, Dynamic Programming is an extension of classical Hamilton–Jacobi field theory from the calculus of variations.

For simplicity, our discussion here will be restricted to the unconstrained problem

$$V^*(x_0) = \min_{u_{[0,\infty)}} \int_0^{\infty} L(x,u)d\tau \tag{2.6a}$$

$$s.t. \quad \dot{x} = f(x,u), \quad x(0) = x_0 \tag{2.6b}$$

with locally Lipschitz dynamics $f(\cdot,\cdot)$. From the Principle of Optimality, it can be seen that (2.6) lends itself to the following recursive definition:

$$V^*(x(t)) = \min_{u[t,t+\Delta t]} \left\{ \int_t^{t+\Delta t} L(x(\tau),u(\tau))d\tau + V^*(x(t+\Delta t)) \right\} \tag{2.7}$$

Assuming that V^* is differentiable, replacing $V^*(x(t+\Delta t)$ with a first-order Taylor-series and the integrand with a Riemannian sum, the limit $\Delta t \rightarrow 0$ yields

$$0 = \min_u \left\{ L(x,u) + \frac{\partial V^*}{\partial x} f(x,u) \right\} \tag{2.8}$$

Equation (2.8) is one particular form of what is known as the Hamilton–Jacobi–Bellman equation. In some cases (such as $L(x,u)$ quadratic in u, and $f(x,u)$ affine in u), equation (2.8) can be simplified to a more standard-looking partial differential equation (PDE) by evaluating the indicated minimization in closed-form.[3] Assuming that a (differentiable) surface $V^* : \mathbb{R}^n \rightarrow \mathbb{R}$ is found (generally by offline numerical solution) which satisfies (2.8), a stabilizing feedback $u = k_{DP}(x)$ can be constructed from the information contained in the surface V^* by simply defining $k_{DP}(x) \triangleq \{u \mid \frac{\partial V^*}{\partial x} f(x,u) = -L(x,u)\}$.[4]

Unfortunately, incorporation of either input or state constraints generally violates the assumed smoothness of $V^*(x)$. While this could be handled by interpreting (2.8)

[3] In fact, for linear dynamics and quadratic cost, (2.8) reduces down to the linear Ricatti equation.
[4] $k_{DP}(\cdot)$ is interpreted to incorporate a deterministic selection in the event of multiple solutions. The existence of such a u is implied by the assumed solvability of (2.8).

in the context of *viscosity solutions* (see Reference 40 for definition), for the purposes of application to MPC it is more typical to simply restrict the domain of $V^* : \Omega \to \mathbb{R}$ such that $\Omega \subset \mathbb{R}^n$ is feasible with respect to the constraints.

2.5.3 *Inverse-optimal control Lyapunov functions*

While knowledge of a surface $V^*(x)$ satisfying (2.8) is clearly ideal, in practice analytical solutions are only available for extremely restrictive classes of systems, and almost never for systems involving state or input constraints. Similarly, numerical solution of (2.8) suffers the so-called "curse of dimensionality" (as named by Bellman) which limits its applicability to systems of restrictively small size.

An alternative design framework, originating in Reference 155, is based on the following definition.

Definition 2.5.3. *A control Lyapunov function (CLF) for* (2.1) *is any* C^1*, proper, positive definite function* $V : \mathbb{R}^n \to \mathbb{R}_{\geq 0}$ *such that, for all* $x \neq 0$*:*

$$\inf_u \frac{\partial V}{\partial x} f(x, u) < 0. \tag{2.9}$$

Design techniques for deriving a feedback $u = k(x)$ from knowledge of $V(\cdot)$ include the well-known "Sontag's controller" of Reference 153, which led to the development of "pointwise min-norm" control of the form in References 64, 65, and 150

$$\min_u \gamma(u) \quad s.t. \quad \frac{\partial V}{\partial x} f(x, u) < -\sigma(x) \tag{2.10}$$

where γ, σ are positive definite, and γ is radially unbounded. As discussed in References 65 and 150, relation (2.9) implies that *there exists* a function $L(x, u)$, derived from γ and σ, for which $V(\cdot)$ satisfies (2.8). Furthermore, if $V(x) \equiv V^*(x)$, then appropriate selection of γ, σ (in particular that of Sontag's controller [153]) results in the feedback $u = k_{clf}(x)$ generated by (2.9) satisfying $k_{clf}(\cdot) \equiv k_{DP}(\cdot)$. Hence this technique is commonly referred to as "inverse-optimal" control design, and can be viewed as a method for approximating the optimal control problem (2.6) by replacing $V^*(x)$ directly.

Chapter 3
Review of nonlinear MPC

The ultimate objective of a model predictive controller is to provide a *closed-loop feedback* $u = \kappa_{mpc}(x)$ that regulates (2.1) to its target set (assumed here $x = 0$) in a fashion that is optimal with respect to the *infinite-time* problem (2.6), while enforcing pointwise constraints of the form $(x, u) \in \mathbb{X} \times \mathbb{U}$ in a constructive manner. However, rather than defining the map $\kappa_{mpc} : \mathbb{X} \to \mathbb{U}$ by solving a PDE of the form (2.8) (i.e., thereby pre-computing knowledge of $\kappa_{mpc}(x)$ for *every* $x \in \mathbb{X}$), the MPC philosophy is to solve for, at time t, the control move $u = \kappa_{mpc}(x(t))$ for the *particular* value $x(t) \in \mathbb{X}$. This makes the online calculations inherently trajectory-based, and therefore closely tied to the results in Section 2.5.1 (with the caveat that the initial conditions are continuously referenced relative to current (t, x)). Since it is not practical to pose (online) trajectory-based calculations over an infinite prediction horizon $\tau \in [t, \infty)$, a truncated prediction $\tau \in [t,\ t + T]$ is used instead. The truncated tail of the integral in (2.6) is replaced by a (designer-specified) terminal penalty $W : \mathbb{X}_f \to \mathbb{R}_{\geq 0}$, defined over any local neighborhood $\mathbb{X}_f \subset \mathbb{X}$ of the target $x = 0$. This results in a feedback of the form

$$u = \kappa_{mpc}(x(t)) \triangleq u^*_{[t,t+T]}(t) \tag{3.1a}$$

where $u^*_{[t,t+T]}$ denotes the solution to the $x(t)$-dependent problem

$$u^*_{[t,t+T]} \triangleq \arg \min_{u^p_{[t,t+T]}} \left(V_T(x(t), u^p_{[t,t+T]}) \triangleq \int_t^{t+T} L(x^p, u^p)\, d\tau + W(x^p(t+T)) \right) \tag{3.1b}$$

$$\text{s.t. } \forall \tau \in [t, t+T] : \tfrac{d}{d\tau} x^p = f(x^p, u^p), \quad x^p(t) = x(t) \tag{3.1c}$$

$$(x^p(\tau), u^p(\tau)) \in \mathbb{X} \times \mathbb{U} \tag{3.1d}$$

$$x^p(t+T) \in \mathbb{X}_f \tag{3.1e}$$

Clearly, if one could define $W(x) \equiv V^*(x)$ globally, then the feedback in (3.1) must satisfy $\kappa_{mpc}(\cdot) \equiv k_{DP}(\cdot)$. While $W(x) \equiv V^*(x)$ is generally unachievable, this motivates the selection of $W(x)$ as a CLF such that $W(x)$ is an inverse-optimal approximation of $V^*(x)$. A more precise characterization of the selection of $W(x)$ is the focus of the next section.

3.1 Sufficient conditions for stability

A very general proof of the closed-loop stability of (3.1), which unifies a variety of earlier, more restrictive, results is presented[1] in the survey [126]. This proof is based upon the following set of sufficient conditions for closed-loop stability:

Criterion 3.1.1. *The function $W : \mathbb{X}_f \to \mathbb{R}_{\geq 0}$ and set $\mathbb{X}_f$ are such that a local feedback $k_f : \mathbb{X}_f \to \mathbb{U}$ exists to satisfy the following conditions:*

1. $0 \in \mathbb{X}_f \subseteq \mathbb{X}$, $\mathbb{X}_f$ *closed (i.e., state constraints satisfied in* $\mathbb{X}_f$*)*
2. $k_f(x) \in \mathbb{U}$, $\forall x \in \mathbb{X}_f$ *(i.e., control constraints satisfied in* $\mathbb{X}_f$*)*
3. $\mathbb{X}_f$ *is positively invariant for* $\dot{x} = f(x, k_f(x))$
4. $L(x, k_f(x)) + \frac{\partial W}{\partial x} f(x, k_f(x)) \leq 0, \quad \forall x \in \mathbb{X}_f$.

Only existence, not knowledge, of $k_f(x)$ is assumed. Thus by comparison with (2.9), it can be seen that C4 essentially requires that $W(x)$ be a CLF over the (local) domain $\mathbb{X}_f$, in a manner consistent with the constraints.

In hindsight, it is nearly obvious that closed-loop stability can be reduced entirely to conditions placed upon only the terminal choices $W(\cdot)$ and $\mathbb{X}_f$. Viewing $V_T(x(t), u^*_{[t,t+T]})$ as a Lyapunov function candidate, it is clear from (2.3) that V_T contains "energy" in both the $\int L\, d\tau$ and terminal W terms. Energy dissipates from the front of the integral at a rate $L(x, u)$ as time t flows, and by the Principle of Optimality one could implement (3.1) on a *shrinking* horizon (i.e., $t + T$ constant), which would imply $\dot{V} = -L(x, u)$. In addition to this, C4 guarantees that the energy transfer from W to the integral (as the point $t + T$ recedes) will be non-increasing, and could even dissipate additional energy as well.

3.2 Sampled-data framework

3.2.1 General nonlinear sampled-data feedback

Within the (non-MPC) nonlinear control literature, the ideas of "sampled-data (SD)" control [72, 130], "piecewise-constant (PWC) control" [37–39], or "sample-and-hold feedback" [86] are all nearly equivalent. The basic idea involves:

Algorithm 3.2.1. *Closed-loop implementation of general SD controller:*

1. *define a* partition, π, *of the time axis, consisting of an infinite collection of sampling points:* $\pi \triangleq \{t_i,\ i \in \mathbb{N} \mid t_0 = 0,\ \ t_i < t_{i+1},\ \ t_i \to \infty \text{ as } i \to \infty\}$
2. *define a feedback* $k(x)$*, or more generally use a parameterized family of feedbacks* $k_T(x, T)$
3. *at time* t_i*, sample the state* $x_i \triangleq x(t_i)$

[1] In the context of both continuous- and discrete-time frameworks.

4. *over the interval* $t \in [t_i, t_{i+1})$ *implement the control via the zero-order hold:* $u(t) = k(x_i)$, *or alternatively* $u(t) = k_T(x_i, t_{i+1} - t_i)$
5. *at time* t_{i+1}, *repeat back to (3).*

Depending upon the design of the feedback $k(x)$, stability of these approaches generally hinges upon the fact that $\overline{\pi} \triangleq \sup_i (t_{i+1} - t_i)$ has a sufficiently small upper bound. Within this context, Fontes [63] demonstrated that the choice $k(x_i) \triangleq \kappa_{mpc}(x_i)$ is stabilizing, where κ_{mpc} is as defined in (3.1) (i.e., minimized over arbitrary signals $u_{[t,t+T]}$). Although *implemented* within a SD framework, the approach of Fontes [63] does not really qualify as "SD-MPC," which we will discuss next.

As highlighted in Reference 58, while the notion of "sample and hold" necessarily applies to the measurement signal of x, there is no fundamental reason why the input signal u necessitates using a hold of any kind. This means that one could easily implement over the interval $t \in [t_i, t_{i+1})$ a time-varying SD feedback of the form $u(t) = k(t, x_i, \pi)$. In other words, the SD framework can be generalized to involve implementing *open-loop* control policies during the inter-sample interval, with the feedback loop being closed intermittently at the sampling times of π. MPC, which inherently involves the generation of open-loop control trajectories, is an ideal choice for the design of such "time-varying" feedback laws.

3.2.2 Sampled-data MPC

The distinguishing characteristic of SD-MPC, in comparison to other frameworks for SD control, is the manner in which the inter-sample behavior is defined. This involves:

Algorithm 3.2.2. *Closed-loop implementation of SD-MPC:*

1. *define* π *as above*
2. *at time* t_i, *sample the state* $x_i \triangleq x(t_i)$
3. *"instantaneously" solve the finite-horizon optimal control problem in* (3.1) *for a prediction interval* $\tau \in [t_i, t_{i+N}]$, *yielding solution* $u^{i*}_{[t_i,t_{i+N}]}$
4. *while* $t \in [t_i, t_{i+1})$, *implement* $u(t) = u^{i*}_{[t_i,t_{i+N}]}(t)$; *i.e., implement, in open loop, the first* $[t_i, t_{i+1})$ *segment of the solution* $u^{i*}_{[t_i,t_{i+N}]}$
5. *at time* t_{i+1}, *repeat back to* (2).

A thorough treatment of this approach is provided in Reference 59. Of fundamental importance is that over any given interval, the *actual* trajectories $(x_{[t_i,t_{i+1})}, u_{[t_i,t_{i+1})})$ of the system "exactly"[2] correspond to the prediction $(x^p_{[t_i,t_{i+1})}, u^{i*}_{[t_i,t_{i+1})})$ generated at time t_i. This means that at time t_{i+1}, the Principle of Optimality (Definition 2.5.1) can be used (together with condition (C4)) to claim that the new N-step optimal cost $V^*_N(x_{i+1}, u^{i+1*}_{[t_{i+1},t_{i+N+1}]})$ must be bounded by the remaining portion of the previous solution: $V^*_N(x_{i+1}, u^{i+1*}_{[t_{i+1},t_{i+N+1}]}) \leq V^*_{N-1}(x_{i+1}, u^{i*}_{[t_{i+1},t_{i+N}]})$, the so-called *monotonicity property*.

[2] Assuming a perfect system model.

While in most situations the time required to solve (3.1) over "arbitrary" trajectories $u_{[t_i, t_{i+N}]}$ is unreasonable, it is clear that one can easily restrict the search in (3.1) to any desired subclass of signals $u_{[t_i,t_{i+N}]} \in \mathcal{U}_{[t_i,t_{i+N}]}(\pi)$ which are locally supported[3] by π. Therefore, approaches such as References 112 and 115 that couple a piecewise constant PWC parameterization of u^p together with a zero-order hold on the implementation $u(t)$ are encompassed by defining $\mathcal{U}_{[t_i,t_{i+N}]}(\pi)$ to be the class of trajectories constant over intervals of π.

3.2.3 Computational delay and forward compensation

In practice, SD implementation of a model predictive controller almost never proceeds as described above. The selection of π is generally based on the computation time required to solve (3.1), so it is unlikely that the solutions are achievable in a faster timescale than the intervals of π.

If the computational lag Δt satisfies $\Delta t \ll t_{i+1} - t_i$, then one might hope to just ignore it and implement the control $u(t + \Delta t) = u^{i*}_{[t_i,t_{i+N}]}(t)$ as soon as it becomes available. It can be shown that the resulting mismatch between (2.1) and (3.1c) can be encompassed into the claim of nominal robustness of the SD control (see Reference 59). This means that the lag will not be destabilizing as long as Δt is "sufficiently small."

For the more typical case where $\Delta t \approx t_{i+1} - t_i$, it is better to use the method of *forward compensation* detailed in References 36 and 57. Assume that a bound $\Delta_i t \geq \Delta t$ is known (often $\Delta_i t \equiv t_{i+1} - t_i$, but this is not required). When posing the optimal control problem (3.1), the additional constraint $u^p_{[t_i,t_i+\Delta_i t]} \equiv u^{i-1*}_{[t_i,t_i+\Delta_i t]}$ is imposed upon the class of signals over which (3.1) is minimized. This means that by construction, the first portion $t \in [t_i, t_i + \Delta_i t]$ of the optimal trajectory $u^{i*}_{[t_i,t_{i+N}]}$ is "known" even before solution of (3.1) is complete. This is equivalent to saying that, based on x_i and the input $u^{i-1*}_{[t_i,t_i+\Delta_i t]}$, the prediction $x^p(t_i + \Delta_i t)$ is used as the initial condition for solving (3.1) over the interval $\tau \in [t_i + \Delta_i t,\ t_{i+N}]$.

3.3 Computational techniques

The last two decades have seen significant development in the area of numerical methods for the online solution of dynamic optimization problems such as (3.1). Early MPC implementations generally made use of "off-the-shelf" sequential quadratic programming (SQP) solvers, developed originally for offline minimization of the optimal control problem (2.3). However, the relatively poor performance of these solvers in online implementation has motivated the development of new, or sometimes modified, methods more suitable for use in (3.1).

Solving (3.1) inherently involves two tasks: the search for the optimal trajectory $u^*_{[t,t+T]}$ and the solution of (3.1c) to generate the corresponding state trajectory $x^p_{[t,t+T]}$.

[3] That is, individual sub-arcs $u_{[t_j,t_{j+1}]}$, $u_{[t_k,t_{k+1}]}$, $j \neq k$, as partitioned by π can be chosen independently of each other in $\mathcal{U}_{[t_i,t_{i+N}]}(\pi)$.

Numerical methods can be classified into two categories, based upon how these tasks are handled.

1. **Sequential approaches:**
   These approaches parameterize the input trajectory $u^p_{[t,t+T]}$, and solve a NLP to minimize $V(x, u_{[t,t+T]})$ over that parameter space. The prediction $x^p_{[t,t+T]}$ is removed from the optimization by cascading the NLP solver with a standard ordinary differential equation (ODE) solver for generating solutions to (3.1c). In doing so, the state constraints are transformed into additional constraints in the parameter space for $u^p_{[t,t+T]}$, which are imposed on the NLP. Examples of this approach include the Newton-type algorithms proposed in References 45, 46, and 107, or for example, those applied to large-scale systems in References 97 and 127.
2. **Simultaneous approaches:**
   These methods parameterize *both* the input and state trajectories $(u^p_{[t,t+T]}, x^p_{[t,t+T]})$, and solve a constrained NLP over this combined parameter space. Approaches for parameterizing $x^p_{[t,t+T]}$ such that it satisfies (3.1c) include
   - ***Orthogonal collocation***. These involve parameterizing $x^p_{[t,t+T]}$ according to linear weighting of a basis function expansion, and defining a collection of time-points in $[t,\ t+T]$ at which the vector field $f(x^p, u^p)$ must be satisfied. This generates very large, but sparse, NLPs. Examples of this approach include References 11, 25, and 26.
   - ***Direct multiple shooting***. These approaches partition the prediction horizon $[t, t+T]$ into N segments, and assign a (n-dimensional) parameter for the value of x^p at each node point. An ODE solver is then applied to each interval *independently*, using the x^p parameters as initial conditions for each interval. Continuity of the $x^p_{[t,t+T]}$ trajectory at the node points is enforced by adding equality constraints into the NLP. Essentially, this has the structure of a constrained NLP (in the combined (x^p, u^p) parameter space) cascaded with a collection of N, parallel-computed, ODE solvers. Examples of this method for real-time application include References 52, 54, 55, 145, and 146.

In all approaches, finding the precise location of the minimizing solution to (3.1) is a challenging task for any solver to perform in a computationally-limited online implementation. Fortunately, as was shown in References 128 and 148, early termination of the search can still result in a stabilizing feedback as long as the resulting parameterization of $u^p_{[t_i,t_{i+N}]}$ and corresponding solution $x^p_{[t_i,t_{i+N}]}$ of (3.1c) are *feasible* (albeit suboptimal), and the cost $V(x(t), u^p_{[t_i,t_{i+N}]})$ decreases in comparison to that of the previous instance. As discussed in Reference 121, a key property of many online solvers is therefore their ability to generate strictly feasible paths, so that termination may occur at any time.

Since large open-loop intervals (i.e., in a SD implementation) are undesirable for robustness considerations, several different approaches have been developed to simplify calculations. These approaches aim to maximize the input–output sampling rate of the feedback, by extending the idea of early termination to the limiting case of an incrementally-evolving search. We briefly present on a few of these approaches below.

3.3.1 Single-step SQP with initial-value embedding

An interesting SQP approach is presented in References 52, 54, and 55. This SD approach aims to avoid the need for forward compensation by attempting to minimize the lag Δt between sampling and implementation. The approach is based upon a simultaneous multiple-shooting method, and allows for the dynamics (3.1c) to be described by very general parameter-dependent differential-algebraic equations. However for consistency of presentation, the approach can be viewed as solving within (3.1) an optimal control problem of the form (with notation following (2.3))

$$P(x_0): \min_{\substack{s_0^u,\dots s_{N-1}^u, \\ s_0^x,\dots s_N^x}} \sum_{i=0}^{N-1} L(s_i^x, s_i^u) + W(s_N^x) \tag{3.2a}$$

$$\text{subject to } \tfrac{d}{d\tau} x_i^p = f(x_i^p(\tau), s_i^u), \quad x_i^p(\tau_i) = s_i^x, \quad \tau \in [\tau_i, \tau_{i+1}], \quad \forall i \neq N \tag{3.2b}$$

$$s_{i+1}^x - x_i^p(\tau_{i+1}) = 0 \qquad i = 0, \dots, N-1 \tag{3.2c}$$

$$s_0^x - x_0 = 0 \tag{3.2d}$$

$$h(s_i^x, s_i^u) \leq 0 \qquad i = 0, \dots, N \tag{3.2e}$$

$$w(s_N^x, s_N^u) \leq 0 \tag{3.2f}$$

The vector s^u contains the parameters used in the (PWC) parameterization of the input $u^p_{[\tau_0, \tau_N]}$, while s^x defines the multiple-shooting parameterization of $x^p_{[\tau_0, \tau_N]}$ over the same time-partition π of the horizon. The *continuity constraints* in (3.2c) ensure that the minimizing solution generates a continuous $x^p_{[\tau_0, \tau_N]}$, although feasibility of $x^p_{[\tau_0, \tau_N]}$ is only tested at the partition points s^x in (3.2e) (note that the constraint $g(x) \leq 0$ of (2.3c) is assumed to be encompassed in (3.2e)).

The NLP in (3.2) can therefore be summarized as follows

$$\min_z F(z) \quad \text{subject to} \begin{cases} G(z) = 0 \\ H(z) \leq 0 \end{cases} \tag{3.3}$$

where $z \triangleq [s_0^x, s_0^u, \dots, s_{N-1}^x, s_{N-1}^u, s_N^x]$, with $G(z)$ containing (3.2c) and (3.2d) and any equalities in (3.2e) and (3.2f), and $H(z)$ the remaining inequalities. Starting from an initial guess z^0, the SQP proceeds by generating iterates

$$z^{k+1} = z^k + \alpha^k \Delta z^k, \quad \alpha^k \in (0, 1) \tag{3.4a}$$

$$\text{where } \Delta z^k \triangleq \arg\min_{\Delta z} \left(\nabla F(z^k)^T \Delta z + \tfrac{1}{2} \Delta z^T A^k \Delta z \right) \tag{3.4b}$$

$$\text{subject to } G(z^k) + \nabla G(z^k)^T \Delta z = 0 \tag{3.4c}$$

$$H(z^k) + \nabla H(z^k)^T \Delta z \leq 0 \tag{3.4d}$$

where A^k denotes any approximation of the hessian $\nabla_z^2 \mathcal{H}$ of the Lagrangian function $\mathcal{H} = F(z) + \mu_G^T G(z) + \mu_H^T H(z)$.

Although (3.2)–(3.4) is a straightforward multiple-shooting-based SQP formulation, the main contribution of this work consists of the following two unique characteristics:

- The update (3.4a) is terminated after a single iteration, meaning the parameters (s^x, s^u) are only updated by a single step per sampling interval of π. Contractivity and convergence of the single-step iteration is shown in Reference 55, with nominal robustness shown in Reference 56.
- In the calculation of the direction vector Δz:
  - by linearizing the equality and active inequality constraints in (3.3), the variables $s_1^x \cdots s_N^x$ are eliminated, and the calculations projected onto the condensed space of $s_0^x, s_0^u, \cdots s_{N-1}^u$. This results in a very dense, low dimensional approximation of (3.4a).
  - the expansion back to the full space of z, and calculation of the next round of gradients/hessians for both the full and condensed versions of (3.4a) are all done prior to the next sample t_{i+1}.
  - upon sampling of x_{i+1}, one only needs to solve for Δs^u (given Δs_0^x) in the condensed version of (3.4a) to generate u_{i+1}, where all necessary matrices have been pre-computed.

Overall, it is suggested that these modifications result in a very fast calculation, which therefore allow the sampling intervals of π to be chosen as small as possible.

3.3.2 Continuation methods

Rather than explicitly posing a NLP to *search* for the optimal trajectory $u^*_{[t,t+T]}$, the approaches in References 132 and 133 instead assume that initially the optimal solution $u^*_{[t_0,t_0+T]}$ is known, and focus on continuously *propagating* $u^*_{[t,t+T]}$ as t evolves. The problem setup is similar to (2.3), except that only equality constraints of the form[4] $h(x,u) = 0$ are included; inequalities require somewhat ad-hoc use of penalties.

The basic idea is to try and propagate the input trajectory $u^*_{[t,t+T]}$ and the multiplier $\mu^*_{[t,t+T]}$ (denoting $\mu \equiv \mu_h$) such that they remain in the kernel of the first-order optimality conditions in (2.5). This is done by discretizing the prediction horizon using a *very fine* partition of N (uniform) intervals, with $u^*_{[t,t+T]}$ and $\mu^*_{[t,t+T]}$ thus described by the discretization

$$U(t) \triangleq [u_0^*(t), \mu_0^*(t), u_1^*(t), \mu_1^*(t), \ldots, u_{N-1}^*(t), \mu_{N-1}^*(t)]^T \in \mathbb{R}^{(m_u+m_h)N} \tag{3.5}$$

As seen by the notation in (3.5), this discretization is fundamentally different than that of a SD partition. Whereas parameterizations in a SD approach are based on a partition π of the *actual* time coordinate $t \in [t_0, \infty)$, the discretization in (3.5) is based on a partition of the horizon *length* $\tau \in [0, T]$, which is treated as an orthogonal coordinate to the actual time t.

[4] The results in Reference 132 actually allow for a known time-dependent parameter vector $p(t)$ in $f(x, u, p(t))$ and $h(x, u, p(t))$, which we omit for clarity.

According to Reference 132, the optimality conditions (2.5e) and (2.5f) can be treated as defining the system of equations

$$F(U(t),x(t)) \triangleq \begin{bmatrix} \nabla_u^T \mathcal{H}(x_0^*(t),\lambda_1^*(t),u_0^*(t),\mu_0^*(t)) \\ h(x_0^*(t),u_0^*(t)) \\ \vdots \\ \nabla_u^T \mathcal{H}(x_{N-1}^*(t),\lambda_N^*(t),u_{N-1}^*(t),\mu_{N-1}^*(t)) \\ h(x_{N-1}^*(t),u_{N-1}^*(t)) \end{bmatrix} = 0 \tag{3.6}$$

where the terms $\{x_i^*,\lambda_i^*\}_{i=0}^N$ are interpreted as $x_i^*(t) \equiv x_i^*(x(t),U(t))$ and $\lambda_i^*(t) \equiv \lambda_i^*(x(t),U(t))$ under the assumption that they are generated by recursively solving the discretized equivalent of (2.5a, b):

$$x_{i+1}^*(t) = x_i^*(t) + f(x_i^*(t),u_i^*(t))\left(\tfrac{T(t)}{N}\right) \tag{3.7a}$$

$$\lambda_i^*(t) = \lambda_{i+1}^*(t) + \mathcal{H}_x^T(x_i^*(t),\lambda_{i+1}^*(t),u_i^*(t)u_i^*(t),\mu_i^*(t))\left(\tfrac{T(t)}{N}\right) \tag{3.7b}$$

subject to $x_0^*(t)=x(t)$ and $\lambda_N^*(t)=\nabla_x W^T(x_N^*(t))$. In other words, $\{x_i^*,\lambda_i^*\}_{i=0}^N$ are assumed to be available from $(x(t),U(t))$ by solving, in a faster timescale than $F(U(t),x(t),t)$, the two-point boundary value problem in (2.5a) and (2.5b) by what is essentially an Euler-based ODE solution technique.

As mentioned, it is assumed that $U(0)$ initially satisfies $F(U(0),x(0))=0$. The continued equality $F(U(t),x(t))=0$ is preserved by selecting a Hurwitz matrix A_s, and determining $\dot{U}$ such that

$$\dot{F}(U,x) = A_s F(U,x) \tag{3.8}$$

from which it is clear that $\dot{U}$ is obtained by

$$\dot{U} = (\nabla_U F)^{-1}(A_s F - \nabla_x F \dot{x}) \tag{3.9}$$

For numerical calculation of $\nabla_U F$ (which essentially consists of the hessian $\nabla_U^2 \mathcal{H}$ and gradient $\nabla_U h$), it is suggested that a set of forward difference equations can be efficiently solved using a particular linear equation solver (i.e., the generalized minimum residual method of Reference 87).

It is therefore proposed that $U(t)$, which defines the control according to $u=u_0^*(t)$, be propagated online by continuous integration of (3.9). The author describes this as a "continuation method," based on its similarity to numerical methods such as Reference 12, which track changes in the root $y(\sigma)$ of an expression $\tilde{F}(y(\sigma),\sigma)=0$ for variations in σ. However unlike true continuation methods, online implementation of (3.9) requires that the solution $U(t)$ be generated incrementally in strictly forward time, which makes it ambiguous in what sense the author claims that this approach is fundamentally different from using an optimization-based approach (assumed to start at a local minimum $F(U(0),x(0))=0$).

Presumably, a rigourous treatment of inequality constraints is prevented by the difficulty in dealing with the discontinuous behavior (2.5c) of the multiplier λ, induced by the active set. Even in the absence of inequality constraints, the optimal trajectory

$u^*_{[0,T]}$ characterized in (2.5) is not guaranteed to be continuous on $[0, T]$, and neither is the optimal solution $u^*_{[0,\infty)}$ of (2.6). Since the states $\{u_i^*(t)\}_{i=0}^N$ parameterizing $u^*_{[0,T]}$ flow continuously in time, and in particular the control is generated by the continuous flow $u(t) = u_0^*(t)$, this means that the ability of the closed-loop control to approximate optimal discontinuous behavior is limited by scaling considerations in (3.9). In particular, this will be the case when aggressive penalty functions are used to tightly approximate the inequalities.

A second point to note is that the proof of stability of this method makes the inherent assumption that $\frac{T}{N} \approx 0^+$. This is compounded by the inability to enforce a terminal inequality $x^p(T) \in \mathbb{X}_f$, which implies that instead T needs to be maintained large enough for $x^p(T) \in \mathbb{X}_f$ to be guaranteed implicitly. Therefore, the underlying dimension N of the calculations may need to be chosen quite large, placing a heavy burden on the calculations of $\nabla_U F$.

3.3.3 Continuous-time adaptation for $\mathcal{L}^2$-stabilized systems

Although it is not really a computational result, and is very limited in its applicability, we briefly review here the approach of References 29 and 30, which solves a constrained predictive control problem using an adaptive (i.e., incrementally updating) approach. The optimal control problem is assumed to be of the form[5]

$$J_Q(x(t), u^p_{[t,\infty)}) \triangleq \int_t^\infty \|e^p\|^2 + \lambda \|u^p\|^2 d\tau, \quad e^p \triangleq h(x^p) \tag{3.10a}$$

$$\text{s.t. } \forall \tau \geq t: \ \dot{x^p} = f(x^p) + g(x^p)u^p, \quad x^p(t) = x(t) \tag{3.10b}$$

$$u^p(\tau) \in \{u \,|\, \underline{u}_0 \leq u \leq \overline{u}_0 \quad \text{and} \quad \underline{u}_1 \leq \dot{u} \leq \overline{u}_1\} \tag{3.10c}$$

$$e^p(\tau) \in \{e \,|\, \underline{e}_0 \leq e \leq \overline{e}_0 \quad \text{and} \quad \underline{e}_1 \leq \dot{e} \leq \overline{e}_1\} \tag{3.10d}$$

The approach differs from most receding-horizon NMPC approaches in that, rather than using a finite-horizon approximation of the cost, it is assumed that a stabilizing feedback $k(x)$ is known, such that $u = k(x)$ generates trajectories whose cost J_Q is $\mathcal{L}^2$-integrable. It is further assumed that the system is passive, such that functions $V_u(t) \equiv V_u(x(t))$ and $V_e(t) \equiv V_e(x(t))$ can be found satisfying

$$V_u(t+\tau) \geq \varepsilon_u \int_t^{\tau - t} \|u^p(\sigma)\|^2 d\sigma \quad \dot{V}_u(t+\tau) \leq u^p(\tau)e^p(\tau) \tag{3.11}$$

$$V_e(t+\tau) \geq \varepsilon_e \int_t^{\tau - t} \|e^p(\sigma)\|^2 d\sigma \quad \dot{V}_e(t+\tau) \leq -e^p(\tau)k(x^p(\tau)) \tag{3.12}$$

[5] The problem presented here is modified substantially for clarity and consistency with the rest of the book, but can be shown equivalent to results in References 29 and 30.

for all $\tau \geq t$. The input is defined as

$$u^p(\tau) = k(x^p(\tau)) + \phi^T(\tau - t)c_\phi(\tau) \tag{3.13}$$

where ϕ represents a vector of exponential basis functions $\phi^T(\sigma) = \phi_0^T e^{A_\phi \sigma}$, with the corresponding (matrix-valued) weights c_ϕ being a state of the controller. This results in closed-loop dynamics of the form

$$\dot{z} \triangleq \begin{bmatrix} \dot{x} \\ \dot{c}_\phi \end{bmatrix} \triangleq \begin{bmatrix} f(x) + g(x)(k(x) + \phi_0^T c_\phi) \\ A_\phi c_\phi + v \end{bmatrix} \tag{3.14}$$

where v is an additional control input. For online calculation, the cost J_Q is replaced by the bound (where the necessary term $\gamma(\varepsilon_e, \varepsilon_u)$ is defined in Reference 29)

$$J(z(t)) \triangleq V_e(x(t)) + V_u(x(t)) + \gamma(\varepsilon_e, \varepsilon_u)c_\phi^T P_c c_\phi \geq J_Q(x(t)) \tag{3.15a}$$

$$P_c = P_c^T \;\text{ s.t. }: P_c A_\phi + A_\phi^T P_c = -\phi_0\, \phi_0^T \tag{3.15b}$$

which, in the absence of constraints, is a strictly decreasing Lyapunov function for (3.14) with $v \equiv 0$[6]. The expression for v used in Reference 29 is of the form

$$v = \text{Proj}\{-\nabla_{c_\phi} J, S_c(x)\} \tag{3.16}$$

where "Proj" denotes a parameter projection law (similar to such standard projections in Reference 99). This projection acts to keep c_ϕ in the set $S_c(x) \triangleq \{c_\phi \mid z \in S\}$, in which S denotes a control-invariant subset of the feasible region.

The expression $\nabla_{c_\phi} J$ is calculable in closed-form (due to the $\mathcal{L}^2$-nature of ϕ), and does not require online predictions. Since *a priori* knowledge of S is highly unlikely, the predictive aspect of the controller lies in the fact that one requires online prediction of the dynamics, "sufficiently far" into the future, to guarantee containment of z in S.

3.4 Robustness considerations

As can be seen in Proposition 2.5.2, the presence of inequality constraints on the state variables poses a challenge for numerical solution of the optimal control problem in (3.1). While locating the times $\{t_i\}$ at which the active set changes can itself be a burdensome task, a significantly more challenging task is trying to guarantee that the tangency condition $N(x(t_{i+1})) = 0$ is met, which involves determining if x lies on (or crosses over) the critical surface beyond which this condition fails.

As highlighted in Reference 70, this critical surface poses more than just a computational concern. Since both the cost function and the feedback $\kappa_{mpc}(x)$ are potentially discontinuous on this surface, there exists the potential for *arbitrarily small* disturbances (or other plant-model mismatch) to compromise closed-loop stability. This situation arises when the optimal solution $u^*_{[t,t+T]}$ in (3.1) switches between

[6] Of course, then the control is essentially just a dissipativity-based design, so it cannot really be classified as "predictive" is any sense.

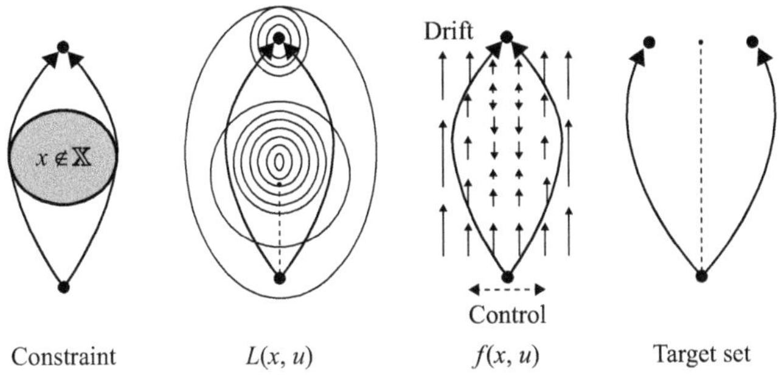

Figure 3.1 Examples of nonconvexities susceptible to measurement error

disconnected minimizers, potentially resulting in invariant limit cycles (e.g., as a very low-cost minimizer alternates between being judged feasible/infeasible.)

A modification suggested in Reference 70 to restore nominal robustness, similar to the idea in Reference 119, is to replace the constraint $x(\tau) \in \mathbb{X}$ of (3.1d) with one of the form $x(\tau) \in \mathbb{X}^o(\tau - t)$, where the function $\mathbb{X}^o : [0, T] \to \mathbb{X}$ satisfies $\mathbb{X}^o(0) = \mathbb{X}$, and the strict containment $\mathbb{X}^o(t_2) \subset \mathbb{X}^o(t_1)$, $t_1 < t_2$. The gradual relaxation of the constraint limit as future predictions move closer to current time provides a safety margin that helps to avoid constraint violation due to small disturbances.

The issue of robustness to measurement error is addressed in Reference 157. On one hand, nominal robustness to measurement noise of an MPC feedback was already established in Reference 69 for discrete-time systems, and in Reference 59 for SD implementations. However, Reference 157 demonstrates that as the sampling frequency becomes arbitrarily fast, the margin of this robustness may approach zero. This stems from the fact that the feedback $\kappa_{mpc}(x)$ of (3.1) is inherently discontinuous in x if the indicated minimization is performed globally on a nonconvex surface, which by References 42 and 77 enables a fast measurement dither to generate flow in any direction contained in the convex hull of the discontinuous closed-loop vector field. In other words, additional attractors or unstable/infeasible modes can be introduced into the closed-loop behavior by arbitrarily small measurement noise.

Although Reference 157 deals specifically with situations of obstacle avoidance or stabilization to a target set containing disconnected points, other examples of problematic nonconvexities are depicted in Figure 3.1. In each of the scenarios depicted in Figure 3.1, measurement dithering could conceivably induce flow along the dashed trajectories, thereby resulting in either constraint violation or convergence to an undesired equilibrium.

Two different techniques were suggested in Reference 157 for restoring robustness to the measurement error, both of which involve adding a hysteresis-type behavior in the optimization to prevent arbitrary switching of the solution between separate minimizers (i.e., making the optimization behavior more decisive).

Chapter 4
A real-time nonlinear MPC technique

4.1 Introduction

The rigourous theoretical underpinnings of NMPC evolved historically within the contexts of both continuous-time [34, 81, 124] and discrete-time [85, 112] systems (see review in Reference 126), despite the fact that the vast majority of applications involve continuous-time physical processes. Although more meaningful from a physical perspective, the implementation of a truly continuous-time framework is generally intractable due to the infinite dimensionality of the required search. While the justification for using discrete-time models is ostensibly to simplify the calculations, it generally comes at the expense of neglecting potentially important inter-sample behavior. As well, the success of discrete-time control designs hinge critically upon appropriate selection of the underlying discretization period to balance losses in model accuracy and performance versus computational complexity, a fact which tends to be neglected in much of the MPC literature.

The SD framework discussed in Section 3.2.2 has the advantage of explicitly considering the continuous-time evolution of the process. This allows for the consideration of important inter-sample behavior, as well as enabling the use of more accurate variable-step ODE solvers than are likely to be inherent in a discretized process model. In the most literal interpretation, imposing a SD framework upon the process does not itself simplify the calculation of the necessary optimal control problem; the search is still infinite-dimensional and "instantaneous," but performed less often. In practice however, computational tractability is achieved in two distinct manners: (1) by restricting the search for the optimal input to some finitely parameterized class of trajectories, locally supported by the sampling partition, and (2) distributing the calculations throughout the sampling interval by way of either the forward compensation [36, 57] or initial-value embedding [52, 54, 55] techniques described in Chapter 3.

The primary downside to any SD implementation (regardless of whether continuous-time or discrete-time predictions are used) is the ensuing interval of open-loop operation that occurs between sampling instants. Reducing the duration of this open-loop operation is the motivation behind the single-step computational approach of References 52, 54, and 55 discussed in Section 3.3.1. While the computational method itself is a relatively efficient SQP algorithm, the achievable sampling frequency is fundamentally limited by the fact that the dimension of the SQP grows with the sampling frequency, thereby increasing the necessary computation time.

As with any SD implementation, achieving a satisfactory balance between closed-loop performance and computational feasibility depends critically upon selection of the sampling period.

While the continuation-based approach of References 132 and 133 in Section 3.3.2 is algorithmically very different from the SD implementation described above, one can draw similar conclusions regarding its computational requirements. The algorithm provides a numerically efficient means of updating the parameterization of the prediction trajectories. However, the liberal use of simplistic first-order approximations throughout the model necessitates a very fine discretization of the prediction-time coordinate that imposes a limitation on how closely the algorithm can approach the assumed "continuous-time" implementation.

In contrast to *either* of the two approaches discussed above, stability of the adaptation-based approach in Reference 29 (Section 3.3.3) is not fundamentally tied to the dimension of the calculations, since the input parameterization does not depend upon the actual sampling rate of the "continuous-time" implementation. Not only is the approach quite limited in the class of systems that can be addressed, but more importantly, the parameterization of the input trajectory is not locally supported across the prediction-time coordinate. Therefore, the exponential basis used in the parameterization may require a very large dimension to provide an acceptable approximation of the nonsmooth behavior of the true optimal input trajectory, or to find feasible parameter values. Such nonsmooth, or even discontinuous, behavior of the trajectories is very typical when dealing with constrained systems.

In this chapter, we provide a framework for implementing real-time predictive control calculations. As in all of the above approaches, the objective is to maximize the rate of input–output sampling frequency such that the controller approaches a continuous-time state feedback. This fast sampling rate is again achieved by simplifying the parameter search to involve only incremental improvements, so that the parameter values evolve incrementally as additional controller states in a dynamic feedback law. In contrast to any of the previous methods, however, our results do not require any assumption on the dimension of the parameterization, other than to assume that an initial set of feasible parameter values exists. The advantage of this approach is that the role of the adaptation mechanism is entirely reduced to that of performance improvement, and thus its performance may be arbitrarily suboptimal in terms of both the update increment and the parameterization basis, while still preserving stability and feasibility. As such, the most important focus and contribution of the result is the manner in which the optimization is *posed* at each instant, rather than the specifics of how it is solved.

4.2 Problem statement and assumptions

We consider the general, continuous-time nonlinear system

$$\dot{x} = f(x, u). \tag{4.1}$$

The primary objective of the control design is assumed to be practical asymptotic regulation of the state variables $x \in \mathbb{R}^n$ to the origin ($x = 0$) using an arbitrary state feedback $u = \kappa_{mpc}(x, \cdot)$. The mapping $f : \mathbb{R}^n \times \mathbb{R}^m \to \mathbb{R}^n$ is assumed to be sufficiently smooth (at most C^{2+} required), and satisfies $f(0, 0) = 0$. The state and input trajectories are required to satisfy the pointwise-in-time constraints $x \in \mathbb{X}$ and $u \in \mathbb{U}$, with $\mathbb{X} \subseteq \mathbb{R}^n$ and $\mathbb{U} \subseteq \mathbb{R}^m$ assumed closed, convex sets such that $\mathring{\mathbb{X}}$ is nonempty.

A secondary objective of the control design is to achieve satisfactory closed-loop performance with respect to a given cost function $J_\infty(x_{[t_0,\infty)}, u_{[t_0,\infty)}) = \int_{t_0}^{\infty} L(x, u)d\tau$. Since online prediction of the cost-to-go $J_\infty(x^p_{[t,\infty)}, u^p_{[t,\infty)})$ is impractical (where x^p, u^p denote arbitrary predictions), J_∞ is approximated over a finite-horizon as

$$J_{rh}(x^p_{[t,t_f]}, u^p_{[t,t_f]}) = \int_t^{t_f} L(x^p, u^p)\, d\tau + W(x^p(t_f)) \tag{4.2}$$

where the horizon length $t_f - t$ is not necessarily fixed. The penalty $W : \mathbb{X}_f \to \mathbb{R}_{\geq 0}$ is assumed (strictly) positive definite and C^{1+} on its convex domain $\mathbb{X}_f \subset \mathring{\mathbb{X}}$, with $\mathring{\mathbb{X}}_f$ nonempty. The function $L : \mathbb{X} \times \mathbb{U} \to \mathbb{R}_{\geq 0}$ is assumed C^{1+}, and for convenience[1] we assume $\underline{\gamma}_L(\|x, u\|) \leq L(x, u) \leq \overline{\gamma}_L(\|x, u\|)$, for some $\underline{\gamma}_L, \overline{\gamma}_L \in \mathcal{K}$. Furthermore, to preclude certain pathological situations, we assume the constraint limits $\mathbb{X} \times \mathbb{U}$ and penalty $L(x, u)$ have been selected to satisfy the following assumption.

Assumption 4.2.1. *The selection of* $\mathbb{X}$, $\mathbb{U}$, *and* $L(x, u)$ *satisfy at least one of the following two conditions. Either*

1. $\mathbb{X}$ *is compact, or*
2. *each of the following hold:*
   i. $\underline{\gamma}_L, \overline{\gamma}_L \in \mathcal{K}_\infty$.
   ii. *the "extended velocity set"* $\{(v, \ell) \in \mathbb{R}^n \times \mathbb{R}_{\geq 0} \mid v = f(x, u),\ \ell \geq L(x, u),\ u \in \mathbb{U}\}$ *is convex for all* $x \in \mathbb{X}$.
   iii. *there exist constants* $c_1,\ c_2 > 0$ *such that*

$$\min_{u \in \mathbb{U}} \left(\frac{L(x, u)}{\|f(x, u)\|} \right) \geq \frac{c_2}{\|x\|} \quad \forall x \in \mathbb{X} \backslash B(0, c_1) \tag{4.3}$$

Assumption 4.2.1 provides a sufficient condition for establishing compactness of optimal closed-loop trajectories, as indicated in the following claim.

Claim 4.2.2. *Let* $J^*_\infty(x)$ *denote the infinite-time optimal cost-to-go function (i.e.,* $J^*_\infty(x) \triangleq \min_{u(\cdot)} \int_0^\infty L(x, u)\, d\tau$, *subject to indicated constraints). Then for every* $c > 0$, *the corresponding set* $\Omega(c) \triangleq \{x \in \mathbb{R}^n \mid J^*_\infty(x) \leq c\} \cap \mathbb{X}$ *is compact.*

Remark 4.2.3. *Assumptions* 4.2.1 2i *and* 2ii *are identical to the assumptions in Reference* 63, *and stem from standard sufficient conditions for global optimality such as discussed in Reference* 159. *Assumption* 4.2.1 2iii *is sufficient for radial*

[1] It is well known, see Reference 126, that more general $L(x, u) \geq 0$ can be used if the system satisfies an appropriate detectability condition with respect to L.

unboundedness of $J^*_\infty(x)$*, and is weaker than the assumed linear growth bound on* $f(x, u)$ *that is typical in optimal control sources such as* [60]. ■

As will be seen in Section 4.4, our approach is based upon minimizing J_{rh} with respect to the restricted class of trajectories $u^p_{[t,t_f]}$ which are PWC functions of the prediction-time $\tau \in [t, t_f]$. (However, this *does not* necessarily imply that the closed-loop implementation of $u = \kappa_{mpc}(x, \cdot)$ should result in PWC behavior as x evolves in *actual* time t.) As a result, the set of conditions in Reference 126 for $W(x)$ and $\mathbb{X}_f$ which are sufficient to guarantee closed-loop stability of a continuous-time NMPC must be modified slightly.

The conditions in Reference 126 essentially require the *existence* of locally stabilizing feedback $u = \kappa(x) \in \mathbb{U}$ with (invariant) domain of attraction $\mathbb{X}_f$, over which $W(x)$ is a CLF. Motivated by such works as References 37, 86, and 130, we will instead assume *knowledge* of a family of locally Lipschitz feedbacks $\kappa(x, T_\kappa)$, parameterized by $T_\kappa \in (0, M_T]$ for some constant $M_T > 0$. The stabilizer $\kappa(x, T_\kappa)$ is designed to guarantee practical-asymptotic stability of the origin when implemented in the sample-and-hold fashion $u(t) = \kappa(x(t_i), T_\kappa)$, $t \in [t_i, \ t_i + T_\kappa)$, a notion made more precise in Assumption 4.2.4 below. The period length T_κ is itself assumed generated by a feedback $T_\kappa = \delta(x)$, with $\delta : \mathbb{X}_f \to (0, M_T]$ a known locally Lipschitz function that is lower bounded by some class $\mathcal{K}$ function $\gamma_\delta(\|x\|)$. The resulting feedback $\kappa_\delta(x) \equiv \kappa(x, \delta(x))$ is therefore a local stabilizer under the sample-and-hold implementation $u_{[t_i,t_{i+1}]} \equiv \kappa_\delta(x(t_i))$, $t_{i+1} \triangleq t_i + \delta(t_i)$, with corresponding solutions of $\dot{x} = f(x, \kappa_\delta(x(t_i)))$ denoted $x^{\kappa_\delta}_{[t_i,t_{i+1}]}$.

In the analysis presented in this chapter, we will focus on stabilization of x to a target set Σ_x, which we assume to be a closed neighborhood of the origin of the form $\Sigma_x = \{x : W(x) \leq c_{\Sigma_x}\}$ for some $c_{\Sigma_x} \geq W(0) = 0$. Furthermore, we will make use of the following inner approximation of Σ_x:

$$\Sigma^\varepsilon_x \triangleq \{x : W(x) \leq (c_{\Sigma_x} - \varepsilon)\} \subset \Sigma_x \tag{4.4}$$

for any $\varepsilon > 0$. Note that $\varepsilon > c_{\Sigma_x}$ is allowed, with the implication then that $\Sigma^\varepsilon_x = \emptyset$. In order to ensure stability of the closed-loop algorithm to be presented in Section 4.4, the selection of the above feedbacks $\kappa(\cdot, \cdot)$ and $\delta(\cdot)$ are assumed to obey the following assumption.[2]

Assumption 4.2.4. *Let* $\Sigma_x \subset \mathbb{X}_f$ *denote the target set. Denoting* $t' \triangleq t + \delta(x)$*, and* $x' \triangleq x^{\kappa_\delta}(t')$*, the functions* $\kappa(\cdot, \cdot)$*,* $\delta(\cdot)$*,* $W(\cdot)$ *and the set* $\mathbb{X}_f$ *are chosen to satisfy:*

1. $\mathbb{X}_f \subset \mathring{\mathbb{X}}$*,* $\mathbb{X}_f$ *closed,* $0 \in \Sigma_x \subset \mathring{\mathbb{X}}_f$.
2. Σ_x *compact,* $\Sigma_x = \{x : W(x) \leq c_{\Sigma_x}\}$*,* $c_{\Sigma_x} \geq 0$.
3. $\kappa_\delta(x) \in \mathbb{U}$ *for all* $x \in \mathbb{X}_f$.
4. $x \in \mathring{\mathbb{X}}_f$ *implies* $x^{\kappa_\delta}_{[t,t']} \in \mathring{\mathbb{X}}_f$ *(point-wise).*
   Specifically, inner approximations $\mathbb{X}^\varepsilon_f \triangleq \overleftarrow{B}(\mathbb{X}_f, \varepsilon)$ *are invariant for* $\varepsilon \in [0, \varepsilon^*]$*, for some* ε^* *sufficiently small.*

[2] Representing a strengthening of the sufficient conditions presented previously in Section 3.1.

5. *For $\varepsilon > 0$ sufficiently small, $\exists\, \gamma(\cdot) \in \mathcal{K}$ such that for all $x \in \mathbb{X}_f \backslash \mathring{\Sigma}_x^\varepsilon$,*

$$W(x') - W(x) + \int_t^{t'} L(x^{\kappa_\delta}, \kappa_\delta(x^{\kappa_\delta}))\, d\tau \leq -\gamma(\|x\|) \tag{4.5}$$

■

By focussing on stabilization to a *neighborhood* Σ_x of the origin, we enable the use of design methods for the local stabilizer $\kappa_\delta(x)$ which achieve only practical stability of the origin. As discussed in References 37 and 130 (and references therein), sampled-and-hold feedbacks are able to (practically) stabilize a much larger class of systems than can be stabilized using continuous feedbacks. For our purposes however, the primary advantage is that from the results in Reference 37 it follows that one potential approach for constructing a practical-stabilizer $\kappa_\delta(x)$ is to simply pass a continuous-time stabilizing feedback through an appropriately short zero-order hold. This motivates the (potentially) variable period $T_\kappa = \delta(x)$, since, as observed in Reference 37, stabilizing arbitrarily small Σ_x may require arbitrarily fast switching near the origin. It is important to note, however, that although the expression $\kappa_\delta(x)$ will ultimately factor into our MPC controller design, this does not imply that the closed-loop behavior of the model predictive feedback will result in PWC-in-time behavior of $u(t)$.

4.3 Preliminary results

4.3.1 Incorporation of state constraints

As a means of enforcing the state constraints $x^p \in \mathbb{X}$ and $x^p(t_f) \in \mathbb{X}_f$, both active-set and interior-point approaches offer well-established alternatives (see References 121, 140, and 161 and references therein). The advantages of selecting an interior-point-based approach include that constraint satisfaction is guaranteed *at every instant* along the continuous-time trajectory, while the primary downside is that the constraint handling is inherently conservative, and may result in performance degradation (with respect to the optimal cost attainable by an active-set approach).

Since the approach in Section 4.4 will emphasize the advantages of using a crudely partitioned prediction horizon, the guarantee of feasibility within partition intervals motivates our preference for an interior-point approach over active sets. To this end, we assume sufficiently differentiable convex barrier functions $\mathcal{B}_x$ and $\mathcal{B}_f$ are defined on the respective domains $\mathring{\mathbb{X}}$ and $\mathring{\mathbb{X}}_f$, satisfying $\lim_{x \to \partial\mathbb{X}_{(f)}} \mathcal{B}_{x(f)} \to \infty$. Following Reference 161, we center the barriers about the origin by defining

$$\mathcal{B}_x^o(x) = \mathcal{B}_x(x) - \mathcal{B}_x(0) - \nabla\mathcal{B}_x(0)^T x. \tag{4.6}$$

(with analogous definition for $\mathcal{B}_f^o(x)$), and incorporate the state constraints into the controller design by augmenting the functions $L(\cdot, u)$ and $W(\cdot)$

$$L^a(x,u) \triangleq L(x,u) + \mu\mathcal{B}_x^o(x) \tag{4.7a}$$

$$W^a(x_f) \triangleq W(x_f) + \mu_f\mathcal{B}_f^o(x_f) \tag{4.7b}$$

As in Reference 161, the values μ, $\mu_f > 0$ are *constant* design parameters, chosen large enough to ensure robustness of the calculations. While this increases the conservatism of the constraint handling, it also simplifies the online controller calculations. To ensure stability, we will require the following assumption.

Assumption 4.3.1. *The barriers $\mathcal{B}_x(\cdot)$, $\mathcal{B}_f(\cdot)$, and weightings μ, μ_f are chosen to satisfy*

$$\mu_f\big(\mathcal{B}_f^o(x') - \mathcal{B}_f^o(x)\big) + \mu \int_t^{t'} \mathcal{B}_x^o(x^{\kappa_\delta}(\tau))\, d\tau \leq \gamma(\|x\|) \tag{4.8}$$

$\forall x \in \mathring{\mathbb{X}}_f \backslash \mathring{\Sigma}_x^{\varepsilon}$, *where x', t', and ε are from Assumption* 4.2.4. ■

Assumption 4.3.1 is admittedly somewhat restrictive on the selection of $\mathcal{B}_x$ and $\mathcal{B}_f$, although a design approach satisfying (4.8) for systems with stabilizable linearizations is given in Reference 172. Alternatively, we motivate the practicality of Assumption 4.3.1 by the following result.

Lemma 4.3.2. *Suppose that Assumption* 4.2.4 *holds with $\mathbb{X}_f$ compact, and $\mathring{\Sigma}_x \neq \emptyset$. Suppose further that there exists a C^{1+} function $V : \mathbb{X}_f \to \mathbb{R}_{\geq 0}$ and constants $0 \leq c_{f1} < c_{f2}$ such that every arbitrary level set $\mathbb{X}_f^{c_f} \triangleq \{x \in \mathbb{X}_f \,|\, V(x) \leq c_f, c_f \in [c_{f1}, c_{f2}]\}$ satisfying the strict containment*

$$\mathbb{X}_f^{c_{f1}} \subset \mathbb{X}_f^{c_f} \subset \mathbb{X}_f^{c_{f2}} \equiv \mathbb{X}_f \tag{4.9}$$

is positive invariant with respect to (4.1) *for $u = \kappa_\delta(x)$ on $t \in [0, \delta(x)]$. Define $\mathcal{B}_f \triangleq \tilde{\mathcal{B}}_f \circ V$, with $\tilde{\mathcal{B}}_f : [0, c_{f2}] \to \mathbb{R}_{\geq 0} \cup \{\infty\}$ a non-decreasing function satisfying $\lim_{s \to c_{f2}} \tilde{\mathcal{B}}_f \to \infty$. Then* (4.8) *is satisfied for sufficiently small μ, $\mu_f > 0$.* ■

If $\mathbb{X}_f$ is a compact level set of $W(x)$, then $V(x) \equiv W(x)$ satisfies Lemma 4.3.2.

4.3.2 Parameterization of the input trajectory

To reduce the optimization of (4.2) to a finite-dimensional nonlinear programming problem, the horizon $[t, t_f]$ is partitioned into N intervals defined by the (ordered) time-support $t^\theta \in \mathbb{R}^N$, with the trajectory $u^p_{[t,t_f]}$ generated by the parameterized mapping $v : \mathbb{R}_{\geq 0} \times \Theta \times \mathbb{R}^N \to \mathbb{R}^m$ defined as

$$u^p(\tau) = v(\tau, \theta, t^\theta) \triangleq \begin{cases} \theta_1, & \tau \in [t,\, t_1^\theta] \\ \theta_i, & \tau \in (t_{i-1}^\theta,\, t_i^\theta], \quad i = \{2 \cdots N\}. \\ 0, & \text{otherwise} \end{cases} \tag{4.10}$$

Definition 4.10 is clearly equivalent to a PWC (a.k.a. zero-order hold) parameterization, where each parameter vector $\theta_i \in \mathbb{R}^m$ defines u^p over the *i'th* interval of t^θ, and $\theta \in \mathbb{R}^{mN}$ contains the $\{\theta_i\}$. However, we choose to refer to the θ_i as "parameters" (rather than "control values") to highlight the fact that selection of a PWC basis for defining $v(\tau, \theta, t^\theta)$ is an entirely arbitrary choice which can be easily generalized, as will be seen in Chapter 5.

Definition 4.3.3. *A* control parameterization *will refer to the pair* $(\theta, t^\theta) \in \mathbb{R}^{mN} \times \mathbb{R}^N$ *which defines the trajectory* $u^p_{[t,t_f]}$ *according to* (4.10). ■

Let $(t_0, x_0) \in \mathbb{R} \times \mathring{\mathbb{X}}$ be an arbitrary initial condition for system (4.1), and let (θ, t^θ) be an arbitrary control parameterization satisfying $t_0 \leq t_1^\theta$. We denote the resulting solution (in the classical sense) to the prediction model (4.1) and (4.10), along the time-coordinate $\tau \in [t_0, t_N^\theta]$ from initial condition $x^p(t_0) = x_0$, by $x^p(\tau, t_0, x_0, \theta, t^\theta)$ and $u^p(\tau, t_0, x_0, \theta, t^\theta)$. At times we will suppress this notation to $x^p(\tau)$, $u^p(\tau)$, when the omitted arguments are considered obvious.

Definition 4.3.4. *A control parameterization* (θ, t^θ) *will be declared* feasible *(with respect to* (t_0, x_0)*), if on the interval* $\tau \in [t_0, t_N^\theta]$*, the solution* $x^p(\tau; t_0, x_0, \theta, t^\theta)$*,* $u^p(\tau; t_0, x_0, \theta, t^\theta)$ *exists and satisfies* $x^p(\tau) \in \mathring{\mathbb{X}}$*,* $u^p(\tau) \in \mathbb{U}$*, and* $x^p(t_N^\theta) \in \mathring{\mathbb{X}}_f$*.* ■

Definition 4.3.5. *For given* (t_0, x_0) *and given* N*, we denote by* $\Phi^N(t_0, x_0) \subseteq \mathbb{R}^{mN} \times \mathbb{R}^N$ *the set of all feasible control parameterizations.* ■

Throughout this chapter, we interpret the statement $\theta \in \mathbb{U}^N$ to imply $\theta_i \in \mathbb{U}$, $\forall i \in \{1, \ldots, N\}$. The following property is important in that it ensures the use of (4.10) does not unduly restrict the stabilizable region, and can essentially be viewed as a corollary of various results from Reference 37, or as a slightly less-restrictive restatement of Reference 154, Proposition 3.7.2.

Corollary 4.3.6. *Let* $\mathcal{X}^0 \subseteq \mathring{\mathbb{X}}$ *denote the set of initial states* x_0 *for which there exists continuous open-loop trajectories* $x_{[t_0,t_f]}$*,* $u_{[t_0,t_f]}$ *solving* (4.1)*, and satisfying the constraints* $x(t) \in \mathring{\mathbb{X}}$*,* $u(t) \in \mathbb{U}$ *and* $x(t_f) \in \mathring{\mathbb{X}}_f$ *for some* $t_f \geq t_0$*. Then, for every* $(t_0, x_0) \in \mathbb{R} \times \mathcal{X}^0$*, there exists* $N^*(x_0)$ *such that* $\Phi^N(t_0, x_0)$ *has positive Lebesgue measure in* $\mathbb{U}^N \times \mathbb{R}^N$ *for all* $N \geq N^*(x_0)$*.* ■

Corollary 4.3.6 ensures that by selecting N sufficiently large, a feasible initial parameterization $(\theta, t^\theta) \in \Phi^N(t_0, x_0)$ is locatable in finite-time (FT).

4.4 General framework for real-time MPC

4.4.1 Description of algorithm

Below we outline the steps involved in calculating our MPC controller, beginning from an arbitrary initial condition $(t_0, x_0) \in \mathbb{R} \times \mathcal{X}^0$. For convenience of notation, we define $z \triangleq [x^T, \theta^T, {t^\theta}^T]^T$, the vector of closed-loop states. To begin, with ε the same as in Assumption 4.2.4 point 5, we define the following smoothed version of an indicator function for (the complement to) Σ_x:

$$\rho(x) \triangleq \begin{cases} 1 & x \notin \Sigma_x \\ \rho_0(W(x)) & x \in \Sigma_x \backslash \Sigma_x^\varepsilon \\ 0 & x \in \Sigma_x^\varepsilon \end{cases} \tag{4.11}$$

where $\rho_0 : [c_{\Sigma_x} - \varepsilon,\ c_{\Sigma_x}] \to [0,\ 1]$ can be any monotone C^{1+} function satisfying $\lim_{W \uparrow c_{\Sigma_x}} \rho_0 = 1$, $\lim_{W \uparrow c_{\Sigma_x}} \frac{d\rho_0}{dW} = 0$, $\lim_{W \downarrow (c_{\Sigma_x} - \varepsilon)} \rho_0 = 0$, and $\lim_{W \downarrow (c_{\Sigma_x} - \varepsilon)} \frac{d\rho_0}{dW} = 0$. The function $\rho : \mathbb{X} \to [0,\ 1]$ is therefore C^{1+} over $\mathbb{R}^n$ (and likewise could be made C^{2+} if necessary). We then use the following modified version of (4.2) as the cost function:

$$J(t,z) \equiv J(t,x,\theta,t^\theta) \triangleq \int_t^{t_N^\theta} L_\rho^a(x^p(\tau,t,z),u^p(\tau,t,z))\,d\tau + W_\rho^a(x^p(t_N^\theta,t,z)) \tag{4.12a}$$

$$L_\rho^a(x,u) \triangleq \rho(x)\left(L(x,u) + \mu \mathcal{B}_x^o(x)\right) \tag{4.12b}$$

$$W_\rho^a(x) \triangleq \rho(x)\left(W(x) + \mu_f \mathcal{B}_f^o(x)\right) \tag{4.12c}$$

Since $W(\cdot)$ is not necessarily a CLF inside Σ_x^ε, replacing $(L^a,\ W^a)$ with $(L_\rho^a,\ W_\rho^a)$ prevents the minimization of J from compromising the forward invariance of Σ_x^ε achieved under $\kappa_\delta(x)$.

The steps of the algorithm are as follows:

Step 1: Initialization of control parameterization

If $\mathcal{X}^0 \subseteq \mathbb{X}_f$, then $(\theta,\ t^\theta)_0$ could be generated by simulation of (4.1) under the local stabilizer $\kappa_\delta(x)$. In general however, a dual programming problem [23] may need to be solved to identify an acceptable $(\theta,\ t^\theta)_0 \in \Phi^N(t_0, x_0)$, with finiteness of the search guaranteed by Corollary 4.3.6. Assuming knowledge of a feasible initial control parameterization is a common starting point for many numerical NMPC approaches, so we omit further details in this regard.

Step 2: Continuous flow under dynamic feedback

At any instant $t \in [t_0,\ t_1^\theta]$, we assume that the current model prediction $x^p_{[t,t_N^\theta]}(\cdot,t,z)$, $u^p_{[t,t_N^\theta]}(\cdot,t,z)$ and predicted cost (4.12) are "instantaneously" available (this assumption is discussed further in Section 4.5.2). Unlike SD approaches, the predictions are continuously updated using continuous online measurement[3] of the current $x(t)$. The closed-loop dynamics (with respect to *ordinary* time t) evolve over the interval $t \in [t_0,\ t_1^\theta]$ as

$$\dot{z} = \begin{bmatrix} \dot{x} \\ \dot{\theta} \\ \dot{t}^\theta \end{bmatrix} = \begin{bmatrix} f(x,\ \theta_1) \\ \Psi(t,z) \\ 0 \end{bmatrix} \tag{4.13}$$

in which θ evolves as a dynamic controller state.

[3] Assumed to be available. In practice, this means that measurements are sampled in a faster timescale than the dynamics of the closed-loop process.

Criterion 4.4.1. *The update law $\Psi(t,z)$ must be chosen to ensure the following:*

1. $\langle \nabla_\theta J,\ \Psi(t,z)\rangle \leq 0$
2. $(\theta(t),\ t^\theta) \in \Phi^N(t,x(t))$, $\forall t \in [t_0,\ t_1^\theta]$, *where* $\theta_{[t_0,t_1^\theta]}$ *and* $x_{[t_0,t_1^\theta]}$ *are prescribed by* (4.13).
3. $\Psi(t,z)$ *is continuous in t and locally Lipschitz in z, for all* $(\theta,\ t^\theta) \in \Phi^N(t,x)$. ■

The term $\nabla_\theta J$ is the gradient of (4.12) with respect to θ, the calculation of which is discussed in Section 4.6. Examples of update laws satisfying Criterion 4.4.1 are discussed in Section 4.5.

Step 3: Discrete transitions at switching times
Upon equality $t = t_1^\theta$, the control parameterization is updated by the "instantaneous" reset:

$$z^+ = \begin{bmatrix} x^+ \\ \theta^+ \\ \\ t^{\theta+} \end{bmatrix} = \begin{bmatrix} x \\ \Upsilon(t,z) \\ \begin{cases} t_{i+1}^\theta & i = \{1,\ldots,N-1\} \\ t_N^\theta + \delta(x^p(t_N^\theta)) & i = N \end{cases} \end{bmatrix} \tag{4.14}$$

where z and z^+ denote pre- and post-reset values, both at time t.

Criterion 4.4.2. *The jump mapping $\Upsilon(t,z)$ is chosen to ensure the following:*

1. $J(t,z^+) - J(t,z) \leq 0$
2. $(\theta^+,\ t^{\theta+}) \in \Phi^N(t,x)$ ■

Step 4: Iteration of Steps 2 and 3
Criterion 4.4.2 point 2 ensures that the new control parameterization is feasible for the current state (t,x). Thus the procedure can iterate back to Step 2.

It is worth noting that the terminal prediction time $t_f \equiv t_N^\theta$ recedes in a discrete fashion during the reset (4.14), while remaining fixed during the flow (4.13). The possibility of assigning more general dynamics $\dot{t}^\theta = \Psi_t(t,z)$ and $t^{\theta+} = \Upsilon_t(t,z)$ in (4.13)–(4.14) is discussed in Chapter 5.

4.4.2 A notion of closed-loop "solutions"

The feedback $u = \kappa_{mpc}(t,z)$ resulting from implementing the above algorithm is a dynamic, time-varying control law which is technically set-valued at the switching instances of Step 3. The notion of a "solution" to (4.1) is therefore unclear, since neither classical nor "sample-and-hold" [37] notions of solution apply, while Filippov solutions are much too unwieldy. To facilitate analysis, we pose our dynamics as a hybrid system, thus adopting a notion of solutions from Reference 67 in which trajectories simultaneously evolve over an orthogonal pair (t, k) of continuous- and discrete-time coordinates,[4] jointly referred to as "hybrid time."

By augmenting z to include time as an additional state (i.e., $z_a \triangleq [z^T, \pi]^T$, $\pi_0 = t_0$, $\dot{\pi} = 1$, $\pi^+ = \pi$), then (4.13) has the form of a continuous flow $\dot{z}_a = \Psi_z(z_a)$ on the "flow domain"

$$S_\Psi \triangleq \left\{z_a \mid \pi \le t_1^\theta \text{ and } (\theta, t^\theta) \in \Phi^N(\pi, x)\right\} \subset \mathbb{R}^{n+mN+N+1}. \tag{4.15a}$$

Similarly, (4.14) has the form $z_a^+ = \Upsilon_z(z_a)$, defined on the "jump domain"

$$S_\Upsilon \triangleq \left\{z_a \mid \pi \ge t_1^\theta \text{ and } (\theta, t^\theta) \in \Phi^N(\pi, x)\right\} \subset \mathbb{R}^{n+mN+N+1}. \tag{4.15b}$$

The dynamics of z_a therefore define a hybrid system of the exact form discussed in References 67 and 111, where (4.15) follows the notations of Reference 67. Since $z_a(t, k)$ and $z_a(t, k+1)$ occur at different instances in hybrid time, the feedback $u(t, k) = \kappa_{mcp}(z_a(t, k))$ is no longer set-valued, and so meaningful guarantees of existence and uniqueness of (hybrid-time) solutions to (4.13), (4.14) follow from Reference 111, Lemma III. 1, 2. This enables the stability analysis of the main result below to be based upon direct application of an invariance principle developed for hybrid systems in Reference 144, which can be shown to be nearly equivalent to that in Reference 111.

Remark 4.4.3. *To be completely correct, an invariance principle cannot be applied to the hybrid system in* (4.13), (4.14) *since the controller state t^θ grows without bound, thus violating the assumption of pre-compactness inherent in any invariance principle. While one could get around this by defining a state transformation in which t^θ is replaced with coordinates defined relative to current time t, this would render the interpretation of many expressions in the analysis (e.g., terms appearing in $\frac{dJ}{dt}$) rather non-intuitive. A simpler approach is to recognize that the resulting feedback $u(t) = \kappa_{mpc}(x, t, z)$ is unchanged if all occurrences of time t in* (4.12) *are replaced with "time since reset" π (where $\dot{\pi} = 1$, $\pi^+ = 0$), and similarly the jump-map $(t^\theta)^+$ is modified by subtracting t_1^θ from all entries (i.e., thereby redefining t^θ as being relative to the time of last reset). Since the boundedness of $\delta(x)$ ensures that a reset must occur at least every M_T time-units, this ensures boundedness of the (re-defined) controller state t^θ. For simplicity, we neglect this technicality in the remainder.*

4.4.3 Main result

We are now ready to present the main result of this chapter. While the theorem itself may appear to be a straightforward consequence of Criteria 4.4.1 and 4.4.2, the usefulness of its generality will become apparent in Section 4.5 when we provide specific examples of $\Psi(t, z)$ and $\Upsilon(t, z)$ which meet these criteria.

[4] In other words, the notation $z(t', k')$ denotes the value of state z after exactly t' units of ordinary time and k' event executions have occurred. If time t' is not in the interval between the switching times of the k' and $(k'+1)$ events, then by the convention of Reference 60 the state $z(t', k')$ is undefined.

Theorem 4.4.4. *Let $L(\cdot,\cdot)$, $\kappa_\delta(\cdot)$, $W(\cdot)$, $\mathbb{X}$, $\mathbb{X}_f$ be chosen to satisfy Assumptions* 4.2.1 *and* 4.2.4 *for given Σ_x, and let $\mathcal{B}_x$, $\mathcal{B}_f$, μ, μ_f satisfy Assumption* 4.3.1. *For any initial conditions $(t_0, x_0) \in \mathbb{R} \times \mathcal{X}^0$ of* (4.1), *and any initial feasible control parameterization $(\theta, t^\theta)_0 \in \Phi^N(t_0, x_0)$, the set $x \in \Sigma_x$ is asymptotically stabilized under* (4.13) *and* (4.14). *Furthermore, the resulting closed-loop trajectories satisfy all point-wise input, state, and terminal constraints.* ■

4.5 Flow and jump mappings

As shown in Theorem 4.4.4, asymptotic convergence to $x \in \Sigma_x$ is guaranteed as long as $\Psi(t,z)$ and $\Upsilon(t,z)$ do not result in increase of $J(t,z)$. In this section, we look at how these mappings can be designed to satisfy Criteria 4.4.1 and 4.4.2, and how either Ψ or Υ can be used to improve the control parameterization.

4.5.1 Improvement by Υ: the SD approach

The most defining characteristic of the hybrid-time framework presented here is the time-varying nature of the trajectory $\theta_{[t_0, t_1^\theta]}$ during Step 2 of Section 4.4.1. However, if we make the following choices for $\Psi(\cdot,\cdot)$ and $\Upsilon(\cdot,\cdot)$

$$\Psi(t,z) \equiv 0 \tag{4.16a}$$

$$\Upsilon(t,z) = \arg\min_{\theta^+ \in \mathbb{U}^N} J(t, x, \theta^+, t^{\theta+}) \tag{4.16b}$$

(where $t^{\theta+}$ is given by (4.14)), then the closed-loop control action $u = \kappa_{mpc}(z)$ exhibits PWC behavior with respect to *actual* time t. In this case, our framework is equivalent to the SD result in Reference 115 (modulo our interior-point approach to constraint handling). It can be observed that the practice in Reference 115 of using a shorter control horizon $N_c < N$ can simply be viewed as imposing upon the optimization in (4.16b) the constraints $\theta_i^+ = \kappa(x^p(t_{i-1}^\theta),\ t_i^\theta - t_{i-1}^\theta)$ for $i > N_c$.

Satisfaction of Criterion 4.4.1 is trivial, as is Criterion 4.4.2 when $x \in \Sigma_x^\varepsilon$. For $x \notin \Sigma_x^\varepsilon$,

$$\begin{aligned}
J(t,z^+) - J(t,z) &= \int_t^{t_N^{\theta+}} L_\rho^a(x^p(\tau,t,z^+), u^p(\tau,t,z^+))\,d\tau + W_\rho^a(x^p(t_N^{\theta^+}, t, z^+)) \\
&\quad - \int_t^{t_N^\theta} L_\rho^a(x^p(\tau,t,z), u^p(\tau,t,z))\,d\tau - W_\rho^a(x^p(t_N^\theta, t, z)) \\
&\le \int_{t_N^\theta}^{t_N^{\theta+}} L_\rho^a(x^p(\tau,t_N^\theta,\bar{z}), u^p(\tau,t_N^\theta,\bar{z}))\,d\tau + W_\rho^a(x^p(t_N^{\theta^+}, t_N^\theta, \bar{z})) - W_\rho^a(x_f^p) \\
&\le \rho(x_f^p)\left[\int_{t_N^\theta}^{t_N^{\theta+}} L^a(x^p(\tau,t_N^\theta,\bar{z}), u^p(\tau,t_N^\theta,\bar{z}))\,d\tau + W^a(x^p(t_N^{\theta^+}, t_N^\theta, \bar{z})) - W^a(x_f^p)\right] \\
&\le 0
\end{aligned} \tag{4.17}$$

where $x_f^p \triangleq x^p(t_N^\theta, t, z)$, and $\bar{z} \triangleq [x_f^p, \theta^+, t^{\theta+}]$. The inequalities hold by virtue of the (suboptimal) choice $\theta_i^+ = \theta_{i+1}, i < N, \theta_N^+ = \kappa_\delta(x_f^p)$, from which both Criterion 4.4.2 point 1 and 2 follow (since violation of Criterion 4.4.2 point 2 would necessitate

$J \to \infty$). It is then clear that given (4.16), Theorem 4.4.4 holds and therefore encompasses most stable results from the SD-MPC literature that are based upon PWC control parameterizations, such as Reference 115.

The chief downside to (4.16) is the required assumption that the optimal control problem posed in (4.16b) is solved "instantaneously," an assumption that is inherent in much of the SD literature. This assumption could be overcome using forward compensation (i.e., by imposing on (4.16b) the constraint $\theta_1^+ = \theta_2$, which makes θ_1^+ available for immediate feedback from knowledge of the previous solution), but at the cost of adding a one-sample delay in the feedback path. In either case, the fact that the feedback path in (4.16) is only closed at the switching instants of t^θ implies that robustness considerations impose an upper bound M_T on the allowable switching period $T = \delta(x)$ used in the control parameterization (thereby requiring N to be large enough to ensure sufficient horizon length to satisfy the terminal constraint).

4.5.2 Improvement by Ψ: a real-time approach

As discussed in Section 4.1, our interest lies in establishing that stability can be preserved when the minimization in (4.16b) is replaced by a gradient-based search involving the simplest possible calculation, allowed to evolve throughout the interval $t \in [t_0, t_1^\theta]$. The motivating benefit of this approach would be the nearly continuous closing of the feedback path, effectively eliminating intervals of open-loop operation and thereby improving disturbance attenuation. Within our framework this implies that the burden of cost improvement is carried by Ψ, allowing Υ to be simplified to an expression more realistically "instantaneous" than (4.16b). The simplest such definition is of the form

$$\Upsilon(t,z) := \theta_i^+ = \begin{cases} \theta_{i+1} & i \in \{1, \ldots, N-1\} \\ \kappa_\delta\big(x^p(t_N^\theta, t, z)\big) & i = N \end{cases} \tag{4.18a}$$

$$\Psi(t,z) = \mathrm{Proj}\big\{\vartheta(t,z),\, \Gamma(t,z),\, \theta,\, \mathbb{U}^N\big\} \tag{4.18b}$$

$$\vartheta(t,z) = -k_\theta \Gamma(t,z) \nabla_\theta J \tag{4.18c}$$

where $\Gamma : \mathbb{R} \times \mathbb{X} \times \Phi^N \to \mathbb{R}^{mN \times mN}$ is a locally Lipschitz matrix-valued function satisfying $\Gamma(t,z) = \Gamma^T(t,z) > 0$, which defines the metric by which the descent direction is selected. The gain $k_\theta > 0$ describes the rate of descent in the chosen direction.

The $\mathrm{Proj}\{\cdot,\cdot,\cdot,\cdot\}$ operator is a locally Lipschitz parameter projection, identical to those discussed in the nonlinear adaptive control literature such as Reference 99, which is designed to ensure $\theta \in \mathbb{U}^N$. For constraints $\mathring{\mathbb{U}} \neq \emptyset$ with smooth boundary $\partial\mathbb{U}$, one such appropriate definition is

$$\mathrm{Proj}\big\{\vartheta, \Gamma, \theta, \mathbb{U}^N\big\} \triangleq$$

$$\begin{cases} \vartheta & \theta \in \mathring{\mathbb{U}}_r^N \quad \text{or} \quad \mathrm{v}_\perp^T \vartheta \le 0 \\ \left(I - \mathrm{sat}_0^1\left(\dfrac{r - \epsilon(\theta)}{r}\right) \Gamma \dfrac{\mathrm{v}_\perp \mathrm{v}_\perp^T}{\mathrm{v}_\perp^T \Gamma \mathrm{v}_\perp}\right) \vartheta & \theta \in \mathbb{U}^N \backslash \mathring{\mathbb{U}}_r^N \quad \text{and} \quad \mathrm{v}_\perp^T \vartheta > 0 \end{cases} \tag{4.19}$$

where $\mathbb{U}_\epsilon^N$, $\epsilon \in [0, r]$, denotes a family of closed *inner* approximations to $\mathbb{U}^N$ satisfying the strict containments $\mathbb{U}_\epsilon^N \subset \mathbb{U}_{\epsilon'}^N$, $\epsilon > \epsilon'$, and where $\partial\mathbb{U}_\epsilon^N$ continuously approaches $\partial\mathbb{U}^N$ as $\epsilon \downarrow 0$. The vector $\text{v}_\perp \equiv \text{v}_\perp(\theta)$ is the outward normal vector to the specific level set $\mathbb{U}_{\epsilon(\theta)}^N$ at the point $\theta \in \partial\mathbb{U}_{\epsilon(\theta)}^N$. The term $\text{sat}_0^1(\cdot)$ implies saturation with respect to the interval [0, 1].

Satisfaction of Criterion 4.4.2 by (4.18a) follows from (4.17). Satisfaction of Criterion 4.4.1 point 1 by (4.18)–(4.19) is obvious for the first case of (4.19), while in the second case

$$\begin{aligned}\langle \nabla_\theta J,\, \Psi(t,z)\rangle &= (\nabla_\theta J)^T \left(I - \text{sat}_0^1\left(\frac{r-\epsilon(\theta)}{r}\right)\Gamma\frac{\text{v}_\perp \text{v}_\perp^T}{\text{v}_\perp^T\Gamma\text{v}_\perp}\right)\vartheta \\ &\leq -\frac{1}{k_\theta}\,\vartheta^T\left(\Gamma^{-1} - \frac{\text{v}_\perp \text{v}_\perp^T}{\text{v}_\perp^T\Gamma\text{v}_\perp}\right)\vartheta \leq 0 \end{aligned} \tag{4.20}$$

By Reference 99, Lemma E.1 (4.19) is a locally Lipschitz operator, and guarantees that $\theta(t) \in \mathbb{U}^N$ for all $t \geq 0$. The remaining requirements of Criterion 4.4.1, that $x^p(\tau, t, z) \in \mathbb{X}$ for all $\tau \in [t, t_N^\theta]$ and $x^p(t_N^\theta, t, z) \in \mathbb{X}_f$, follow from the fact that Criterion 4.4.1 point 1 ensures $\dot{J} < 0$ (see Theorem 4.4.4 proof), and that $J(t,z) \to \infty$ continuously as $x^p(t_f)$ approaches $\partial\mathbb{X}_f$ or any point $x^p(\tau)$ approaches $\partial\mathbb{X}$.

Remark 4.5.1. *The operator* (4.19) *is essentially an active-set for the constraint* $\theta \in \mathbb{U}^N$. *While this constraint could alternatively be enforced by a barrier function (as recommended by Reference* 161*), in many practical situations the computation of* (4.19) *is acceptably simple, and less conservative. In particular, for hypercubic* $\mathbb{U}$ *the projection* (4.19) *can be implemented as element-wise saturation within the (eventually discretized) implementation of the update of* θ *in* (4.13).

Remark 4.5.2. *While* (4.18b) *will admittedly require at least some computation time, our purpose here is to demonstrate the limiting behavior of using a single-step, gradient-based solution method for minimizing* J. *Since the complexity of computing* any *gradient-based update (with* (4.18b) *being one example) obviously scales strongly with dimension* mN *of* θ *(in a manner depending on* Γ*), it is therefore critical to recognize that the* only *restriction on* N *in Theorem* 4.4.4 *is that a feasible initialization* $(\theta, t^\theta)_0$ *exists. Beyond this, the partition of the prediction horizon by* t^θ *may be arbitrarily coarse. "Instantaneous" computability of an incremental step for* (4.18b), *an approximation shared with other works such as References* 132 *and* 133, *is therefore justified here more on the basis of achievable dimensional reduction of the calculations than by any requirement of efficiency of the calculations themselves.*

Remark 4.5.3. *While a coarse partition* t^θ *would obviously result in significant loss of performance with respect to* (4.12), *it will be shown in Chapter* 5 *that by changing the underlying basis of parameterization* (4.10), *this performance degradation can be substantially reduced.*

4.5.3 Other possible definitions for Ψ and Υ

The above two sections present perspectives in which the burden of performance improvement is transferred entirely to either Ψ or to Υ. Other intermediate options include:

- Defining Υ to involve any type of (crude, quasi-) global search, most likely making use of forward compensation, which can be used to overcome the local nature of Ψ by occasional resetting of the local search.
- Defining Υ (together with an appropriately-defined Υ_t) as a mechanism for further partitioning the intervals of t^θ during a reset, in order to increase the "resolution" of the parameterization defining $u^p(\tau)$ as the prediction time τ nears t (i.e., such that $u^p_{[t,t^\theta_N]}$ is always parameterized more coarsely for distant parts of the prediction). This essentially generalizes the idea of using different control and prediction horizons advocated in Reference 115, which can be viewed as a lumping of the parameters $\{\theta_i\}$ for $i \in \{N_c, \ldots, N\}$.

4.6 Computing the real-time update law

4.6.1 Calculating gradients

In addition to solving the model predictions $x^p(\tau)$ and $u^p(\tau)$, incremental propagation of (4.18b) requires calculation of the gradient vector $\nabla_\theta J$. By definition, $\nabla_\theta J$ is given by

$$\nabla_\theta J = \int_t^{t^\theta_N} \left(\frac{\partial L^a_\rho}{\partial x}\frac{\partial x^p}{\partial \theta} + \frac{\partial L^a_\rho}{\partial u}\frac{\partial u^p}{\partial \theta} d\tau \right) + \frac{dW^{a\,T}_\rho}{dx}\frac{\partial x^p}{\partial \theta}(t^\theta_N) \tag{4.21}$$

where $\frac{\partial L^a}{\partial x}$, $\frac{\partial L^a}{\partial u}$, and $\frac{dW^a}{dx}$ are evaluated along the prediction trajectory $x^p(\tau)$, $u^p(\tau)$. If the dimensions satisfy $n < m(N-1)$, then instead of propagating the full state sensitivity matrix $\frac{\partial x^p}{\partial \theta_i}$ along the prediction horizon, the computation can be more efficiently decomposed by propagating within each interval $\tau \in [t^\theta_{i-1}, t^\theta_i]$ (where $t^\theta_0 \equiv t$) the sensitivities $S_u \in \mathbb{R}^{n\times m}$, $S_x \in \mathbb{R}^{n\times n}$

$$\dot{S}_u = \frac{\partial f}{\partial x}S_u + \frac{\partial f}{\partial u} \quad S_u(t^\theta_{i-1}) = 0 \tag{4.22a}$$

$$\dot{S}_x = \frac{\partial f}{\partial x}S_x \quad S_x(t^\theta_{i-1}) = I \tag{4.22b}$$

where the expressions for $\frac{\partial f}{\partial x}$ and $\frac{\partial f}{\partial u}$ are evaluated along the arguments $x^p_{[t^\theta_{i-1},t^\theta_i]}$ and $u^p_{[t^\theta_{i-1},t^\theta_i]}$. Then for each θ_i, the sensitivity $\frac{\partial x^p}{\partial \theta_i}(\tau)$ is calculated over $\tau \in [t, t^\theta_N]$ as

$$\frac{\partial x^p}{\partial \theta_i}(\tau) \triangleq \begin{cases} 0 & \tau < t^\theta_{i-1} \\ S_u(\tau) & \tau \in [t^\theta_{i-1}, t^\theta_i) \\ S_x(\tau)\frac{\partial x^p}{\partial \theta_i}(t^\theta_{j-1}) & \tau \in [t^\theta_{j-1}, t^\theta_j], \quad j > i \end{cases} \tag{4.23}$$

from which it can be seen that the drift dynamics corresponding to all previous intervals are tracked as linear combinations of S_x (since the $\frac{\partial x^p}{\partial \theta_i}(t^\theta_{j-1})$ terms are effectively known constants). Calculating the integral portion of (4.21) can be similarly decomposed, which is equivalent to augmenting the definition of the state x^p in (4.22)–(4.23) to include a scalar accumulator for $L^a_\rho(x, u)$.

Numerical solution of (4.22) is a well-studied problem, for which numerous efficient techniques exist ([105, 106] and references therein). While solving (4.22)–(4.23) may appear challenging, we emphasize that calculating $\nabla_\theta J$ for (4.18c) is comparable to performing a single gradient evaluation within an iterative gradient-based solver (such as an SQP) applied to (4.16b).

4.6.2 Selecting the descent metric

The matrix $\Gamma(t, z)$ in (4.18) defines the metric used for selecting the descent direction (i.e., viewing (4.18c) as a Newton-like update, $\Gamma = \Gamma^T > 0$ approximates the inverted hessian). Depending on computational resources available, this approximation could be selected from any number of standard forms, some examples being:

4.6.2.1 (Scaled) Steepest descent

Setting $\Gamma(t, z) \equiv I$ generates steepest-descent behavior, where the diagonal elements could be weighted (specified offline) to help improve scaling.

4.6.2.2 Approximate second-order hessian

Computing the full hessian $\nabla_{\theta^2} J$ is almost always impractical, in particular since the second-order sensitivities cannot be decomposed as efficiently as (4.23) due to cross terms. However, neglecting interaction terms between intervals (i.e., $\nabla^2_{\theta_i\theta_j} J,\ i \neq j$) allows for the block-diagonal approximation $\Gamma(t, z) = diag\{\Gamma_i\}$, $i = \{1 \ldots N\}$, with

$$\Gamma_i \triangleq \left[\nabla^2_{\theta_i^2} J + \varepsilon_i(t, z) I\right]^{-1} \tag{4.24}$$

where $\varepsilon_i(t, z) > 0$ is a convexification term satisfying $\varepsilon_i(t, z) > -\min\{0,\ \lambda_{\min}(\nabla^2_{\theta_i^2} J)\}$. The second-order derivatives can be generated in analogous fashion to (4.22)–(4.23) by solving over successive intervals $\tau \in [t^\theta_{j-1}, t^\theta_j]$

$$\dot{S}^k_{uu} = \frac{\partial^2 f_k}{\partial u \partial x} S_u + S_u^T \frac{\partial^2 f_k}{\partial x^2} S_u + \frac{\partial^2 f_k}{\partial u^2} + \sum_{\ell=1}^{n} \frac{\partial f_k}{\partial x_\ell} S^\ell_{uu}, \quad S_{uu}(t^\theta_{j-1}) = 0 \tag{4.25a}$$

$$\dot{S}^k_{xx} = S_x^T \frac{\partial^2 f_k}{\partial x^2} S_x + \sum_{\ell=1}^{n} \frac{\partial f_k}{\partial x_\ell} S^\ell_{xx}, \quad S_{xx}(t^\theta_{j-1}) = 0 \tag{4.25b}$$

$$\frac{\partial^2 x^p_k}{\partial \theta_i^2}(\tau) \triangleq \begin{cases} 0 & i > j \\ S_{uu} & i = j \\ S^k_x \frac{\partial^2 x^p_k}{\partial \theta_i^2}(t^\theta_{j-1}) + \frac{\partial^2 x^p_k}{\partial \theta_i^2}(t^\theta_{j-1})\ S^k_{xx}\ \frac{\partial^2 x^p_k}{\partial \theta_i^2}(t^\theta_{j-1}) & \tau \in [t^\theta_{j-1}, t^\theta_j], \quad i > j \end{cases} \tag{4.25c}$$

where $S_{xx} \in \mathbb{R}^{n \times n \times n}$, $S_{uu} \in \mathbb{R}^{n \times m \times m}$, and $\frac{\partial^2 x^p}{\partial \theta_i^2} \in \mathbb{R}^{n \times m \times m}$, and the $k, \ell \in \{1 \dots n\}$ denote the indexing of the first coordinate. The integral portions of the individual $\nabla^2_{\theta_i^2} J$ in (4.24) can again be calculated in similar manner to (4.25), or by simply appending to the definition of x^p. The complexity of these calculations scale very poorly with the system dimensions n and m, but only super-linearly with the number of partitions N due to the omission of cross-terms in (4.24).

4.6.2.3 Approximate Gauss–Newton

Motivated by a similar approach in Reference 54, $L(x, u)$ and $W(x)$ can be taken to be of the form $L(x, u) = \|l(x, u)\|^2$, $W(x) = \|w(x)\|^2$ so that (4.12) has the form $L^a_\rho = \|l^a_\rho(x, u)\|^2$, $W^a_\rho = \|w^a_\rho(x)\|^2$, where $l^a_\rho(x, u) \triangleq \rho^{\frac{1}{2}}[l^T, (\mu \mathcal{B}^o_x)^{\frac{1}{2}}]^T$ and $w^a_\rho(x) \triangleq \rho^{\frac{1}{2}}[w^T, (\mu_f \mathcal{B}^o_f)^{\frac{1}{2}}]^T$. Neglecting interval-interaction terms as above, the individual interval hessians $\nabla^2_{\theta_i^2} J$ in (4.24) can be replaced with Gauss–Newton approximations of the form

$$\nabla^2_{\theta_i^2} J \approx G_i \triangleq 2 \int_{t^\theta_{i-1}}^{t^\theta_N} \mathbb{L}_i(\tau)^T \mathbb{L}_i(\tau)\, d\tau + 2\,\mathbb{W}_i^T \mathbb{W}_i \tag{4.26a}$$

$$\mathbb{L}_i \triangleq \begin{cases} \dfrac{\partial l^a_\rho}{\partial x} \dfrac{\partial x^p}{\partial \theta_i} + \rho^{\frac{1}{2}} \dfrac{\partial l}{u} \tau \in [t^\theta_{i-1}, t^\theta_i) \\ \dfrac{\partial l^a_\rho}{\partial x} \dfrac{\partial x^p}{\partial \theta_i} \tau \geq t^\theta_i \end{cases} \qquad \mathbb{W}_i \triangleq \frac{\partial w^a_\rho}{\partial x} \frac{\partial x^p}{\partial \theta_i}(t^\theta_N) \tag{4.26b}$$

with $\frac{\partial x^p}{\partial \theta_i}$ given by (4.23). As long as $\limsup_{x \to \Sigma_x} \|\frac{\partial l}{\partial x}\|$ and $\limsup_{x \to \Sigma_x} \|\frac{\partial w}{\partial x}\|$ are bounded, then $\rho(x)$ and $B(x)$ can always be chosen such that $\frac{\partial l^a_\rho}{\partial x}$ and $\frac{\partial w^a_\rho}{\partial x}$ are bounded on $\mathring{\mathbb{X}}$ and $\mathring{\mathbb{X}}_f$, respectively (and in particular in a neighborhood of Σ_x). Otherwise, an element-wise saturation can be added into (4.26b) to ensure G_i remains bounded as $x^p \to \Sigma_x$. Since $G_i = G_i^T \geq 0$ is guaranteed by (4.26a), then $\varepsilon_i(t, z)$ in (4.24) can be taken as a small constant $\varepsilon_i > 0$. From (4.23) and (4.26b), it can be seen that over the interval $\tau \in [t^\theta_{j-1}, t^\theta_j]$, the quadrature associated with each G_i, $i < j$, can be reconstructed from a common accumulator of the form $\frac{\partial x^p}{\partial \theta_i}^T \left(\int_{t^\theta_{j-1}}^{t^\theta_j} S_x^T \frac{\partial l^a_\rho}{\partial x}^T \frac{\partial l^a_\rho}{\partial x} S_x \, d\tau \right) \frac{\partial x^p}{\partial \theta_i}$.

4.7 Simulation examples

4.7.1 Example 4.1

To illustrate the concept of real-time optimization (RTO) proposed in this book, we consider the nonlinear system given in Reference 34, with constraint $\mathbb{U} = [-2, 2]$ and cost $L(x, u) = 0.5\|x\|^2 + u^2$,

$$\dot{x}_1 = x_2 + (0.5 + 0.5x_1)u \quad \dot{x}_2 = x_1 + (0.5 - 2x_2)u$$

The stabilizer $\kappa_\delta(x)$ was designed by exact discretization of the linearized process with constant $\delta = 0.5$, yielding the optimal feedback $\kappa_\delta(x) = [0.1402,\ 0.1402]x$. The terminal penalty

$$W(x) = x^T \begin{bmatrix} 3.6988 & 2.8287 \\ 2.8287 & 3.6988 \end{bmatrix} x$$

was obtained from a Lyapunov equation, and the corresponding terminal region $\mathbb{X}_f = \{x : W(x) \leq 0.141\}$ was enforced using a logarithmic barrier.

The system was simulated from the initial condition $x_0 = [-0.6830,\ -0.8640]$ using a standard SD controller based on the full solution of (4.16), as well as four different "real-time" controllers (RT-1 through RT-4) based on (4.18). All four real-time controllers used steepest-descent updates, starting from $(t^\theta, \theta)_0$ values corresponding to the first 1.5 seconds of the RT-1 trajectory. Controller parameters and their accumulated costs are given in Table 4.1, with the resulting trajectories depicted in Figure 4.1. As one would expect, the performance of the real-time controller approaches the SD control as k_θ is increased, since the convergence of (4.18b) approaches the instantaneous behavior of (4.16b). Although negligible in this case, the controller RT-4 slightly outperforms the SD controller. This results from the fact that (4.18b) continually readjusts θ as the interval $[t, t_1^\theta]$ shrinks.

4.7.2 Example 4.2

To better illustrate the generality of our framework, we consider regulation of the following two-state system, with cost function $L(x,u) = \|x\|^2 + 0.1u^2$, and input constraint $\mathbb{U} = [-4,\ 4]$. For clarity, we express the (x_1, x_2) dynamics in standard polar coordinates (r, ϕ) as

$$\dot{r} = -\frac{1}{2} r\cos(2\phi) + u\sin(\phi) \qquad \dot{\phi} = \frac{1}{r}$$

The trajectories of this system rotate around the origin in the (x_1, x_2)-plane, and the sign of the gain between u and r changes between the upper and lower (x_1, x_2)-half-planes. The unforced flows are open-loop unstable, and the linearization at $(x_1, x_2) = 0$ is neither controllable nor stabilizable. Since the angular rate of rotation increases

Table 4.1 Definition of controllers used in Example 4.1

Name	Υ	Ψ	Γ	δ	N	k_θ	Cost[1]
RT-1	(4.18a)	(4.18b)	0	0.5	3	0	6.329
RT-2	(4.18a)	(4.18b)	I	0.5	3	1	5.134
RT-3	(4.18a)	(4.18b)	I	0.5	3	10	4.822
RT-4	(4.18a)	(4.18b)	I	0.5	3	100	4.800
SD	(4.16b)	(4.16a)	–	0.5	3	–	4.807

[1] $\int_0^7 L\,d\tau + W(x(7))$, although in all cases $W(x(7)) \approx 0$

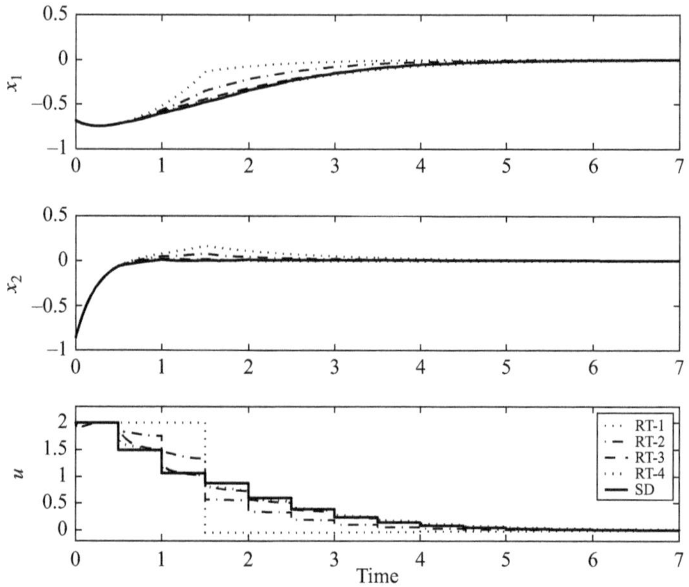

Figure 4.1 Closed-loop response of different controllers for Example 4.1

as the origin is approached, true convergence to the origin requires arbitrarily fast switching of the input.

It can be shown that the origin $x = 0$ is globally stabilized by the feedback $u = -r\cos(\phi) = -x_2$, which motivates our selections $\kappa(x, T_\kappa) = -x_2$, $\mathbb{X}_f \equiv \mathbb{R}^2$, $\Sigma_x \equiv \{0\}$, with the corresponding CLF $W(x) = 1.65\|x\|^2$ satisfying $\dot{W} \leq -1.5L(x, -x_2)$. Rather than solving explicitly for a function $\delta(x)$ satisfying Assumption 4.2.4 point 5, $\delta(x)$ is instead defined implicitly as the largest $\delta \leq 2$ over which $W(x^p(t+\delta)) - W(x^p(t)) \leq 1.25 \int_t^{t+\delta} L(x, -x_2(t))d\tau$ holds, the testing of which can be easily incorporated as a termination criterion of the first model prediction following a reset of (4.14).

Figure 4.2 depicts the closed-loop response of several different controllers, including two standard SD controllers (SD-1, SD-2) based on the full solution of (4.16), two real-time controller (RT-1, RT-2) based on the proposed (4.18), and a continuous (C) feedback $u = -x_2$. The controllers are listed in Table 4.2. Controller RT-1 is essentially an un-optimized sample-and-hold implementation of $\kappa_\delta(x)$, which still outperforms "C" by virtue of $\delta(x)$. The significant performance difference between RT-2 (based on steepest-descent) and SD-1 stems mostly from the ability of RT-1 to continually readjust throughout the first interval in particular, providing a major advantage considering the coarseness of the interval spanned a sign change in the input gain. The final controller SD-2 used a fixed $\delta = 0.5$, and although difficult to distinguish on Figure 4.2, it was the only controller which failed to have converged to $x = 0$ by time $t = 50$, instead exhibiting a stable periodic limit cycle. The final cost of each controller is shown in Table 4.2.

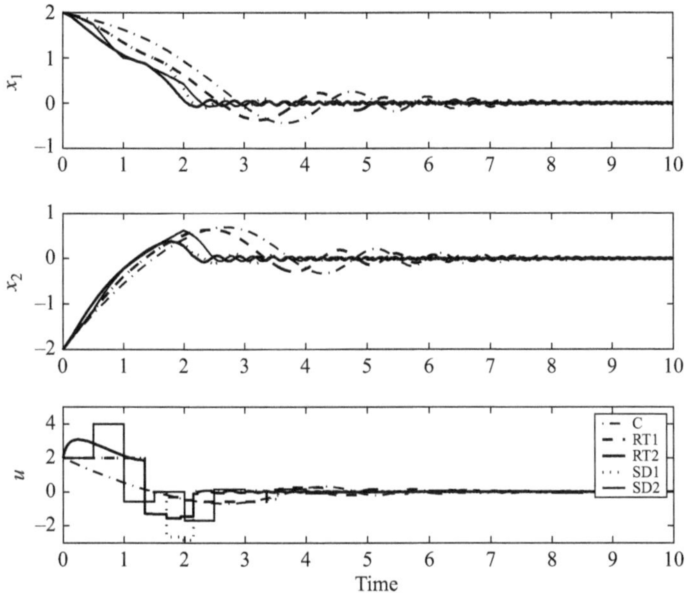

Figure 4.2 Closed-loop response of different controllers for Example 4.2

Table 4.2 Controllers used in Example 4.2

Name	Υ	Ψ	Γ	δ	N	k_θ	Cost[1]
C	–	–	–	0	0	$u = -x_2$	8.23
RT-1	(4.18a)	(4.18b)	–	$\delta(x)$	4	0	6.87
RT-2	(4.18a)	(4.18b)	I	$\delta(x)$	4	10	5.41
SD-1	(4.16b)	(4.16a)	–	$\delta(x)$	4	–	6.51
SD-2	(4.16b)	(4.16a)	–	0.5	4	–	6.22

[1] $\int_0^{10} L\,d\tau + W(x(10))$, where $W(x(10)) \approx 0$ for all but SD-2

4.8 Summary

In this chapter, we have studied a limiting case involving the application of fast, incrementally-improving gradient methods for MPC of continuous-time nonlinear systems. Unlike existing results involving real-time calculations, our study emphasizes the equal importance of computational simplifications not only in how the optimal control problem is *solved*, but also in how it is *posed*. We have shown that for applications in which computational speed is of high priority, a fast gradient-based solution approach can be embedded within a framework that allows for crude, low-dimensional parameterization of the input trajectory. Of key importance is that this input-parameterization involves a time-support which is independent of the controller input–output cycle-time (treated here as a faster timescale), allowing for control calculations to be continually fed back to the process at the fastest possible rate. Ultimately, we have demonstrated that the philosophy from Reference 148 that "feasibility implies stability," exploited by many real-time computational methods to reduce calculation time, is applicable not only to arbitrarily-suboptimal *termination* of the search, but also to arbitrarily-suboptimal restriction of the class of input trajectories over which the search is posed.

4.9 Proofs for Chapter 4

4.9.1 Proof of Claim 4.2.2

If $\mathbb{X}$ is compact then Claim 4.2.2 is obvious, so we focus on sufficiency of the second condition in Assumption 4.2.1. From Reference 159, conditions *2i,ii* are sufficient to ensure global solvability of the analogous unconstrained, infinite-time optimal control problem (whose cost-to-go we denote J_∞^{UC*}). The presence of constraints imposes

$$J_\infty^*(x_0) \geq J_\infty^{UC*}(x_0) \tag{4.27}$$

from which it is clear that proving radial unboundedness of J_∞^{UC*} is sufficient to prove Claim 4.2.2. This we do next.

Let c_1 and c_2 be arbitrary constants satisfying (4.3), and define

$$a_1 \triangleq \min_{x \in \partial B(\Sigma_x, c_1)} J_\infty^{UC*}(x) > 0. \tag{4.28}$$

which is well-defined by the global solvability of the optimal control problem for J_∞^{UC*}. Then, for arbitrary $x_0 \notin B(\Sigma_x, c_1)$, one has

$$J_\infty^{UC*}(x_0) \geq a_1 + \int_{t_0}^{t_1} L(x^{\infty*}, u^{\infty*})\, d\tau \tag{4.29}$$

where $(x^{\infty*}, u^{\infty*})$ denote any minimizing pair (with respect to J_∞^{UC*}) satisfying $x^{\infty*}(t_0) = x_0$, and t_2 denotes the (first) time that $x^{\infty*}$ intersects $B(\Sigma_x, c_1)$.

By the time-invariance of the vector field (4.1), it follows that for any x_0, there must exist[5] a lower bound $c_f \equiv c_f(x_0) > 0$ such that

$$\min_{\tau \in [t_0, t_1]} \|f(x^{\infty*}(\tau), u^{\infty*}(\tau))\| \geq c_f(x_0) > 0 \tag{4.30}$$

This implies the existence of a diffeomorphism $\pi : \mathbb{R}_{\geq 0} \to \mathbb{R}_{\geq 0}$ that defines the length $s = \pi(\tau)$ of the optimal arc $x^{\infty*}_{[t_0,\tau]}$ connecting x_0 and $x^{\infty*}(\tau)$; i.e., satisfying $s(t_0) = 0$. This implies that the integral cost in (4.29) can be expressed as (denoting $s_1 \triangleq \pi^{-1}(t_1)$,

[5] Note that a constant $c_f > 0$ does not necessarily hold uniformly for all $x_0 \in \mathbb{R}^n$.

and the abuse of notation $x^{\infty*}(s) \equiv x^{\infty*}(\pi^{-1}(s))$ and $u^{\infty*}(s) \equiv u^{\infty*}(\pi^{-1}(s)))$

$$\begin{aligned}\int_{t_0}^{t_1} L(x^{\infty*}(\tau), u^{\infty*}(\tau))\, d\tau &= \int_0^{s_1} \frac{L(x^{\infty*}(s), u^{\infty*}(s))}{\|f(x^{\infty*}(s), u^{\infty*}(s))\|} ds \\ &\geq \int_0^{s_1} \frac{c_2}{\|x^{\infty*}(s)\|} ds \\ &\geq \int_0^{s_1} \frac{c_2}{c_1 + s_1 - s} ds\end{aligned}$$

It then follows from (4.29) that

$$J_{\infty}^{UC*}(x_0) \geq a_1 + \ln\left(\frac{c_1 + s_1(x_0)}{c_1}\right). \tag{4.31}$$

Since $\|x_0\| \to \infty$ implies $s_1(x_0) \to \infty$, this proves radial unboundedness of the surface $J_{\infty}^{UC*}(x)$. Compactness of level sets of J_{∞}^* follows from (4.27), completing the claim.

4.9.2 *Proof of Lemma 4.3.2*

From (4.6), we have that $\forall x \in \mathbb{X}_f$ (with x' as per Assumption 4.2.4)

$$\begin{aligned}\mathcal{B}_f^o(x') - \mathcal{B}_f^o(x) &= \mathcal{B}_f(x') - \mathcal{B}_f(x) + \nabla\mathcal{B}_f(0)^T\left(x' - x\right) \\ &\leq \tilde{\mathcal{B}}_f(V(x')) - \tilde{\mathcal{B}}_f(V(x)) + \|\nabla\mathcal{B}_f(0)\| \max_{x,x'\in\mathbb{X}_f}\left(\|x' - x\|\right) \\ &\leq \tilde{\mathcal{B}}_f(c_{f1}) + \|\nabla\mathcal{B}_f(0)\| \max_{x,x'\in\mathbb{X}_f}\left(\|x' - x\|\right) \triangleq b_1\end{aligned}$$

where $b_1 < \infty$ is well defined by the compactness of $\mathbb{X}_f$ and the strict inequality $c_{f1} < c_{f2}$ (recalling that $\lim_{c_f \to c_{f2}} \tilde{\mathcal{B}}_f(c_f) \to +\infty$). The second inequality follows from $x \in \mathbb{X}_f^{c_{f1}}$ being the worst-case, since otherwise $x \in \mathbb{X}_f^{c_{f2}} \backslash \mathbb{X}_f^{c_{f1}}$ would imply $\mathcal{B}_f(x') - \mathcal{B}_f(x) < 0$ by invariance of the level sets of V. Furthermore,

$$\int_t^{t'} \mathcal{B}_x^o(x^{\kappa_\delta}(\tau))\, d\tau \leq M_T \max_{x\in\mathbb{X}_f}(\mathcal{B}_x^o(x)) \triangleq b_2$$

where $\max_{x\in\mathbb{X}_f}\left(\mathcal{B}_x^o(x)\right) < \infty$ exists by the compactness of $\mathbb{X}_f$ and strict containment $\mathbb{X}_f \subset \mathring{\mathbb{X}}$. Finally, the nonemptiness of $\mathring{\Sigma}_x$ guarantees that $\varepsilon > 0$ can be chosen such that

$$b_3 \triangleq \min_{x\in\mathbb{X}_f\backslash\mathring{\Sigma}_x^{\varepsilon}} \gamma(\|x\|) > 0$$

exists. The result follows for any $\mu,\ \mu_f \in (0,\ \mu^*]$, $\mu^* \triangleq \frac{b_3}{b_1 + b_2}$, as seen by substitution into (4.8).

4.9.3 *Proof of Corollary 4.3.6*

Let $x^o_{[t_0,t_f]}$, $u^o_{[t_0,t_f]}$ denote a continuous feasible solution to (4.1) from $x_0 \in \mathcal{X}^0$ such that $\min_{t\in[t_0,t_f]} \min_{x'\in\partial\mathbb{X}} \|x' - x^o(t)\| \geq c$ and $\min_{x'\in\partial\mathbb{X}_f} \|x' - x^o(t_f)\| \geq c$, for some

$c \equiv c(x_0) > 0$. Define the compact sets $\mathbb{T}_x \triangleq \cup_{t\in[t_0,t_f]} x^o$, $\mathbb{T}_x^c \triangleq \{x \mid \min_{x' \in \mathbb{T}_x} \|x - x'\| \leq c\}$ and $\mathbb{T}_u \triangleq \cup_{t\in[t_0,t_f]} u^p$, and let $K_\mathbb{T}$ denote the Lipschitz constant (w.r.t x) of $f(x,u)$ over $\mathbb{T}_x^c \times \mathbb{T}_u$.

For an arbitrary N, define $t^\theta \equiv t^\theta(N) \triangleq \{t_i^\theta, i = 1 \dots N \mid t_i^\theta = t_0 + \frac{i}{N}(t_f - t_0)\}$ with corresponding parameters $\theta \equiv \theta(N) \triangleq \{\theta_i, i = 1 \dots N \mid \theta_i = u^o(t_{i-1}^\theta)\}$. Let $w_{[t_0,t_f]}$, $u_{[t_0,t_f]}^w$ denote a solution to (4.1) with input $u^w(t) = v(t,\theta,t^\theta)$ from (4.10), whose dynamics can be written

$$\begin{aligned} \dot{w} &= f(w, u^o(t)) + g(t,w), \quad w(t_0) = x_0 \\ g(t,w) &\triangleq f(w, u^w(t)) - f(w, u^o(t)) \end{aligned}$$

By the continuity of the (w-parameterized) time-function $f \circ u^o : \mathbb{R} \to \mathbb{R}^n$, there exists an integer $N^* \equiv N^*(x_0)$ such that the bound

$$\max_{(t,w)\in[t_0,t_f]\times\mathbb{T}_x^c} \|g(t,w)\| < \frac{c\, K_\mathbb{T}}{\exp\big(K_\mathbb{T}(t_f - t_0)\big) - 1} \tag{4.32}$$

holds $\forall N \geq N^*$, since the left-hand side can be made arbitrarily small. This implies, by standard results on continuity of solutions with respect to parameters [94, Therom 3.4], the (pointwise) bound $\|x_{[t_0,t_f]} - w_{[t_0,t_f]}\| < c$, and hence $\Phi^N(t_0,x_0)$ contains the particular (θ, t^θ) defined above, as well as arbitrarily close neighbors satisfying (4.32).

4.9.4 Proof of Theorem 4.4.4

As is standard in MPC approaches, we will prove stability by using the finite horizon cost (4.12) as a Lyapunov function. Since our closed-loop trajectory $z_a(t,k) = [x^T, t^{\theta^T}, \theta^T, \pi]^T(t,k)$ evolves in hybrid time (as per Section 4.4.2), and Criterion 4.4.2 only guarantees nonincrease of the (hybrid-time) cost J during the reset (4.14), we rely upon a hybrid-systems version of the invariance principle given by [144, Theorem 4.1, Corollary 4.2].

By assumption, the initial parameterization $(\theta, t^\theta)_0$ selected during the initialization step corresponds to a bounded prediction trajectory with bounded initial cost $J_0 \triangleq J(0, z_0)$ given by (4.12). It then follows by Claim 4.2.2 that this trajectory is contained within the compact set $\Omega(J_0)$, since the optimal cost-to-go (whose level sets define Ω) underbounds the cost in (4.12). If the cost J of (4.12) can be shown non-increasing in hybrid time, then compactness of the state x follows. This we demonstrate next.

Ordinary-time evolution

We start by noting that Criterion 4.4.1 point 3, together with the discussion in Section 4.4.2, guarantees that the continuous trajectory $z(t,k)$ exists over some nonzero interval. We use the notation $\dot{J}_k$ to denote $\frac{d}{dt} J_k(t)$ on a (closed) interval of constant k (i.e., between switches). Then from (4.12)

$$\dot{J}_k = \nabla_t J + \langle \nabla_x J, f(x(t,k), u(t,k)) \rangle + \langle \nabla_\theta J, \dot{\theta} \rangle \tag{4.33}$$

where $u(t,k) \triangleq v(t,\theta(t,k),t^\theta(k))$. Differentiating (4.12), the term $\nabla_t J$ includes the effects of both the integration limit and Lie derivatives associated with the initial condition. This implies $\nabla_t J = -L^a_\rho(x(t,k),u(t,k)) - \langle \nabla_x J, f(x(t,k),u(t,k))\rangle$, in which the second term cancels the middle term in (4.33) (necessarily so, since uniqueness of solutions implies that the locus $x^p_{[t,t_f]}$ remains invariant if $x(t)$ flows along the locus). This implies

$$\begin{aligned} \dot{J}_k &= -L^a_\rho(x(t,k),u(t,k)) + \langle \nabla_\theta J, \dot{\theta}\rangle \\ &\leq -L(x(t,k),u(t,k)) + \langle \nabla_\theta J, \dot{\theta}\rangle, \quad \forall x \in \mathbb{X}\backslash\mathring{\Sigma}_x \\ &\leq -\gamma_L(\|x(t,k),u(t,k)\|), \qquad\qquad \forall x \in \mathbb{X}\backslash\mathring{\Sigma}_x \end{aligned} \tag{4.34}$$

where the second inequality follows from Criterion 4.4.1 point 1. From (4.11), the first line of (4.34) implies that $\dot{J}_k \leq 0$ when $x \in \mathring{\Sigma}_x$.

Event-time evolution
Defining $S \triangleq \{z_a \mid (\theta, t^\theta) \in \Phi^N(\pi, x)\}$ as the set of feasible states, we have $S_\Upsilon = S\backslash\mathring{S}_\Psi$, guaranteeing that the jump mapping $\Upsilon_z(z_a)$ is defined whenever continuous evolution is not possible. Criterion 4.4.2 point 1 then directly gives that $J_{k+1}(t) \leq J_k(t)$ under mapping $\Upsilon_z(\cdot)$.

This demonstrates compactness of the state trajectory $x_{[0,\infty)}$. Since the interval-lengths of the time-support t^θ are underbounded by the class-$\mathcal{K}$ function $\gamma_\delta(\|x\|)$, it follows that the parameter vector θ must also remain bounded (otherwise $x_{[0,\infty)}$ could not remain compact). Boundedness of t^θ is discussed in Remark 4.4.3. Hence, the trajectories of the closed-loop states z_a are guaranteed to remain compact. From the invariance principle in Reference 144, Theorem 4.1, Corollary 4.3, the above implies that z_a converges asymptotically to the invariant set $M = \{z_a : J_{k+1} - J_k = 0$ under $\Upsilon_z\} \cap \{z_a : \dot{J}_k = 0$ under $\Psi_z\}$. From (4.34), it follows $z_a \in M \implies x \in \Sigma_x$. Feasibility of all the pointwise constraints follows directly from Criteria 4.4.1 points 1 and 2 and Definitions 4.3.4 and 4.3.5.

Chapter 5
Extensions for performance improvement

5.1 General input parameterizations, and optimizing time support

The vast majority of MPC implementations are based upon the approximation of $u^p_{[t,t_f]}$ as being PWC in time, making use of a zero-order hold both in the input implementation as well as in the model predictions (potentially in the form of a model discretization). From a theoretical perspective, there is nothing preventing the use of more general parameterizations to describe sub-arcs $u^p_{[t_i,t_{i+1}]}$ when implemented within a SD framework; for example, the SD result in Reference 58 allows for arbitrary arcs $u^p : [t_i, t_{i+1}] \to \mathbb{R}^m$. In general, higher-order parameterizations involving increased number of parameters (per interval) are able to describe u^p over significantly longer intervals $[t_i, t_{i+1}]$, resulting in an overall decrease in the number of parameters required to comparably describe $u^p_{[t,t_f]}$. However, the open-loop nature of the SD intervals requires that their duration be kept reasonably short, thereby eliminating most of the benefits associated with increasing the order of the parameterization. As such, it is rarely practical for SD control designs to make use of any parameterizations beyond the basic PWC selection.

In contrast to a SD framework, in the approach of Chapter 4 the input–output sampling behavior is completely decoupled from the partition t^θ used in parameterizing the trajectory $u^p_{[t,t_f]}$ via (4.10). Since there are no longer any robustness or stability concerns associated with selecting t^θ arbitrarily coarse, redefining (4.10) creates the potential for a substantial reduction in the total number of parameters used to describe $u^p_{[t,t_f]}$, without significantly impacting the overall performance (i.e., with respect to the cost J). This section will briefly demonstrate how the results of Chapter 4 can be extended using a more general version of (4.10). A second contribution of this chapter is to demonstrate that the time support vector t^θ, viewed simply as a parameter used to define $u^p_{[t,t_f]}$, is amenable to online optimization just as is θ. This could be of particular interest for systems with restrictive state constraints, where optimizing the timing of discontinuities in $u_{[t,t_f]}$ using a crude partition t^θ could yield better performance than a computationally-comparable increase in the number of intervals N.

5.1.1 Revised problem setup

The problem of interest is essentially the same as that described in Section 4.2; that is, the creation of a stabilizing continuous-time model-predictive feedback based upon solutions of finite horizon optimal control problems of the nominal form

$$\min_{u^p_{[t,t+T]}} \left\{ \int_t^{t+T} L(x^p, u^p)d\tau + W(x^p(t+T)) \right\} \tag{5.1a}$$

$$\text{s.t. } \dot{x}^p = f(x^p, u^p), \qquad x^p(t) = x \tag{5.1b}$$

$$(x^p_{[t,t+T]}, u^p_{[t,t+T]}) \in \mathbb{X} \times \mathbb{U} \tag{5.1c}$$

$$x^p(t+T) \in \mathbb{X}_f. \tag{5.1d}$$

However, to simplify the presentation and to slightly generalize the result, we will re-define some of the assumptions underlying the function $L(x, u)$.

The control objective is to regulate x to any arbitrary compact set $\Sigma_x \subset \mathbb{R}^n$ (i.e., not necessarily a ball around the origin), which is control-invariant for dynamics (4.1) using inputs in some compact set[1] $u \in \Sigma_u$. Defining $\Sigma \triangleq \Sigma_x \times \Sigma_u$, the mapping $L : \mathbb{X} \times \mathbb{U} \to \mathbb{R}_{\geq 0}$ is assumed to satisfy $\underline{\gamma}_L(\|x, u\|_\Sigma) \leq L(x, u) \leq \overline{\gamma}_L(\|x, u\|_\Sigma)$ for some $\underline{\gamma}_L, \gamma_H \in \mathcal{K}_\infty$, which, in particular, implies[2] $L(\Sigma_x, \Sigma_u) \equiv 0$. Similarly, Assumption 4.2.1 is assumed to hold, with (4.3) interpreted as

$$\min_{u \in \mathbb{U}} \left(\frac{L(x, u)}{\|f(x, u)\|} \right) \geq \frac{c_2}{\|x\|_{\Sigma_x}} \qquad \forall x \in \mathbb{X} \setminus B(\Sigma_x, c_1) \tag{5.2}$$

For reasons that will become apparent shortly, it is convenient to enforce the constraint $\mathbb{U}$ in (5.1c) using a barrier function. We therefore assume non-emptiness of $\mathring{\mathbb{U}}$ in addition to that of $\mathring{\mathbb{X}}$ and $\mathring{\mathbb{X}}_f$, and furthermore assume knowledge of generic barrier functions B_u, B_x, B_f satisfying the following.

Criterion 5.1.1. *Denoting $(s, \mathbb{S})$ a placeholder for any pair $\{(u, \mathbb{U}), (x, \mathbb{X}), (x_f, \mathbb{X}_f)\}$, each barrier B_s is assumed to satisfy*

1. *B_s: $\mathbb{S} \to \mathbb{R}_{\geq 0} \cup \{\infty\}$, and B_s is C^{1+} on the open set $\mathring{\mathbb{S}}$.*
2. *$s \to \partial\mathbb{S}$ (from within) implies $B_s(s) \to \infty$.*
3. *$B_s \equiv 0$ on $s \in \Sigma_s$, and $B_s \geq 0$ on $s \in \mathbb{S} \setminus \Sigma_s$. (With interpretation $\Sigma_{x_f} \equiv \Sigma_x$)* ■

The constraints in (5.1) will therefore be replaced using the following version of (4.7)

$$L^a(x, u) \triangleq L(x, u) + \mu(B_x(x) + B_u(u)) \tag{5.3a}$$

$$W^a(x_f) \triangleq W(x_f) + \mu_f B_f(x_f) \tag{5.3b}$$

[1] Note the results of this section can be easily modified to allow $\Sigma_u \equiv \Sigma_u(x)$, a dependence we will omit for clarity of presentation.
[2] In practice this property could be achieved using an expression $\rho(x, u)$ similar to (4.11) as was done in Chapter 4, the details of which will not be pursued here.

Since the convexity of the constraint sets previously assumed in Section 4.2 was only ever used in the recentering (4.6) and projection algorithm (4.19), satisfaction of Criterion (5.1.1) relaxes the need for $\mathbb{U}$, $\mathbb{X}$, $\mathbb{X}_f$ to be convex sets, or for the barriers B_u, B_x, B_{x_f} to be convex functions.

Remark 5.1.2. *While nonconvexity of the barriers or constraint sets will admittedly add non-convexities into the optimal control problem, this has no impact on the* stability *result due to the fact that the optimal control problem is already generically non-convex (resulting from nonconvexities in any of $f(x,u)$, $L(x,u)$, or $W(x)$). The problem of improving* performance *in the presence of such nonconvexities is discussed in Section 5.2.*

5.1.2 General input parameterizations

Rather than assuming that the control input remains constant within the intervals of the partition t^θ, as was the case in (4.10), we will assume here that the behavior within intervals is described by any general, smooth set of basis functions $\phi : [0, T] \times \Theta \to \mathbb{R}^m$. In particular, we emphasize that it is not necessary for $\phi(\tau, \theta_i)$ to be linearly weighted in $\theta_i \in \Theta$, where $\Theta \subset \mathbb{R}^p$ denotes the associated parameter space. Typical examples might include such definitions as polynomials, exponentials, or radial-basis functions (see examples in Sections 5.1.6 and 5.1.7).

Assumption 5.1.3. *The mapping $\phi : \mathbb{R}_{\geq 0} \times \Theta \to \mathbb{R}^m$ and the set Θ are selected such that* (1) *Θ is compact and convex,* (2) *ϕ is C^{1+} on an open cover of $\mathbb{R}_{\geq 0} \times \Theta$, and* (3) *the image of Θ under ϕ satisfies $\mathbb{U} \subseteq \phi(0, \Theta)$.* ■

We note that Assumption 5.1.3 is not particularly restrictive, as it is unrelated to *feasibility* with respect to $\mathbb{U}$. In practice, Assumption 5.1.3 simply helps avoid problematic singularities or degeneracies of ϕ by appropriately defining Θ.

For reasons which become clear in subsequent sections, the (ordered) vector t^θ used in parameterizing the trajectory $u^p_{[t,t_f]}$ is augmented with an additional element t^θ_0, such that $t^\theta \in \mathbb{R}^{N+1}$. We can then define the following replacement for (4.10)

$$u^p(\tau) = v(\tau, \theta, t^\theta) \triangleq \begin{cases} \phi(\tau - t^\theta_0, \theta_1) & \tau \in [t^\theta_0, t^\theta_1] \\ \phi(\tau - t^\theta_{i-1}, \theta_i) & \tau \in (t^\theta_{i-1}, t^\theta_i], \quad i = \{2 \cdots N\} \\ 0 & \text{otherwise} \end{cases} \tag{5.4}$$

Using Assumption 5.1.3, it can be easily shown that existence of feasible parameter sets is still guaranteed by Corollary 4.3.6, modulo the fact that now $\Phi(t_0, x_0)$ is interpreted to take values in $\Theta^N \times \mathbb{R}^{N+1}$ rather than $\mathbb{U}^N \times \mathbb{R}^N$.

5.1.3 Requirements for the local stabilizer

One advantage to using the PWC parameterizations in Chapter 4 is that significant research focus on the properties of sample-and-hold feedback has resulted in a well-developed body of theory, complete with various constructive approaches for designing such a control policy. In Chapter 4, a local stabilizer $u = \kappa(x, T_\kappa)$, with associated period $T_\kappa = \delta(x)$, could be readily designed according to any of those methods in order to initialize the parameter vector θ_N.

In the context of (5.4) the parameter vector θ no longer has the nice interpretation of containing "values of the input," and so it is admittedly less clear in what sense one can use a "local stabilizer" to initialize θ. To clarify this notion, we will first present an analog to Assumption 4.2.4 which defines exactly the conditions such a stabilizer must satisfy. Following that, we discuss the practicality of obtaining such a feedback.

5.1.3.1 Stability requirements

Similarly to Chapter 4, we assume knowledge of a pair of feedbacks $\kappa_\delta : \mathbb{X}_f \to \Theta$ (defining the input) and $\delta : \mathbb{X}_f \to (0, M_T]$ (defining the associated period), designed to be implemented in a sampled framework of the form: $u = u^{\kappa_\delta}(\tau) = \phi(\tau, \kappa_\delta(x(t_i)))$ for $\tau \in [t_i, t_{i+1}]$, with $t_{i+1} \triangleq t_i + \delta(x(t_i))$. As before, solutions corresponding to $\dot{x} = f(x, u^{\kappa_\delta}(\tau))$ on $\tau \in [t_i, t_{i+1}]$ are denoted $x^{\kappa_\delta}_{[t_i, t_{i+1}]}$.

Assumption 5.1.4. *The penalty $W : \mathbb{X}_f \to \mathbb{R}_{\geq 0}$, the sets $\mathbb{X}_f$ and Σ, the mapping ϕ, and the feedbacks $\delta : \mathbb{X}_f \to \mathbb{R}_{>0}$ and $\kappa_\delta : \mathbb{X}_f \to \Theta$ are all chosen such that*

1. $\Sigma \triangleq \Sigma_x \times \Sigma_u$ *and $\mathbb{X}_f$ are both compact, satisfying $\Sigma_x \subset \mathring{\mathbb{X}}_f$ and $\mathbb{X}_f \subset \mathring{\mathbb{X}}$.*
2. *there exists a constant $\varepsilon_\delta > 0$ such that $\delta(x_0) \geq \varepsilon_\delta$ for all $x_0 \in \mathbb{X}_f$.*
3. $x_0 \in \mathring{\mathbb{X}}_f$ *implies $(x^{\kappa_\delta}_{[0,\delta(x_0)]}, u^{\kappa_\delta}_{[0,\delta(x_0)]}) \in \mathring{\mathbb{X}}_f \times \mathring{\mathbb{U}}$ (specifically, sufficiently small inner approximations of $\mathbb{X}_f \times \mathbb{U}$ are positively invariant)*
4. $x_0 \in \Sigma_x$ *implies $(x^{\kappa_\delta}_{[0,\delta(x_0)]}, u^{\kappa_\delta}_{[0,\delta(x_0)]}) \in \Sigma$ (pointwise)*
5. *there exists $\gamma \in \mathcal{K}$ such that for all $x_0 \in \mathbb{X}_f$ (with $x_f \triangleq x^{\kappa_\delta}(\delta(x_0))$),*

$$W(x_f) - W(x_0) + \int_0^{\delta(x_0)} L(x^{\kappa_\delta}, \phi(\tau, \kappa_\delta(x_0))) \leq -\gamma(\|x\|_{\Sigma_x}) \tag{5.5}$$

■

Similarly, it is assumed that the general barrier functions B_x, B_u, B_f satisfy the following analog of Assumption 4.3.1.

Assumption 5.1.5. *For given choices of $\kappa_\delta(\cdot)$, $\delta(\cdot)$, and $\phi(\cdot,\cdot)$, it follows that the barriers B_u, B_x, B_f, and weightings μ, μ_f are chosen to satisfy*

$$\mu_f\big(B_f(x_f) - B_f(x_0)\big) + \mu \int_0^{\delta(x_0)} B_x(x^{\kappa_\delta}(\tau)) + B_u(\phi(\tau, \kappa_\delta(x_0)))\, d\tau \leq \gamma(\|x\|_{\Sigma_x}) \tag{5.6}$$

$\forall\, x_0 \in \mathring{\mathbb{X}}_f$, *where $x_f \triangleq x^{\kappa_\delta}(\delta(x_0))$.* ■

Just as in Chapter 4, the easiest way to satisfy Assumption 5.1.5 is to ensure that (i) level curves of B_f are invariant, for example, aligning with level curves of W; (ii) the growth rates of B_x and $B_u \circ \phi \circ \kappa_\delta$ are less than that of γ near Σ; and (iii) μ and μ_f are selected sufficiently small.

5.1.3.2 Design considerations for $\kappa_\delta(x)$ and $\delta(x)$

For the purposes of the results here, any pair $\kappa_\delta(x)$, $\delta(x)$ satisfying Assumptions 5.1.4 and 5.1.5 can be used. As mentioned in Chapter 4, for the special case $\phi(\tau, \theta_i) \equiv \theta_i$ there are multiple approaches in the literature for designing κ_δ and δ. As one possible means of constructing κ_δ and δ for more general ϕ, we present here a simple modification of the design approach[3] in [37].

1. Assume that a known feedback $u = k_f(x)$ and associated CLF $W(x)$ satisfy

$$\frac{\partial W}{\partial x} f(x, k_f(x)) + L(x, k_f(x)) \le -\gamma_k(\|x\|_{\Sigma_x}) \quad \forall x \in \mathbb{X}_f \tag{5.7}$$

   for some $\gamma_k \in \mathcal{K}$, with $\mathring{\Sigma} \neq \emptyset$ (if necessary, take Σ as a small neighborhood of the true target). Furthermore, let $\mathbb{X}_f^\varepsilon$ and Σ^ε denote families of strictly nested inner approximations of $\mathbb{X}_f$ and Σ (i.e., satisfying $\mathbb{X}_f^0 \equiv \mathbb{X}_f$ and $\Sigma^0 \equiv \Sigma$). Then for some $\varepsilon^* > 0$, all sets $\mathbb{X}_f^\varepsilon$ and Σ^ε, $\varepsilon \in [0, \varepsilon^*]$, are assumed to be strictly forward-invariant with respect to the dynamics $\dot{x} = f(x, k_f(x))$.
2. Without loss of generality, assume a number $r \in \{0, 1, \ldots, \text{floor}(n_\theta/m) - 1\}$ is known such that $k_f \in C^{r+}$, and

$$\text{span}_{\theta_i \in \Theta} \begin{bmatrix} \phi(0, \theta_i) \\ \vdots \\ \frac{\partial^r \phi}{\partial \tau^r}(0, \theta_i) \end{bmatrix} = \mathbb{U} \oplus \mathbb{R}^{rm}. \tag{5.8}$$

   Select any C^{1+} mapping $\kappa(x) : \mathbb{X}_f \to \{\varpi \in \Theta \mid \varpi \text{ satisfies (5.9) for given } x\}$, whose range is guaranteed to be nonempty by (5.8) and Assumption 5.1.3. In other words, find ϖ by inverting (non-uniquely) the $\mathbb{R}^{(r+1)m}$ equations of (5.9).

$$\begin{bmatrix} k_f(x) \\ \dfrac{\partial k_f}{\partial x} f(x, k_f(x)) \\ \vdots \\ L_f^r k_f \end{bmatrix} = \begin{bmatrix} \phi(0, \varpi) \\ \dfrac{\partial \phi}{\partial \tau}(0, \varpi) \\ \vdots \\ \dfrac{\partial^r \phi}{\partial \tau^r}(0, \varpi) \end{bmatrix} \tag{5.9}$$

   (where $L_f^r k_f$ denotes a Lie derivative (of order r) to the function $k_f(x)$ along the vector field $f(x, k_f(x))$).
3. Using the definition $\gamma(\|x\|_{\Sigma_x}) \triangleq \int_0^{\delta(x)} \frac{1}{2}\gamma_k(\|x^{\kappa_\delta}\|_{\Sigma_x})\, d\tau$, simulate the dynamics of $x_{[0,\tau]}^{\kappa_\delta}$ forward from $x^{\kappa_\delta}(0) = x$ using control $u^{\kappa_\delta} = \phi(\tau, \varpi)$ until one of the conditions in Assumption 5.1.4 fails, at a time $\tau = \delta^*$. Set $\delta(x) = c_\delta \delta^*$, for any $c_\delta \in (0, 1)$.

This approach effectively assigns $\kappa(x)$ by fitting a series approximation of order r, centered at time $\tau = 0$, to the input trajectory generated by $u = k_f(x)$. By the invariance (and compactness) of the inner approximations $\mathbb{X}_f^\varepsilon$ and Σ^ε for some $\varepsilon^* > 0$,

[3] It should be noted that the contribution of Reference 37 extends much beyond simply proposing the (somewhat obvious) design approach for which we have given it credit.

there exists a sufficiently small constant ε_δ which is a lower bound for the function $\delta(x)$ generated by this approach.

An alternative approach for initializing input trajectories is used in Reference 58, where forward simulation of the closed-loop dynamics $\dot{x} = f(x, k_f(x))$ is used to directly generate $u^p_{[0,\Delta t]}$ over any desired interval. A key distinction however, is that Reference 58 does not consider the effects of finitely parameterizing the input trajectory $u^p_{[0,\Delta t]}$. In our context, a trajectory $u^p_{[0,\Delta t]}$ generated by forward simulation of $u = k_f(x)$ would require projection onto the space of time-functions spanned by $\phi(\cdot, \theta_i)$ (i.e., by solving an appropriate min-norm problem to identify the θ_i that provides the closest fit to $u^p_{[0,\Delta t]}$). However, this would necessitate the min-norm calculation for θ_i being part of an inner loop nested within the search for $\delta(x)$, and thus (in our context) this approach could be numerically challenging for online implementation. However, since (5.8) guarantees that the finitely parameterized basis $\phi(\tau, \theta_i)$ can approximate $u^p_{[0,\Delta t]}$ to within arbitrary precision over a sufficiently short interval, this type of approach may be practical if a valid $\delta(x)$ can be generated from a suboptimal lower-bound, rather than performing a search.

5.1.4 Closed-loop hybrid dynamics

Despite superficial appearances, it was relatively easy to justify that the underlying hybrid dynamics of the closed-loop behavior in Chapter 4 are autonomous, and amenable to an invariance principle. While the use of a more general parameterization (5.4) does not really violate the arguments of Remark 4.4.3, the introduction of a non-trivial update law for $\dot{t}^\theta$ does, since it is no longer obvious that the time between resets (i.e., executions of (4.14)) will be finite.

To this end, the vector of closed-loop states z is defined in this chapter as $z \triangleq [x^T\ \theta^T\ t^{\theta^T}\ \pi]^T \in \mathbb{R}^n \oplus \Theta^N \oplus \mathbb{R}^{N+1} \oplus \mathbb{R}$, where π represents "time since last reset," and likewise t^θ is interpreted as being relative to the time of last reset. The cost function is therefore interpreted

$$J(z) = \int_{\pi}^{t_N^\theta} L^a(x^p(\tau, z), u^p(\tau, z))d\tau + W^a(x^p(t_N^\theta, z)) \tag{5.10}$$

where $x^p(\tau, z)$ and $u^p(\tau, z)$ denote solutions on the interval $\tau \in [\pi, t_N^\theta]$ to the system

$$\dot{x}^p = f(x^p, v(\tau, \theta, t^\theta)), \quad x^p(\pi) = x, \tag{5.11}$$

5.1.4.1 Evolution of continuous flows

Similarly to Chapter 4, the continuous dynamics have the form $\dot{z} = \Psi_z(z)$ on the flow domain $z \in S_\Psi \triangleq \{z \mid \pi \le t_1^\theta \text{ and } (\theta, t^\theta) \in \Phi^N(\pi, x)\}$, where $\Psi_z(z)$ is of the form

$$\dot{z} = \begin{bmatrix} \dot{x} \\ \dot{\theta} \\ \dot{t}^\theta \\ \dot{\pi} \end{bmatrix} \triangleq \begin{bmatrix} f(x,\ v(\pi, \theta, t^\theta)) \\ \text{Proj}\{-k_\theta \Gamma_\theta(z) \nabla_\theta J(z),\ \ \Gamma_\theta,\ \theta,\ \Theta^N\} \\ \text{Proj}\{-k_{t^\theta}(z) \Gamma_{t^\theta}(z) \nabla_{t^\theta} J(z),\ \ \Gamma_{t^\theta},\ t^\theta,\ \Xi\} \\ 1 \end{bmatrix} \tag{5.12}$$

$$k_{t^\theta}(z) \triangleq k_{t^\theta}\ \text{sat}_0^1\left(\min\left\{\tfrac{t_1^\theta - \pi}{\varepsilon_k},\ \tfrac{\pi - t_0^\theta}{\varepsilon_k}\right\}\right), \qquad k_{t^\theta}, \varepsilon_k > 0$$

Definition of $v(\pi, \theta, t^\theta)$ is given by (5.4), and the projection algorithm for θ is identical to that in (4.19), which simply maintains $\theta \in \Theta^N$ by projecting $\dot{\theta}$ onto the boundary $\partial\Theta^N$ (using a notion of orthogonality defined by Γ_θ).

The definition of $k_{t^\theta}(z)$ ensures both that the inequality $\pi \geq t_0^\theta$ is preserved, and that any intersection $t_1^\theta = \pi$ occurs transversally (included primarily for the convenience of implying deterministic uniqueness of the trajectories). Note that if Γ_{t^θ} is diagonal, then the terms $\text{sat}_0^1\left(\frac{\pi - t_0^\theta}{\varepsilon_k}\right)$ and $\text{sat}_0^1\left(\frac{t_1^\theta - \pi}{\varepsilon_k}\right)$ could be applied individually to $\dot{t}_0^\theta$ and $\dot{t}_1^\theta$, respectively (i.e., after projection). If desired, θ and t^θ could be appended into a single update law, to allow for $\theta_i - t_i^\theta$ cross-terms in the definition of a common $\Gamma(z)$.

The projection for $\dot{t}^\theta$ ensures ordering of t^θ by preserving inclusion in the convex region $\Xi \triangleq \{t^\theta \mid t_i^\theta \geq t_{i-1}^\theta,\ i = 1 \cdots N, \text{ and } t_N^\theta - t_0^\theta \leq T\}$. Since the region Ξ is a convex linear polytope (and hence has a smooth boundary), the projection operator defined in (4.19) does not technically apply. However, more applicable definitions of the operator can be found in the adaptive control literature (e.g., an appropriate modification of the hypercubic version in Reference 88 would suffice).

Using the hybrid-time notation described in Chapter 4, the ordinary-time evolution of $J_k(t) \triangleq J(z(t,k))$ for $z(t,k) \in S_\Psi$ therefore satisfies

$$\begin{aligned}
\dot{J}_k &= \nabla_\pi J + \langle \nabla_x J,\ \dot{x}\rangle + \langle \nabla_\theta J,\ \dot{\theta}\rangle + \langle \nabla_{t^\theta} J,\ \dot{t}^\theta\rangle && \text{(5.13a)}\\
&= -L(x(t,k), u(t,k)) + \langle \nabla_\theta J,\ \dot{\theta}\rangle + \langle \nabla_{t^\theta} J,\ \dot{t}^\theta\rangle && \text{(5.13b)}\\
&\leq -\underline{\gamma}_L(\|x, u\|_\Sigma). && \text{(5.13c)}
\end{aligned}$$

5.1.4.2 Discrete evolution

On the jump domain $z \in S_\Upsilon \triangleq \{z \mid \pi \geq t_1^\theta \text{ and } (\theta,\ t^\theta) \in \Phi^N(\pi, x)\}$, the discrete reset dynamics $z^+ = \Upsilon_z(z)$ are given by

$$z^+ = \begin{bmatrix} x^+ \\ \theta^+ \\ (t^\theta)^+ \\ \pi^+ \end{bmatrix} \triangleq \begin{bmatrix} x \\ \begin{cases} \theta_{i+1} & i = 1 \cdots N-1 \\ \kappa_{\tilde{\delta}}(x^p(t_N^\theta)) & i = N \end{cases} \\ \begin{cases} t_{i+1}^\theta - t_1^\theta & i = 0 \cdots N-1 \\ t_N^\theta - t_1^\theta + \tilde{\delta}(x^p(t_N^\theta)) & i = N \end{cases} \\ 0 \end{bmatrix} \tag{5.14a}$$

$$\tilde{\delta}(x) = \min\{\delta(x),\ \ T - t_N^\theta + t_1^\theta\} \tag{5.14b}$$

from which it can be seen that the elements of t^θ are reset relative to the instant at which the jump occurs. Recognizing that $J(z)$ in (5.10) is invariant with respect to any uniform translation of the states π and t^θ, it follows that (defining $\bar{x}_0 \triangleq x^p(t_N^\theta, z)$)

$$J(z^+) - J(z) = W^a(x^{\kappa_\delta}(\bar{x}_0)) - W^a(\bar{x}_0) + \int_0^{\delta(\bar{x}_0)} L^a(x^{\kappa_\delta}, \phi(\tau, \kappa_\delta(\bar{x}_0)))\, d\tau \leq 0. \tag{5.15}$$

5.1.5 Stability results

The main intention of this chapter has been to show that the claims of Remark 4.4.3 remain valid for the proposed modifications to the controller design, and therefore the analysis is concluded with the following restatement of Theorem 4.4.4. Despite (5.13) and (5.15), the proof is not yet completely obvious due to the fact that the required boundedness of $z(t,k)$ has yet to be established.

Theorem 5.1.6. *Let $L(\cdot,\cdot)$, $\kappa_\delta(\cdot)$, $W(\cdot)$, $\mathbb{X}$, $\mathbb{X}_f$ be chosen to satisfy Assumptions* 4.2.1 *and* 4.2.4.

Let $L(\cdot,\cdot)$, $\phi(\cdot,\cdot)$, $\delta(\cdot)$, $\kappa_\delta(\cdot)$, $W(\cdot)$, $\mathbb{X}$, $\mathbb{X}_f$ be chosen to satisfy Assumptions 4.2.1 *and* 5.1.3–5.1.4 *(for given Σ), and let B_x, B_u, B_f, μ, μ_f satisfy Assumption* 5.1.5. *For any initial condition $x_0 \in \mathcal{X}^0$ (as defined in Corollary* 4.3.6*) of the dynamics* (4.1)*, and any initial feasible control parameterization $(\theta, t^\theta)_0 \in \Phi^N(\pi_0, x_0)$, the target $x \in \Sigma_x$ is asymptotically stabilized under the closed-loop dynamics* (5.12) *and* (5.14) *using the control parameterization* (5.4). *Furthermore, the resulting closed-loop trajectories satisfy all point-wise input, state, and terminal constraints.* ■

Proof of Theorem 5.1.6

The main property to prove is the boundedness of the "time since reset" state π with respect to both coordinates of hybrid time (i.e., boundedness of $\pi(t,k)$). This comes down to disproving that the adaptation of t^θ could result in t_1^θ perpetually "keeping ahead" of π (which grows at the constant rate $\dot{\pi} = 1$). Boundedness of all remaining states will then follow by the same arguments as used in proving Theorem 4.4.4. To this end, we begin with a (contradictory) assumption:

CA1 Let $\Omega \subseteq \mathcal{X}^0 \oplus \Theta^N \oplus \Xi \oplus \{0\}$ be a compact set such that $\pi = t_0^\theta = 0$ and $(\theta, t^\theta) \in \Phi^N(0,x)$ for every $z \in \Omega$. Then for some $z^*(0,0) \in \Omega$, there exists a constant $k^* \in \{0,1,2\ldots\}$ and a corresponding $t_{k^*} \in \mathbb{R}_{\geq 0}$ such that $\pi^*_{k^*}(t) \equiv \pi^*(t,k^*)$ is defined (and thus radially unbounded) on $t \in [t_{k^*}, \infty)$.

Then, since no resets occur for $t \geq t_{k^*}$, the state $z^*(t_{k^*}, k^*)$ can be viewed as the initial condition of a (non-hybrid) continuous-time flow on $t \in [t_{k^*}, \infty)$, generated by (5.12). From standard results, (5.13) implies $\lim_{t\to\infty} J^*_{k^*}(t) = 0$, and thus $\lim_{t\to\infty} x^*_{k^*}(t) \to \Sigma_x$. More useful is the fact that by defining t_{k^*} sufficiently large (but finite), the "initial condition" $x^*(t_{k^*}, k^*)$ can be assumed within any arbitrarily small neighborhood of Σ_x.

It can be seen that the expression

$$\begin{aligned}\nabla_{t_1^\theta} J^*_{k^*} &= L^a(x^p(t_1^\theta), \phi(t_1^\theta - t_0^\theta, \theta_1)) - L^a(x^p(t_1^\theta), \phi(0, \theta_2)) \\ &\quad + \int_{t_1^\theta}^{t_N^\theta} \frac{\partial L^a}{\partial x}\frac{\partial x^p}{\partial t_1^\theta}\, d\sigma - \int_{t_1^\theta}^{t_2^\theta} \frac{\partial L^a}{\partial u}\frac{\partial \phi}{\partial \tau}\, d\sigma + \frac{\partial W^a}{\partial x_f}\frac{\partial x_f^p}{\partial t_1^\theta}\end{aligned} \tag{5.16}$$

must continuously approach zero as $x^*(t_{k^*}, k^*) \to \Sigma_x$, since the terms L^a, $\frac{\partial L^a}{\partial x}$, $\frac{\partial L^a}{\partial u}$, and $\frac{\partial W^a}{\partial x}$ all continuously approach zero, and the remaining terms are bounded. This implies that $\nabla_{t_1^\theta} J^*_{k^*}(t_{k^*})$, and thus $\dot{t}_1^\theta$, can be assumed arbitrarily small $\forall t \geq t_{k^*}$. However, the fact that $\dot{\pi}^*_{k^*} \equiv 1$ then violates the inherent assumption in **CA1** that the condition $\pi^*_{k^*} \leq t_1^\theta$ holds indefinitely.

This proves boundedness of $\pi(t,k)$. The boundedness of $t^\theta(t,k)$ follows by the definition of Ξ and the fact that $\pi \in [t_0^\theta, t_1^\theta]$. Boundedness of $\theta(t,k)$ comes from compactness of Θ. The boundedness of x follows as before from Claim 4.2.2. Having established the boundedness of $z(t,k)$, the result follows from (5.13) and (5.15) by the same invariance principle [144, Theorem 4.1, Corollary 4.2] used to prove Theorem 4.4.4. ■

5.1.6 *Simulation Example 5.1*

We consider regulation of the (constant level) stirred tank reactor from Reference 112, with exothermic reaction $A \longrightarrow B$ resulting in dynamics

$$\dot{C}_A = \frac{v}{V}(C_{Ain} - C_A) - k_0 \exp\left(\frac{-E}{R\,T_r}\right) C_A$$

$$\dot{T}_r = \frac{v}{V}(T_{in} - T_r) - \frac{\Delta H}{\rho\, c_p} k_0 \exp\left(\frac{-E}{R\,T_r}\right) C_A + \frac{UA}{\rho\, c_p\, V}(T_c - T_r)$$

Constants are taken from Reference 112: $v = 100\,\ell/\text{min}$, $V = 100\,\ell$, $\rho\, c_p = 239\,\text{J}/\ell\,\text{K}$, $\text{E/R} = 8750\,\text{K}$, $k_0 = 7.2 \times 10^{10}\,\text{min}^{-1}$, $UA = 5\times 10^4\,\text{J/min K}$, $\Delta H = -5\times 10^4\,\text{J/mol}$, $C_{Ain} = 1\,\text{mol}/\ell$, and $T_{in} = 350\,\text{K}$. The objective is to regulate the unstable equilibrium $C_A^{eq} = 0.5\,\text{mol}/\ell$, $T_r^{eq} = 350\,\text{K}$, $T_c^{eq} = 300\,\text{K}$, using the coolant temperature T_c as the input, subject to the constraints $0 \leq C_A \leq 1$, $280 \leq T_r \leq 370$, and $280 \leq T_c \leq 370$.

We use the cost function[4] $L(x,u) = x'Qx + u'Ru$, with $x = [C_A - C_A^{eq},\ T_r - T_r^{eq}]'$, $u = (T_c - T_c^{eq})$, $R = 1/300$, and $Q = diag(2,\ 1/350)$, where "*diag*" denotes a diagonal matrix containing the indicated values. By linearizing around $x = 0$, the local controller $k_f(x) = [109.1,\ 3.3242]x$ and terminal penalty function $W(x) = x'Px$, $P = [17.53,\ 0.3475;\ 0.3475,\ 0.0106]$ were chosen according to a Ricatti equation. Four different choices of the basis function $\phi : \mathbb{R}_{\geq 0} \times \Theta \to \mathbb{R}^m$ defining (5.4) were tested:

$$\phi_C(s,\theta_i) = \theta_{i1} \qquad \phi_L(s,\theta_i) = \theta_{i1} + \theta_{i2}s$$

$$\phi_Q(s,\theta_i) = \theta_{i1} + \theta_{i2}s + \theta_{i3}s^2 \qquad \phi_E(s,\theta_i) = \theta_{i1}\exp(-\theta_{i2}s)\,.$$

The piecewise-exponential parameterization is of particular interest, since it has the potential to efficiently approximate the optimal input trajectories for systems which exhibit linear-like response over large intervals. In each case, the gains $k_\theta = 0.1$ and

[4] Values for Q and R taken from Reference 112.

Table 5.1 Definition and performance of different controllers in Example 5.1

Controller	**Linear quadratic regulator (LQR)**	ϕ_C	ϕ_L	ϕ_Q	ϕ_E
N	–	8	4	3	4
$\dim(\theta \oplus t^\theta)$	–	16	13	13	12
Worst CPU time[1] (ms)	–	0.8	1.1	0.7	0.6
$(C_A, T_r)_0$	Accumulated Cost[2]				
(0.3, 363)	0.285	0.310	0.281	0.278	0.279
(0.3, 335)	1.74	1.80	1.55	1.42	1.41
(0.6, 335)	0.596	0.723	0.570	0.567	0.558

[1]For one incremental evaluation of $\dot{\theta}$ and $\dot{t}^\theta$
[2]For $x(10) \approx 0$, so $J_\infty \approx \int_0^{10} L(x,u)\,d\tau + W(x(10))$

$k_{t^\theta} = 0.5$ were used in the update laws, with $\Gamma_t \equiv I$ and Γ_θ a diagonally scaled identity matrix.

The feedbacks $\kappa(x)$ were derived by analytically solving (5.9), while $\delta(x)$ was chosen using forward simulation as described in Section 5.1.3. In all cases, the initial conditions $(\theta, t^\theta)_0$ were chosen offline such that the initial parameterized input trajectory $u^p_{[0,1.5]}$ best approximates (in an integral least-squares sense) the closed-loop trajectory $u_{[0,1.5]} = T_{c\,[0,1.5]} - T_c^{eq}$ that results under the Linear quadratic (LQ) feedback $u = k_f(x)$. In each case, the parameter N defining the number of intervals in t^θ was specified such that the total number of optimization variables in θ and t^θ, and thus the computational requirements[5] were comparable for all controllers, as seen in Table 5.1.

Three different initial conditions were tested, with closed-loop state profiles depicted in Figures 5.1–5.4, and corresponding closed-loop costs reported in Table 5.1. Using higher-order parameterizations such as ϕ_E and ϕ_Q over coarse time-intervals generally resulted in better performance than the low-order ϕ_C, despite the fact that ϕ_C used substantially more intervals in t^θ and was allotted more optimization variables. Although the equilibrium of this system is open-loop unstable, large interval-lengths were not problematic since (5.12) does not involve open-loop operation.

5.1.7 Simulation Example 5.2

Consider the problem of state-feedback regulation of a jacketed non-isothermal reactor with van de Vusse kinetics. The reaction mechanism is

$$A \xrightarrow{k_1} B \xrightarrow{k_2} C \quad 2A \xrightarrow{k_3} D$$

[5] Gradient calculations were performed on an AthlonXP 2000+, in Fortran (called from within MATLAB®), using the sensitivity-ODE solver ODESSA [105]. However, limited effort was devoted to optimizing code efficiency.

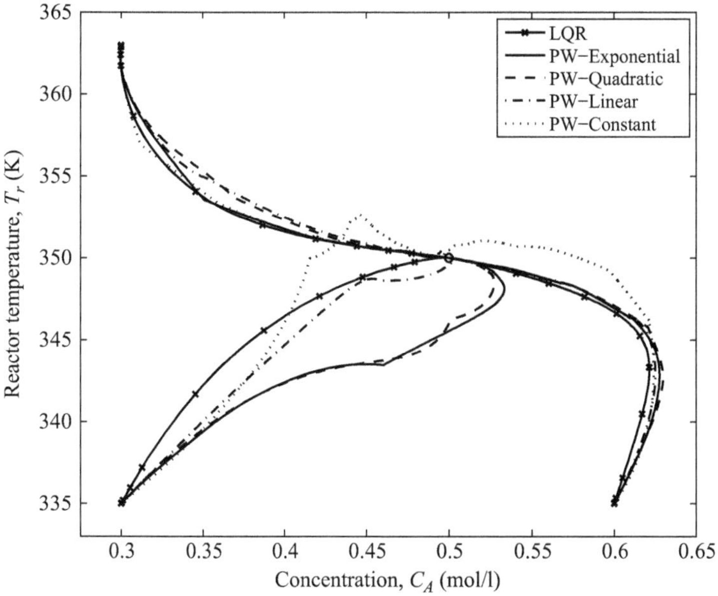

Figure 5.1 Closed-loop profiles from different initial conditions in Example 5.1

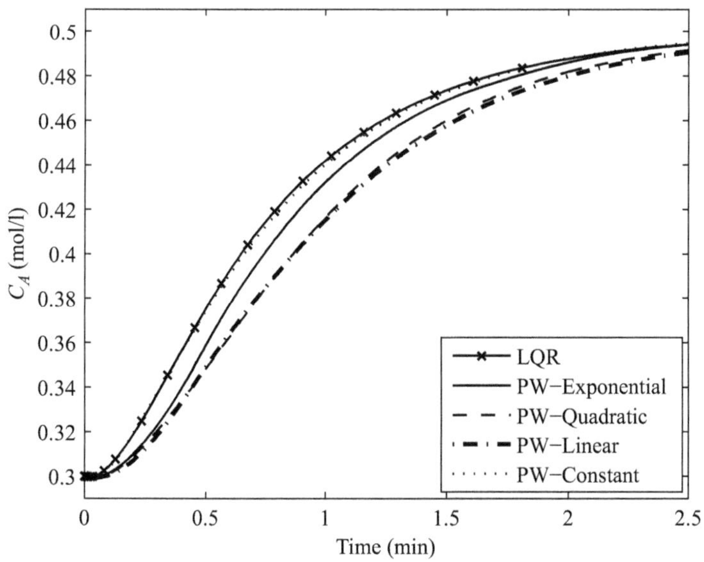

Figure 5.2 Closed-loop concentration profiles from $(C_A, T) = (0.3, 335)$ in Example 5.1

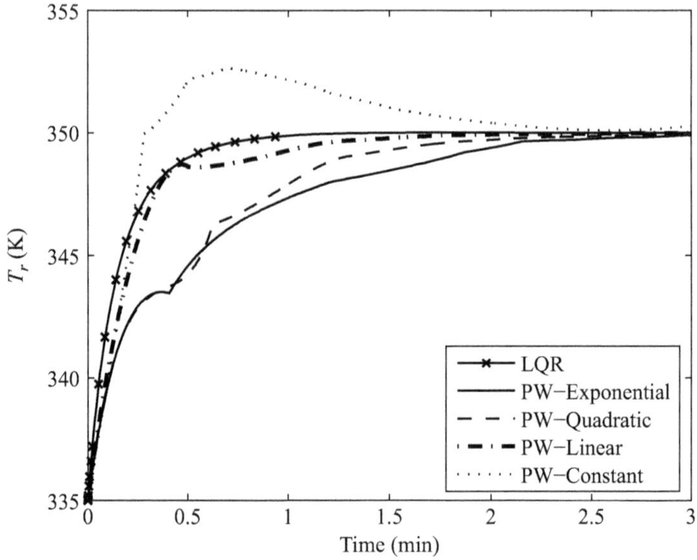

Figure 5.3 Closed-loop temperature profiles from (C_A, T) = (0.3, 335) in Example 5.1

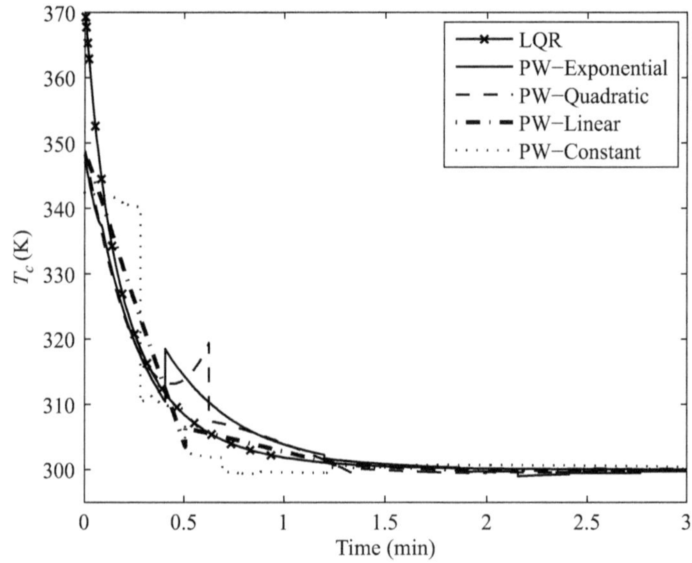

Figure 5.4 Closed-loop input profiles from (C_A, T) = (0.3, 335) in Example 5.1

The states of the system consist of concentrations of components A and B, as well as the temperatures T and T_K occurring in the reactor and cooling jacket, respectively. The manipulated variables consist of the dilution rate $\frac{\dot{V}}{V_R}$ and the rate of heat removal from the jacket. All four states are assumed to be measured, and evolve according to

$$\dot{C}_A = \frac{\dot{V}}{V_R}(C_{A_0} - C_A) - k_1(T)C_A - k_3(T)C_A^2$$

$$\dot{C}_B = -\frac{\dot{V}}{V_R}C_B + k_1(T)C_A - k_2(T)C_B$$

$$\dot{T} = \frac{\dot{V}}{V_R}(T_0 - T) + \frac{k_w A_R}{\rho c_p V_R}(T_k - T) - \frac{1}{\rho c_p}\big(k_1(T)C_A \Delta H_{R_{AB}} + k_2(T)C_B \Delta H_{R_{BC}} + k_3(T)C_A^2 \Delta H_{R_{AD}}\big)$$

$$\dot{T}_k = \frac{1}{m_k c_{pk}}\big(\dot{Q}_k + k_w A_R(T - T_k)\big)$$

where the rate constants $k_i(T)$ follow the Arrhenius law, and the values of all necessary system parameters are reported in Reference 93. The steady state to be regulated is given by $x_r = [C_A, C_B, T, T_k]_r = [2.14\,\frac{\text{mol}}{\text{L}}, 1.09\,\frac{\text{mol}}{\text{L}},\ 114.2°\text{C},\ 112.9°\text{C}]$ beginning from the initial conditions $(x_0 + x_r) = [1\,\frac{\text{mol}}{\text{L}}, 0.5\,\frac{\text{mol}}{\text{L}}, 100°\text{C}, 100°\text{C}]$ using inputs $u = [\frac{\dot{V}}{V_R}, \dot{Q}_k]$. At steady state, $[\frac{\dot{V}}{V_R}, \dot{Q}_k]_r = [14.19\,\text{hr}^{-1}, -1118\,\frac{\text{kJ}}{\text{hr}}]$. The input constraints are given by $3 \le \frac{\dot{V}}{V_R} \le 35\,\text{hr}^{-1}$ and $-9000 \le \dot{Q}_k \le 0\,\frac{\text{kJ}}{\text{hr}}$, enforced using logarithmic barrier functions.

The cost function is taken to be $L(x,u) = x^T Qx + u^T Ru$, with the diagonal matrices $Q = \text{diag}(0.2, 1, 0.5, 0.20)$ and $R = \text{diag}(0.5, 5 \times 10^{-7})$. The terminal cost $W = x^T Px$ and nominal local controller $k(x) = \text{sat}(Kx, \mathbb{U})$ were derived from the algebraic Ricatti equation for the linearized system, and are given by

$$K = \begin{bmatrix} -0.0381 & -0.0405 & -0.1004 & -0.0244 \\ 12.7532 & 6.2581 & 5.9558 & 3.6523 \end{bmatrix},$$

$$P = \begin{bmatrix} 70.1 & 36.6 & 20.1 & 6.4 \\ 36.6 & 33.6 & 9.9 & 3.1 \\ 20.1 & 9.9 & 10.6 & 3.0 \\ 6.4 & 3.1 & 3.0 & 1.8 \end{bmatrix}.$$

The operator $\text{sat}(\cdot, \mathbb{U})$ denotes componentwise saturation to the input constraints. Although $k(x)$ is not necessarily globally asymptotically stabilizing, the above

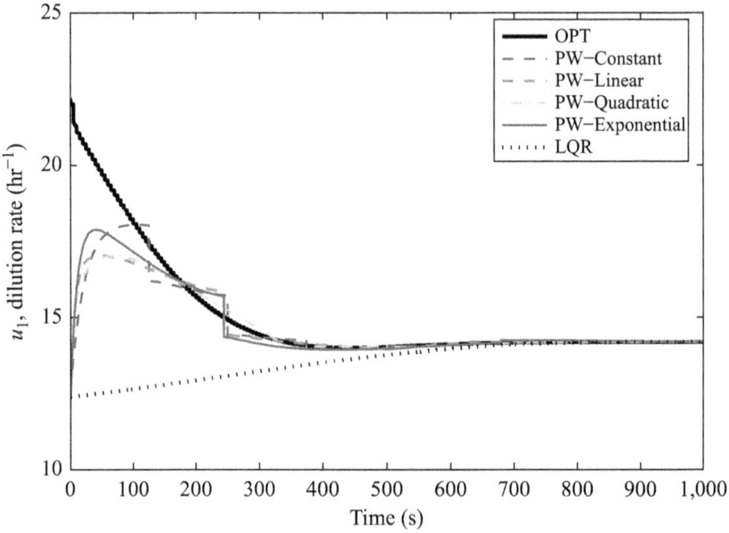

Figure 5.5 Closed-loop dilution rate trajectories for Example 5.2

initial conditions are within its domain of attraction and thus the solution trajectory $u(t) = k(x(t))$ solving $\dot{x} = f(x, k(x))$ was used as the basis for all parameter initializations.

The same four input parameterizations as in the previous example were used:

$$\phi_C(\tau, \theta_i) \triangleq [\theta_{i1}, \theta_{i2}]^T \qquad \phi_L(\tau, \theta_i) \triangleq \begin{bmatrix} \theta_{i1} + \theta_{i2}\tau \\ \theta_{i3} + \theta_{i4}\tau \end{bmatrix}$$

$$\phi_Q(\tau, \theta_i) \triangleq \begin{bmatrix} \theta_{i1} + \theta_{i2}\tau + \theta_{i3}\tau^2 \\ \theta_{i4} + \theta_{i5}\tau + \theta_{i6}\tau^2 \end{bmatrix} \qquad \phi_E(\tau, \theta_i) \triangleq \begin{bmatrix} \theta_{i1}e^{-\theta_{i2}\tau} \\ \theta_{i3}e^{-\theta_{i4}\tau} \end{bmatrix}$$

Closed-loop simulation results for each of the four controllers, as well as those of the nominal controller $u = \text{sat}(Kx, \mathbb{U})$ and the "optimal" solution (solved using 200 uniform 5 s intervals), are depicted in Figures 5.5–5.8. In all cases, a (diagonally-scaled) steepest-descent definition was used for both Γ_θ and Γ_{t^θ}, and the adaptation gains were specified as $k_\theta = 5$, $k_{t^\theta} = 0.1$. The feedback $\kappa_\delta(x)$ was calculated from $k(x)$ using the approach in Section 5.1.3, where the saturation operation $k(x)$ was first smoothed over a small interior approximation of $\mathbb{U}$. The function $\delta(x)$ was nominally specified as a constant $\delta = 1000/N$, reduced as necessary to validate (5.5).

It can be seen from the figures, and the values reported in Table 5.2, that the performance of all of the real-time controllers were essentially comparable, with a

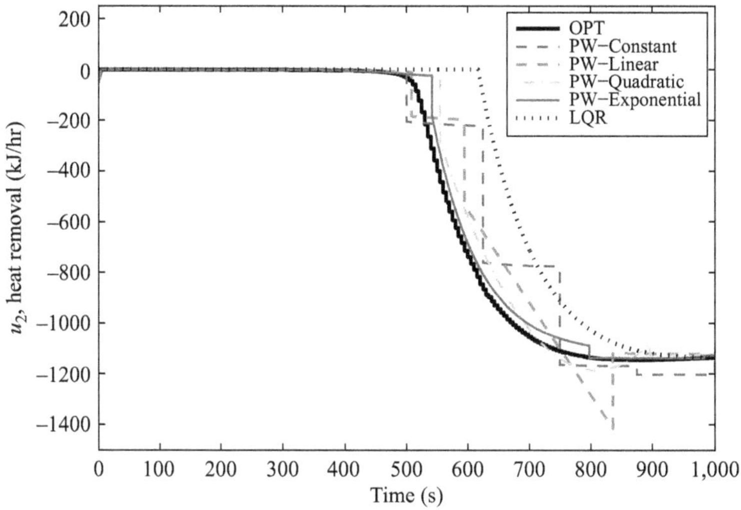

Figure 5.6 Closed-loop heat removal trajectories for Example 5.2

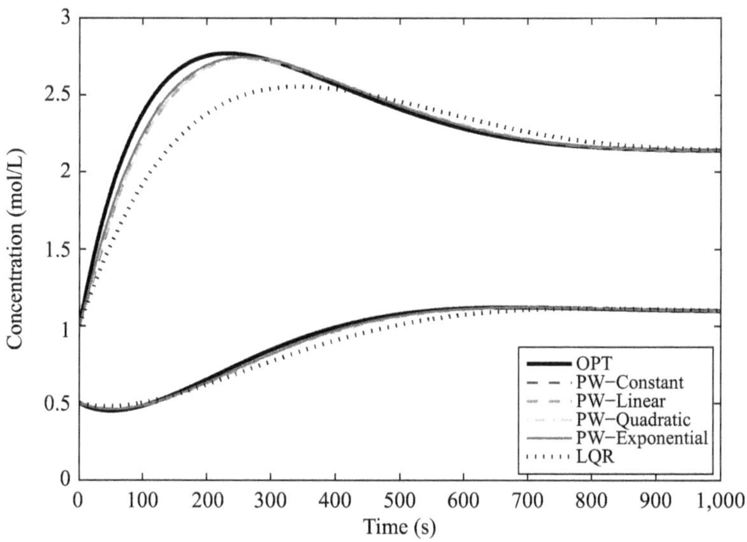

Figure 5.7 Closed-loop concentration trajectories for Example 5.2

slight advantage demonstrated by the higher-order parameterizations. As expected, each of these controllers was somewhat suboptimal (due to the suboptimality of $(\theta, t^{\theta})_0$), but significantly out performed the (saturated) LQ controller upon which the initializations $(\theta, t^{\theta})_0$ and $\kappa(x)$ were based. Calculations were performed in the same computing environment as indicated in Example 5.1.

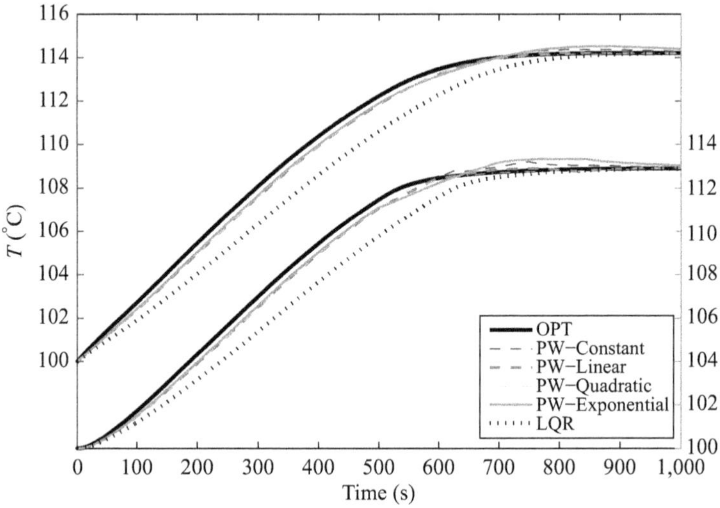

Figure 5.8 Closed-loop temperature trajectories for Example 5.2

Table 5.2 Definition and performance of different controllers in Example 5.2

Controller:	**LQR**	**OPT**	ϕ_C	ϕ_L	ϕ_Q	ϕ_E
N	–	200	8	4	3	4
Nominal[1] $\delta(x)$ (s)	–	5	125	250	333	250
$\dim(\theta \oplus t^\theta)$	–	–	24	21	22	21
Worst CPU time[2] (ms)	–	–	4.5	4.5	5	5
Accumulated cost[3] ($\times 10^3$)	8.91	7.48	7.72	7.68	7.65	7.60

[1] Subject to validation of 5.5
[2] For one incremental evaluation of $\dot{\theta}$ and $\dot{t}^\theta$
[3] For $x(1800) \approx 0$, so $J_\infty \approx \int_0^{1800} L(x,u)\,d\tau + W(x(1800))$

5.2 Robustness properties in overcoming locality

5.2.1 Robustness properties of the real-time approach

As was discussed in Section 3.4, the generically defined MPC feedback of (3.1) has the potential to exhibit a lack of even nominal robustness to arbitrarily small errors in either the prediction model or measurement signal. In the case of model error/disturbance $d(t)$, the underlying problem lies with constraint-induced discontinuities in the (infinite-horizon) optimal value function $V^* : \mathbb{X} \to \mathbb{R}$, which end up being passed along to both the (finite-horizon) optimal cost $J^* : \mathbb{X} \to \mathbb{R}$ and the corresponding feedback map $\kappa_{mpc} : \mathbb{X} \to \mathbb{U}$. In contrast, the susceptibility to measurement error $e(t)$ stems from the fact that the $\arg\min_{u(\cdot)}$ in (3.1) generically involves a global optimization. If the problem is nonconvex (i.e., nonconvexity in any of $L(x,u)$, $W(x)$, $f(x,u)$, $\mathbb{X}$, $\mathbb{X}_f$ or $\mathbb{U}$), this optimization can potentially inject discontinuities into $\kappa_{mpc}(x)$ by switching between disconnected minimizers.

In contrast, the robustness properties of the real-time approach of Section 5.1 are summarized in the following claim. With some abuse of the notation from previous sections, we will here represent the continuous flow-field of (5.12) in the condensed form

$$\dot{x} = f(x, v(\omega)) \quad \dot{\omega} = \Psi_\omega(x, \omega) \tag{5.17}$$

where $\omega = [\pi; \theta; t^\theta]$ contains the controller states, and $v(\omega)$ is given by (5.4); that is, the closed-loop dynamics consist of the vector $z = [x;\ \omega]$. Note that for the purposes of the claim, it is not necessary to consider the discrete evolution $\omega^+ = \Upsilon_\omega(x, \omega)$ of (5.14), since its evolution along event-coordinate k is orthogonal to the time-evolution of $d(t)$ and $e(t)$ (where the assumption that $e(t,\ k+1) \equiv e(t,\ k)$ is justified on the basis that x does not need to be resampled between multiple event-executions, since $x(t, k_{i+1}) \equiv x(t, k_i) \equiv x(t)$).

Definition 5.2.1 (Input-to-state stability (ISS) [90, Definition 4.7]). *The system with dynamics $\dot{y} = g(y, d)$ is said to be ISS if there exist a class $\mathcal{KL}$ function β and a class $\mathcal{K}$ function γ such that for any initial state $y(t_0)$ and any bounded input $d(t)$, the solution $y(t)$ exists for all $t \geq t_0$ and satisfies*

$$\|y(t)\|_{\mathbb{Y}} \leq \beta(\|y(t_0)\|_{\mathbb{Y}},\ t - t_0) + \gamma\left(\sup_{t_0 \leq \tau \leq t} \|d(\tau)\|\right).$$

The following claim demonstrates that the closed-loop dynamics exhibit nominal robustness to disturbance and measurement error, in the sense that they are ISS with respect to perturbations of "sufficiently small" magnitude. Not surprisingly, the robustness margin depends upon the initial condition of the entire closed-loop state $z = [x;\ \omega]$, implying that appropriately-conservative initialization of the control parameters ω is required to guarantee robustness.

Claim 5.2.2. *Given any initial condition $z_0 \in \mathcal{Z} \triangleq \{z \mid (\theta, t^\theta) \in \Phi^N(\pi, x)\}$, there exist constants $M_d \triangleq M_d(z_0) > 0$ and $M_e \triangleq M_e(z_0) > 0$ such that under the conditions of Theorem 5.1.6, the target $x \in \Sigma_x$ under perturbed dynamics*

$$\dot{x} = f(x, v(\omega)) + d \quad \dot{\omega} = \Psi_\omega(x + e,\ \omega) \tag{5.18}$$

(i.e., a perturbation of (5.12)) is ISS with respect to arbitrary signals $d(t)$ and $e(t)$ satisfying pointwise bounds $\|d\| \leq M_d$ and $\|e\| \leq M_e$. ■

Proof of Claim 5.2.2

By the assumed differentiability of relevant functions, it follows by standard results on continuity of solutions [90, Theorem 3.5][6] that at any time $t \in [t_i, t_{i+1}]$ between resets:

- the prediction $x^p_{[\pi,t_N^\theta]}(z)$ varies continuously with respect to perturbation e in its initial condition.
- $\nabla_\theta J$ and $\nabla_{t^\theta} J$ in (5.12) are locally Lipschitz in $x^p_{[\pi,t_N^\theta]}$, and thus $\Psi_\omega(x+e,\, \omega)$ varies continuously with e.
- the closed-loop solution $z_{[t_i,t_{i+1}]}$ of (5.12) varies continuously with respect to the additive disturbances d and $\tilde{\Psi}_\omega \triangleq \Psi_\omega(x+e,\, \omega) - \Psi_\omega(x, \omega)$ entering (5.17).

It then follows that under perturbation, (5.13) has the form

$$\dot{J}_k \leq -\underline{\gamma}_L(\|x, u\|_\Sigma) + \langle \nabla_x J,\; d\rangle + \langle \nabla_\omega J,\; \tilde{\Psi}_\omega\rangle$$

By definition of $\mathcal{Z}$, the unperturbed trajectories satisfy $\bar{z}_0 \in \mathcal{Z} \implies z(t) \in \mathcal{Z}$, $\forall t \geq t_0$. Taking any sufficiently small, strictly feasible compact tube $\mathbb{T}_{[t_0,\infty)}$ surrounding the (unperturbed) trajectory $\bar{z}_{[t_0,\infty)}$, it follows that both $\nabla_x J$ and $\nabla_\omega J$ have a finite upper bound M_∇ over $\mathbb{T}_{[t_0,\infty)}$. The claim then follows from [90, Theorem 4.19]. ■

Claim 5.2.2 is neither particularly surprising, nor is it particularly strong. The robustness to measurement error is the more useful result, which is a consequence of the fact that $e(t)$ enters (5.17) via the update law Ψ_ω rather than directly in the feedback $u = \kappa_{mpc}(x)$ (as was the case in Section 3.4). The robustness stems from the fact that Ψ_ω is only a local search, and is aided by the fact that ω provides an "inertia" to filter out high frequencies in $e(t)$.

In contrast, the robustness to $d(t)$ in Claim 5.2.2 is relatively weak, since it depends entirely upon the conservativeness of the interior-point barrier to provide robustness. In practice, the approach in Section 3.4 of increasing the conservativeness of the constraint handling along the prediction horizon $\tau \in [t, t_N^\theta]$ would significantly improve the robustness to disturbance (i.e., increasing the value of M_d in Claim 5.2.2). One approach would be to replace the constraints $\mathbb{X}$ and $\mathbb{X}_f$ with strictly nested functions of the form $\mathbb{X}^\tau : [0, T] \to \overline{\text{cov}}\,\{\mathbb{X}\}$, as discussed in Section 3.4, hence resulting in barrier functions in (5.3) of the form $B_x(\tau - \pi,\, x)$, $B_f(\tau - \pi,\, x)$. However, a much simpler approach is to simply define μ and μ_f to be of the form $\mu(\tau, \omega) = \mu_0 + \mu_\tau(\tau - \pi)$, where $\mu_0 > 0$ and $\mu_\tau \in \mathcal{K}$. Clearly, either approach will add an additional negative-definite term of the form $-\left(\int_\pi^{t_N^\theta} \frac{d}{d\tau}(\mu B_x)\, d\sigma + \frac{d}{d\tau}(\mu_f B_f)\right)$ into the right-hand side of (5.13), which provides additional robustness to perturbation.

[6] While the piecewise continuity in τ of (5.4) does not quite meet the continuity condition of [90, Theorem 3.5], the theorem can easily be applied successively over the N intervals of t^θ.

5.2.2 Robustly incorporating global optimization methods

To avoid the optimization variables θ and t^θ from becoming trapped by local minima, it was hinted in Section 4.5.3 that one could potentially make use of the reset mapping Υ. However, as discussed in Section 3.4, there are potential robustness issues (with respect to measurement error) associated with the incorporation of global search methods, in particular when an excessively coarse partition t^θ makes it undesirable to wait for the next occurrence of $\pi = t_1^\theta$ before applying such a reset. Ironically, one approach recommended in Reference 157 to address this robustness issue is to augment the controller (in a discrete-time framework) with a set of "memory variables" to retain knowledge of the previous solution for comparison purposes, thus providing a sense of inertia to controller decision-making. As highlighted in Claim 5.2.2, the gradient-based approach presented here already exhibits this property by its very nature.

A simple, but effective method for addressing nonconvexity is to pose the controller as being a selector between multiple (potentially cooperative) agents. For example, assume that resources allow for the parallel computation of multiple parameter sets of the form[7] $\omega^j, j \in \{1, \dots, \bar{j}\} \triangleq \mathcal{J}$ (where each ω^j may involve unique choices $N^j, \phi^j, \kappa_\delta^j, \delta^j$ defining v^j in (5.4)). Furthermore, let $j^* \in \mathcal{J}$ record the particular index selected to be "active" (i.e., j^* is an integer-valued state of the system). For the remainder, let z be interpreted to include *all* of the closed loop states: system x, parameter sets ω^j, $\forall j$, and the selector j^*.

Let $\eta \in (0, 1)$ be a chosen discount factor, and let $J(x, \omega^j)$ denote the obvious interpretation of (5.10) for any $j \in \mathcal{J}$. Furthermore, we denote by $\mathcal{J}_\Phi \equiv \mathcal{J}_\Phi(z)$ the set of feasible indices $\mathcal{J}_\Phi \triangleq \{j \in \mathcal{J} \mid (\theta, t^\theta)^j \in \Phi^N(\pi, x)^j\}$. Then on the domain

$$S_{\Upsilon_z} \triangleq \Big\{ z \mid (\theta, t^\theta)^{j^*} \in \Phi^N(\pi, x)^{j^*} \textbf{ and } \\ \big(\exists j \in \mathcal{J} \;\; s.t. \;\; \pi^j \geq (t_1^\theta)^j \quad \textbf{or} \quad \exists j \in \mathcal{J}_\Phi \;\; s.t. \;\; J(x, \omega^j) \leq \eta\, J(x, \omega^{j^*})\big)\Big\}$$

the reset action $z^+ = \Upsilon_z(z)$ is defined to have the form

$$x^+ = x \tag{5.19a}$$

$$(\omega^j)^+ = \begin{cases} \Upsilon_\omega^j(x, \omega^j) & \pi^j = (t_1^\theta)^j \\ \omega^j & \text{otherwise} \end{cases} \qquad \forall j \in \mathcal{J} \tag{5.19b}$$

$$(j^*)^+ = \begin{cases} \text{choose}\{\arg\min_{j \in \mathcal{J}_\Phi} J(x, \omega^j)\} & \exists j \in \mathcal{J}_\Phi \; s.t. \; J(x, \omega^j) \leq \eta\, J(x, \omega^{j^*}) \\ j^* & \text{otherwise} \end{cases} \tag{5.19c}$$

where $\Upsilon_\omega^j(x, \omega^j)$ denotes the reset for (π, θ, t^θ) defined in (5.14), using the particular $\delta^j(x)$ and $\kappa_\delta^j(x)$. The "choose" operator denotes an arbitrary, deterministic selection rule for the case of multiple minimizers.

[7] Using notation from (5.17).

Similarly, the continuous flow-field $\dot{z} = \Psi_z(z)$ is defined on the domain

$$S_{\Psi_z} \triangleq \Big\{ z \mid ((\theta,\, t^\theta)^{j^*} \in \Phi^N(\pi, x)^{j^*}) \quad \textbf{and} \quad (\pi^j \leq (t_1^\theta)^j, \quad \forall j \in \mathcal{J}) \\ \textbf{and} \quad (J^j(x, \omega^j) \geq \eta\, J(x, \omega^{j^*}), \quad \forall j \in \mathcal{J}_\Phi \setminus \{j^*\}) \Big\}$$

with the vector fields given by

$$\dot{x} = f(x,\, v^{j^*}(\omega^{j^*})) \tag{5.20a}$$

$$\dot{\omega}^j = \begin{cases} \Psi_\omega^{j^*}(x, \omega^{j^*}) & j = j^* \\ \Lambda^j(z) & \text{otherwise} \end{cases} \qquad \forall j \in \mathcal{J} \tag{5.20b}$$

$$\dot{j}^* \equiv 0 \tag{5.20c}$$

The expression $\Psi_\omega^{j^*}(x, \omega^{j^*})$ denotes the feasibility-preserving, descent-based update law of (5.12), which is used to update the active parameter set. In contrast, the update laws $\Lambda^j(z)$ applied to the non-active parameter sets need not satisfy any of these properties; it need not preserve feasibility or generate a descent-direction, and it is allowed to cooperate with other agents.

The obvious reason why $\Lambda^j(z)$ does not need to satisfy any conditions for stability is that it does not feed back into the control law, and is simply excluded from consideration in (5.19c) if it is infeasible. Thus the most fundamental property of the real-time approach, that is, the guaranteed invariance of the feasibility-inclusion $z(t, k) \in \mathcal{Z}$ along evolution of both t and k, is not impacted by the modifications in (5.19) or (5.20). Stability of the switched controller therefore follows directly from Theorem 5.1.6, given the fact that the hysteresis effect introduced by η prevents zeno-like switching between multiple minimizing agents. This hysteresis therefore also provides some nominal robustness to measurement errors $e(t)$, by essentially eliminating noise-induced dithering behavior (a.k.a. zeno-like switching), as long as $e(t)$ is “sufficiently small.” Thus, the system exhibits nominal robustness against the type of noise-induced instability or infeasibility discussed in Reference 157.

We specifically distinguish between separate update behavior for the active and inactive update laws not just for the sake of generality, but also because there are significant potential benefits in doing so. Without any attempt to be rigorous, there is the potential to incorporate any of the following ideas into the design of $\Lambda^j(z)$.

5.2.2.1 Infeasible-point handling

Methods such as References 21, 22, and 158 combine the use of interior-point barriers with exterior-point penalties to allow a descent-based NLP to temporarily pass through infeasible regions. By alternately relaxing and re-tightening feasibility constraints, it is claimed that the solver is not only less likely to become trapped by local minima (specifically, those which result from nonconvexities in the constraint), but also follows a shorter search path to the feasible minimizer.

Whether by these particular methods or otherwise, the ability of $\Lambda^j(z)$ to remain defined for infeasible ω^j is necessary since, for inactive parameterizations, the vector field for the true state x *does not* lie in the tangent space of $x^p(\pi^j, x, \omega^j)$. In other words, it may be impossible to avoid the condition $(\theta, t^{\theta})^j \in \Phi(\pi^j, x)^j, j \neq j^*$, from becoming infeasible if x evolves in a very different direction under ω^{j^*} than it would have under the input move defined by ω^j.

5.2.2.2 Quasi-global "roaming" of the surface

Various methods in the global optimization literature attempt to modify local search methods such that over time, the search will visit many different (local) minima. One example is the approach in Reference 82, which toggles between ascent and descent modes of operation. A second example is the so-called "heavy ball" method of Reference 16, which augments a typical NLP with additional "velocity" states, whose (dissipative) inertia helps the search to escape from the basin of attraction surrounding shallow local minima.

5.2.2.3 Cooperative behavior

As mentioned above, the difference between the actual flow of x and that predicted under inactive ω^j acts as a disturbance to the search of inactive agents. However, since the future behavior of x under the current ω^{j^*} is predictable, there is potential to introduce a feedforward term into $\Lambda^j(z)$ to counteract the effects of ω^{j^*}.

Another benefit of cooperation is the ability to prevent the individual searches from clustering in a common region of the parameter-space, for example, by penalizing parameterizations that become "too similar," or by posing the different $\Lambda^j(z)$ as multi-objective searches which simultaneously try to maximize some measure of distance[8] between the individual ω^j.

Ultimately, whether or not it is beneficial or computationally realistic to incorporate any of these approaches will depend on the nature of the system of interest. For systems in which the non-convexity is not too excessive, it may be more advantageous to devote all available CPU resources to improving a single local search, rather than splitting resources between multiple searches. However, it should be noted that one could easily modify the definition of the individual Υ^j_ω in (5.19) such that the discretization-level N^j of a particular parameterization $(\theta, t^{\theta})^j$ changes when ω^j transitions between active/inactive status. This implies that one could devote the majority of computational resources to the active search, while still parallel-computing some very coarsely parameterized additional searches, which could still yield beneficial results if the active search becomes trapped by an excessively suboptimal local minimum.

5.2.3 Simulation Example 5.3

In order to illustrate the basic idea of this section, we consider a simple exothermic reaction $A \rightarrow B$ taking place in a non-isothermal, gas-phase continuously stirred tank reactor (CSTR). The system is comprised of three states; although many equivalent coordinate systems can be used, the equations are most clearly expressed for the choice: n (total moles of gas in reactor), n_A (moles of A), and T (reactor temperature).

[8] For example, measures of the form $\int \|u^p(\tau, \omega^j) - u^p(\tau, \omega^{j2})\|^2 d\tau$, which allows $\phi^{j1} \neq \phi^{j2}$.

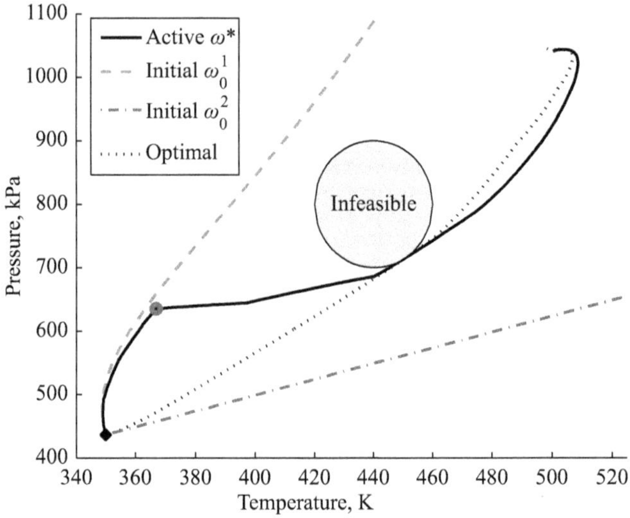

Figure 5.9 System trajectories in the $P - T$ plane for Example 5.3. Small circle indicates switch in active ω^j

The control objective is specified as regulation to the target $(n, n_A, T)_{ss} = (2.5\,\text{kmol}, 0.25\,\text{kmol}, 500\,\text{K})$, corresponding to a system pressure of 1040 kPa, from the initial state $(n, n_A, T)_0 = [2,\ 0.5,\ 350]$. Manipulated variables are the outlet molar flow F_{out}, and rate of heat removal $\dot{Q}$. Using several simplifying assumptions, the system equations are

$$\dot{n} = F_{in} - F_{out} \tag{5.21a}$$

$$\dot{n}_A = F_{in} - \frac{n_A}{n} F_{out} - k(T) n_A \tag{5.21b}$$

$$\dot{T} = F_{in} \frac{T_{in} - T}{n} - k(T) \frac{\Delta H_r}{\overline{c_p}} \frac{n_A}{n} + \frac{\dot{Q}}{\overline{c_p}\, n} \tag{5.21c}$$

where $k(T) = k_0 e^{-\frac{E}{RT}}$. System parameters are

$$\Delta H_r = -5000\,\frac{\text{kJ}}{\text{kmol}} \quad R = 8.314\,\frac{\text{m}^3\,\text{kPa}}{\text{kmol K}} \quad E = 8000\,\frac{\text{m}^3\,\text{kPa}}{\text{kmol}} \quad \overline{c_p} = 10\,\frac{\text{kJ}}{\text{kmol K}}$$

$$V = 10\,\text{m}^3 \quad T_{in} = 300\,\text{K} \quad k_0 = 6.2\,\text{s}^{-1} \quad F_{in} = 0.25\,\text{kmol s}^{-1}$$

An important objective of the controller is to ensure the system trajectories avoid passing through the shaded region in the $P - T$ plane shown in Figure 5.9, for example, to avoid undesirable thermodynamic behavior of other components in the gas stream occurring in that region. This constraint in the $P - T$ space was transformed into a

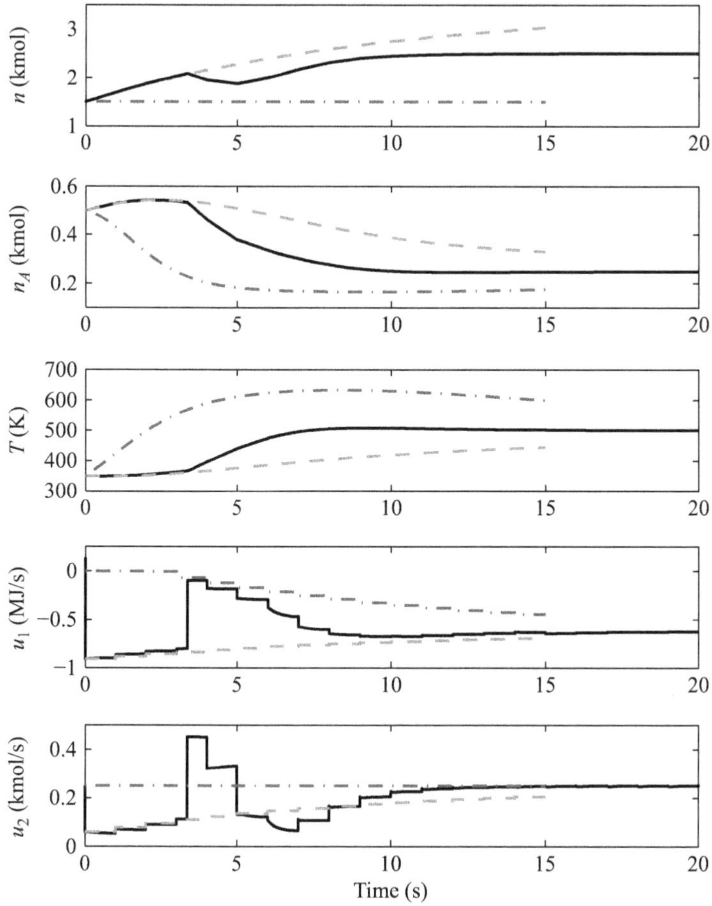

Figure 5.10 Closed-loop system trajectories for Example 5.3

constraint $\mathbb{X}$ in the statespace (n, n_A, T) using the algebraic ideal gas law $P(n, T) \triangleq n R T \frac{1}{V}$.

The cost function was $L(x, u) = x^T \mathcal{Q} x + u^T \mathcal{R} u$, $\mathcal{Q} = \text{diag}(10, 10, 10^{-3})$, $\mathcal{R} = \text{diag}(0.2, 0.5)$, where $x^T = [n, n_A, T]_{dev}$ and $u^T = [\dot{Q}, F_{out}]_{dev}$ are deviations from the indicated steady state. The terminal penalty $W(x)$ is the quadratic solution to an algebraic Riccati equation for the linearized (5.21), and the local control law $\kappa(x)$ was derived from the optimal linear controller using the method in Section 5.1.3.

Using simple PWC parameterizations, two sets of parameters ω^1 and ω^2 (both with $N = 15$, $\delta(x) = 1$, shown in Figure 5.10) were adapted online. The initializations ω_0^1 and ω_0^2 corresponded to different paths around the constraint region in the P–T plane (whose image $\mathbb{X}$ is a skewed infinite cylinder in the actual x-space). For simplicity, the mapping $\Lambda^j(z)$ in (5.20) for the inactive parameterization was calculated in similar fashion to the active gradient-based update $\Psi^{j^*}(x, \omega^{j^*})$, except that a quadratic was used to penalize constraint violation for the inactive parameterization, while a logarithmic barrier was used for the active parameterization.

As can be seen in Figure 5.9, neither initialization started very close to the (infinite-horizon) optimal solution. The controller initially selected the active parameterization ω^1 (i.e., passing "over" the constraint in Figure 5.9), but as the dynamics and adaptation progressed, the active parameterization switched to ω^2 in time to feasibly pass under the constraint region.

Chapter 6

Introduction to adaptive robust MPC

In this book, we focus on the more typical role of adaptation as a means of coping with uncertainties in the system model. A standard implementation of MPC using a nominal model of the system dynamics can, with slight modification, exhibit nominal robustness to disturbances and modeling error. However in practical situations, the system model is only approximately known, so a guarantee of robustness which covers only "sufficiently small" errors may be unacceptable. In order to achieve a more solid robustness guarantee, it becomes necessary to account (either explicitly, or implicitly) for *all* possible trajectories which could be realized by the uncertain system, in order to guarantee feasible stability. The obvious numerical complexity of this task has resulted in an array of different control approaches, which lie at various locations on the spectrum between simple, conservative approximations versus complex, high-performance calculations. Ultimately, selecting an appropriate approach involves assessing, for the particular system in question, what is an acceptable balance between computational requirements and closed-loop performance.

6.1 Review of NMPC for uncertain systems

While a vast majority of the robust-MPC literature has been developed within the framework of discrete-time systems,[1] for consistency with the rest of this book most of the discussion will be based in terms of their continuous-time analogs. The uncertain system model is therefore described by the general form

$$\dot{x} = f(x, u, d) \tag{6.1}$$

where $d(t)$ represents any arbitrary $\mathcal{L}_\infty$-bounded disturbance signal, which takes point-wise[2] values $d \in \mathcal{D}$. Equivalently, (6.1) can be represented as the differential inclusion model $\dot{x} \in F(x, u) \triangleq f(x, u, \mathcal{D})$.

In the next two sections, we will discuss approaches for accounting explicitly for the disturbance in the online MPC calculations. We note that significant effort has also been directed toward various means of increasing the inherent robustness of the controller without requiring explicit online calculations. This includes the suggestion

[1] Presumably for numerical tractability, as well as providing a more intuitive link to game theory.
[2] The abuse of notation $d_{[t_1,t_2]} \in \mathcal{D}$ is likewise interpreted pointwise.

in Reference 116 (with a similar discrete-time idea in Reference 44) to use a modified stage cost $\overline{L}(x,u) \triangleq L(x,u) + \langle \nabla_x V_T^*(x),\ f(x,u)\rangle$ to increase the robustness of a nominal-model implementation, or the suggestion in Reference 95 to use an prestabilizer, optimized offline, of the form $u = Kx + v$ to reduced online computational burden. Ultimately, these approaches can be considered encompassed by the banner of nominal-model implementation.

6.1.1 Explicit robust MPC using open-loop models

As seen in Chapters 3, 4, and 5, essentially all MPC approaches depend critically upon the Principle of Optimality (Def 2.2.1) to establish a proof of stability. This argument depends inherently upon the assumption that the predicted trajectory $x^p_{[t,t+T]}$ is an invariant set under open-loop implementation of the corresponding $u^p_{[t,t+T]}$; that is, that the prediction model is "perfect." Since this is no longer the case in the presence of plant-model mismatch, it becomes necessary to associate with $u^p_{[t,t+T]}$ a *cone* of trajectories $\{x^p_{[t,t+T]}\}_{\mathcal{D}}$ emanating from $x(t)$, as generated by (6.1).

Not surprisingly, establishing stability requires a strengthening of the conditions imposed on the selection of the terminal cost W and domain $\mathbb{X}_f$. As such, W and $\mathbb{X}_f$ are assumed to satisfy Criterion 3.1.1, but with the revised conditions:

C3a. $\mathbb{X}_f$ is strongly positively invariant for $\dot{x} \in f(x, k_f(x), \mathcal{D})$.

C4a. $L(x, k_f(x)) + \frac{\partial W}{\partial x} f(x, k_f(x), d) \leq 0, \quad \forall (x,d) \in \mathbb{X}_f \times \mathcal{D}$.

While the original C4 had the interpretation of requiring W to be a CLF for the nominal system, so the revised C4a can be interpreted to imply that W should be a robust-CLF like those developed in Reference 65.

Given such an appropriately defined pair $(W, \mathbb{X}_f)$, the model predictive controller explicitly considers all trajectories $\{x^p_{[t,t+T]}\}_{\mathcal{D}}$ by posing the modified problem

$$u = \kappa_{mpc}(x(t)) \triangleq u^*_{[t,t+T]}(t) \tag{6.2a}$$

where the trajectory $u^*_{[t,t+T]}$ denotes the solution to

$$u^*_{[t,t+T]} \triangleq \arg \min_{\substack{u^p_{[t,t+T]} \\ T\in[0,T_{\max}]}} \left(\max_{d_{[t,t+T]}\in\mathcal{D}} V_T(x(t), u^p_{[t,t+T]}, d_{[t,t+T]}) \right) \tag{6.2b}$$

The function $V_T(x(t), u^p_{[t,t+T]}, d_{[t,t+T]})$ appearing in (6.2) is as defined in (3.1), but with (3.1c) replaced by (6.1). Variations of this type of design are given in References 35, 104, 123, 128, and 139, differing predominantly in the manner by which they select $W(\cdot)$ and $\mathbb{X}_f$.

If one interprets the word "optimal" in Definition 2.2.1 in terms of the worst-case trajectory in the optimal cone $\{x^p_{[t,t+T]}\}^*_{\mathcal{D}}$, then at time $\tau \in [t, t+T]$ there are only two possibilities:

- The actual $x_{[t,\tau]}$ matches the subarc from a worst-case element of $\{x^p_{[t,t+T]}\}^*_{\mathcal{D}}$, in which case the Principle of Optimality holds as stated.

- The actual $x_{[t,\tau]}$ matches the subarc from an element in $\{x^p_{[t,t+T]}\}^*_{\mathcal{D}}$ which was *not* the worst case, so implementing the remaining $u^*_{[\tau,t+T]}$ will achieve overall less cost than the worst-case estimate at time t.

One will note however, that the bound guaranteed by the Principle of Optimality applies only to the remaining subarc $[\tau, t+T]$, and says nothing about the ability to extend the horizon. For the nominal-model results of Chapter 3, the ability to extend the horizon followed from C4 of Criterion 3.1.1. In the present case, C4a guarantees that for *each* terminal value $\{x^p_{[t,t+T]}(t+T)\}^*_{\mathcal{D}}$ there exists a value of u rendering W decreasing, but not necessarily a single such value satisfying C4a for *every* $\{x^p_{[t,t+T]}(t+T)\}^*_{\mathcal{D}}$. Hence, receding of the horizon can only occur at the discretion of the optimizer. In the *worst* case, T could contract (i.e., $t+T$ remains fixed) until eventually $T=0$, at which point $\{x^p_{[t,t+T]}(t+T)\}^*_{\mathcal{D}} \equiv x(t)$, and therefore by C4a an appropriate extension of the "trajectory" $u^*_{[t,t]}$ exists.

Although it is not an explicit min–max type result, the approach in Reference 119 makes use of global Lipschitz constants to determine a bound on the worst-case distance between a solution of the uncertain model (6.1), and that of the underlying nominal model estimate. This Lipschitz-based uncertainty cone expands at the fastest possible rate, necessarily containing the actual uncertainty cone $\{x^p_{[t,t+T]}\}_{\mathcal{D}}$. Although ultimately just a nominal-model approach, it is relevant to note that it can be viewed as replacing the "max" in (6.2) with a simple worst-case upper bound.

Finally, we note that many similar results [30, 94] in the linear robust-MPC literature are relevant, since nonlinear dynamics can often be approximated using uncertain linear models. In particular, linear systems with polytopic descriptions of uncertainty are one of the few classes that can be realistically solved numerically, since the calculations reduce to simply evaluating each node of the polytope.

6.1.2 *Explicit robust MPC using feedback models*

Given that robust control design is closely tied to game theory, one can envision (6.2) as representing a player's decision-making process throughout the evolution of a strategic game. However, it is unlikely that a player even moderately skilled at such a game would restrict themselves to preparing only a single *sequence of moves* to be executed in the future. Instead, a skilled player is more likely to prepare a *strategy* for future game-play, consisting of several "backup plans" contingent upon future responses of their adversary.

To be as least-conservative as possible, an ideal (in a worst-case sense) decision-making process would more properly resemble

$$u = \kappa_{mpc}(x(t)) \triangleq u^*_t \tag{6.3a}$$

where $u^*_t \in \mathbb{R}^m$ is the constant value satisfying

$$u^*_t \triangleq \arg\min_{u_t} \left(\max_{d_{[t,t+T]} \in \mathcal{D}} \; \min_{u^p_{[t,t+T]} \in \mathcal{U}(u_t)} \; V_T(x(t), u^p_{[t,t+T]}, d_{[t,t+T]}) \right) \tag{6.3b}$$

with the definition $\mathcal{U}(u_t) \triangleq \{u^p_{[t,t+T]} \mid u^p(t) = u_t\}$. Clearly, the "least conservative" property follows from the fact that a separate response is optimized for every possible sequence the adversary could play. This is analogous to the philosophy in Reference 149, for system $x^+ = Ax + Bu + d$, in which polytopic $\mathcal{D}$ allows the max to be reduced to selecting the worst index from a finitely indexed collection of responses; this equivalently replaces the innermost minimization with an augmented search in the outermost loop over *all* input responses in the collection.

While (6.3) is useful as a definition, a more useful (equivalent) representation involves minimizing over *feedback policies* $k : [t, t+T] \times \mathbb{X} \rightarrow \mathbb{U}$ rather than trajectories:

$$u = \kappa_{mpc}(x(t)) \triangleq k^*(t, x(t)) \tag{6.4a}$$

$$k^*(\cdot,\cdot) \triangleq \arg\min_{k(\cdot,\cdot)} \max_{d_{[t,t+T]}\in\mathcal{D}} \Big(V_T(x(t), k(\cdot,\cdot), d_{[t,t+T]}) \Big) \tag{6.4b}$$

$$V_T(x(t), k(\cdot,\cdot), d_{[t,t+T]}) \triangleq \int_t^{t+T} L(x^p, k(\tau, x^p(\tau)))\, d\tau + W(x^p(t+T)) \tag{6.4c}$$

$$\text{s.t. } \forall \tau \in [t, t+T]: \ \frac{d}{d\tau}x^p = f(x^p, k(\tau, x^p(\tau)), d), \quad x^p(t) = x(t) \tag{6.4d}$$

$$(x^p(\tau), k(\tau, x^p(\tau))) \in \mathbb{X} \times \mathbb{U} \tag{6.4e}$$

$$x^p(t+T) \in \mathbb{X}_f \tag{6.4f}$$

There is a recursive-like elegance to (6.4), in that $\kappa_{mpc}(x)$ is essentially defined as a search over future candidates of itself. Whereas (6.3) explicitly involves *optimization-based* future feedbacks, the search in (6.4) can actually be (suboptimally) restricted to *any* arbitrary sub-class of feedbacks $k : [t, t+T] \times \mathbb{X} \rightarrow \mathbb{U}$. For example, this type of approach first appeared in References 94, 104, and 123, where the cost functional was minimized by restricting the search to the class of linear feedback $u = Kx$ (or $u = K(t)x$).

The error cone $\{x^p_{[t,t+T]}\}^*_{\mathcal{D}}$ associated with (6.4) is typically *much* less conservative than that of (6.2). This is due to the fact that (6.4d) accounts for future disturbance attenuation resulting from $k(\tau, x^p(\tau))$, an effect ignored in the open-loop predictions of (6.2). In the case of (6.3) and (6.4) it is no longer necessary to include T as an optimization variable, since by condition C4a one can now envision extending the horizon by appending an increment $k(T + \delta t, \cdot) = k_f(\cdot)$.

This notion of feedback MPC has been applied in References 113 and 114 to solve $\mathcal{H}_\infty$ disturbance attenuation problems. This approach avoids the need to solve a difficult Hamilton–Jacobi–Isaacs (HJI) equation, by combining a specially selected stage cost $L(x, u)$ with a local HJI approximation $W(x)$ (designed generally by solving an $\mathcal{H}_\infty$ problem for the linearized system). An alternative perspective of the implementation of (6.4) is developed in Reference 101, with particular focus on obstacle-avoidance in Reference 138. In this work, a set-invariance philosophy is used to propagate the uncertainty cone $\{x^p_{[t,t+T]}\}_{\mathcal{D}}$ for (6.4d) in the form of a control-invariant tube. This enables the use of efficient methods for constructing control invariant sets based on approximations such as polytopes or ellipsoids.

6.1.3 Adaptive approaches to MPC

Despite the fact that the ability to adjust to changing process conditions was one of the earliest industrial motivators for developing predictive control techniques, the progress in this area has been negligible. The small amount of progress that has been made is restricted to systems which do not involve constraints on the state, and which are affine in the unknown parameters. We will briefly describe two such results.

6.1.3.1 Certainty-equivalence implementation

The result in Reference 125 implements a certainty equivalence nominal-model[3] MPC feedback of the form $u(t) = \kappa_{mpc}(x(t), \hat{\theta}(t))$, to stabilize the uncertain system

$$\dot{x} = f(x, u, \theta) \triangleq f_0(x, u) + g(x, u)\theta \tag{6.5}$$

subject to an input constraint $u \in \mathbb{U}$. The vector $\theta \in \mathbb{R}^p$ represents a set of unknown constant parameters, with $\hat{\theta} \in \mathbb{R}^p$ denoting an identifier. Certainty equivalence implies that the nominal prediction model (3.1c) is of the same form as (6.5), but with $\hat{\theta}$ used in place of θ.

At any time $t \geq 0$, the identifier $\hat{\theta}(t)$ is defined to be a (min-norm) solution of

$$\int_0^t g(x(s), u(s))^T \big(\dot{x}(s) - f_0(x(s), u(s))\big) ds = \int_0^t g(x(s), u(s))^T g(x(s), u(s)) ds \; \hat{\theta} \tag{6.6}$$

which is solved over the window of all past history, under the assumption that $\dot{x}$ is measured (or computable). If necessary, an additional search is performed along the nullspace of $\int_0^t g(x, u)^T g(x, u) ds$ in order to guarantee $\hat{\theta}(t)$ yields a controllable certainty-equivalence model (since (6.6) is controllable by assumption).

The final result simply shows that there must exist a time $0 < t_a < \infty$ such that the regressor $\int_0^t g(x, u)^T g(x, u) ds$ achieves full rank, and thus $\hat{\theta}(t) \equiv \theta$ for all $t \geq t_a$. However, it is only by assumption that the state $x(t)$ does not escape the stabilizable region during the identification phase $t \in [0, t_a]$; moreover, there is no mechanism to decrease t_a in any way, such as by injecting excitation.

6.1.3.2 Stability-enforced approach

One of the early stability results for nominal-model MPC by Primbs [136, 137] involved the use of a global CLF $V(x)$ *instead* of a terminal penalty. Stability was enforced by constraining the optimization such that $V(x)$ is decreasing, and performance achieved by requiring the predicted cost to be less than that accumulated by simulation of pointwise min-norm control.

This idea was extended in Reference 1 to stabilize unconstrained systems of the form

$$\dot{x} = f(x, u, \theta) \triangleq f_0(x) + g_\theta(x)\theta + g_u(x)u \tag{6.7}$$

[3] Since this result arose early in the development of NMPC, it happens to be based upon a terminal-constrained controller (i.e., $\mathbb{X}_f \equiv \{0\}$); however, this is not critical to the adaptation.

Using ideas from robust stabilization, it is assumed that a global ISS-CLF[4] is known for the nominal system. Constraining $V(x)$ to decrease ensures convergence to a neighborhood of the origin, which gradually contracts as the identification proceeds. Of course, the restrictiveness of this approach lies in the assumption that $V(x)$ is known.

6.2 An adaptive approach to robust MPC

Both the theoretical and practical merits of model-based predictive control strategies for nonlinear systems are well established, as reviewed in Chapter 3. To date, the vast majority of implementations involve an "accurate model" assumption, in which the control action is computed on the basis of predictions generated by an approximate nominal process model, and implemented (unaltered) on the actual process. In other words, the effects of plant-model mismatch are completely ignored in the control calculation, and closed-loop stability hinges upon the critical assumption that the nominal model is a "sufficiently close" approximation of the actual plant. Clearly, this approach is only acceptable for processes whose dynamics can be modeled *a priori* to within a high degree of precision.

For systems whose true dynamics can only be approximated to within a large margin of uncertainty, it becomes necessary to directly account for the plant-model mismatch. To date, the most general and rigourous means for doing this involves explicitly accounting for the error in the online calculation, using the robust-MPC approaches discussed in Section 6.1. While the theoretical foundations and guarantees of stability for these tools are well established, it remains problematic in most cases to find an appropriate approach yielding a satisfactory balance between computational complexity, and conservatism of the error calculations. For example, the framework of min–max feedback-MPC [113, 149] provides the least-conservative control by accounting for the effects of future feedback actions, but is in most cases computationally intractable. In contrast, computationally simple approaches such as the open-loop method of Reference 119 yield such conservatively large error estimates, that a feasible solution to the optimal control problem often fails to exist.

For systems involving primarily *static* uncertainties, expressible in the form of unknown (constant) model parameters $\theta \in \Theta \subset \mathbb{R}^p$, it would be more desirable to approach the problem in the framework of adaptive control than that of robust control. Ideally, an adaptive mechanism enables the controller to improve its performance over time by employing a process model which asymptotically approaches that of the true system. Within the context of predictive control, however, the transient effects of parametric estimation error have proven problematic toward developing anything beyond the limited results discussed in Section 6.1.3. In short, the development of a general "robust adaptive-MPC" remains at present an open problem.

[4] That is, a CLF guaranteeing robust stabilization to a neighborhood of the origin, where the size of the neighborhood scales with the $\mathcal{L}_\infty$ bound of the disturbance signal.

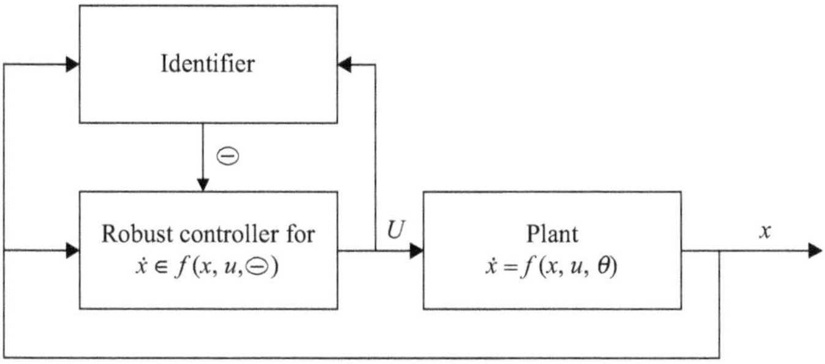

Figure 6.1 Adaptive robust feedback structure

In the following two chapters, we make no attempt to construct such a "robust adaptive" controller; instead we propose an approach more properly referred to as "adaptive robust" control. The approach differs from typical adaptive control techniques, in that the adaptation mechanism does not directly involve a parameter identifier $\hat{\theta} \in \mathbb{R}^p$. Instead, a set-valued description of the parametric uncertainty, Θ, is adapted online by an identification mechanism. By gradually eliminating values from Θ that are identified as being inconsistent with the observed trajectories, Θ gradually contracts upon θ in a nested fashion. By virtue of this nested evolution of Θ, it is clear that an adaptive feedback structure of the form in Figure 6.1 would retain the stability properties of any underlying robust control design.

The idea of arranging an identifier and robust controller in the configuration of Figure 6.1 is itself not entirely new. For example, the robust control design of Reference 41, appropriate for nonlinear systems affine in u whose disturbances are bounded and satisfy the so-called "matching condition," has been used by various authors [27, 41, 156] in conjunction with different identifier designs for estimating the disturbance bound resulting from parametric uncertainty. A similar concept for linear systems is given in Reference 91.

However, to the best of our knowledge this idea has not been well explored in the situation where the underlying robust controller is designed by robust-MPC methods. The advantage of such an approach is that one could then potentially embed an internal model of the identification mechanism into the predictive controller, as shown in Figure 6.2. In doing so the effects of future identification are accounted for within the optimal control problem, the benefits of which are discussed in Chapter 6.3.

In order to demonstrate that this methodology is not tied to any one particular robust-MPC technique, Chapters 6 and 7 each develop the idea within robust-MPC frameworks that represent opposing extremes with respect to computational complexity. The results in Chapter 6 focus upon demonstrating how this approach can further improve upon even the best-performing robust-MPC design, if computational limitations are not restrictive. In contrast, Chapter 7 seeks to show that at least some of these same benefits are realizable within even the most computationally simplistic framework for robust-MPC.

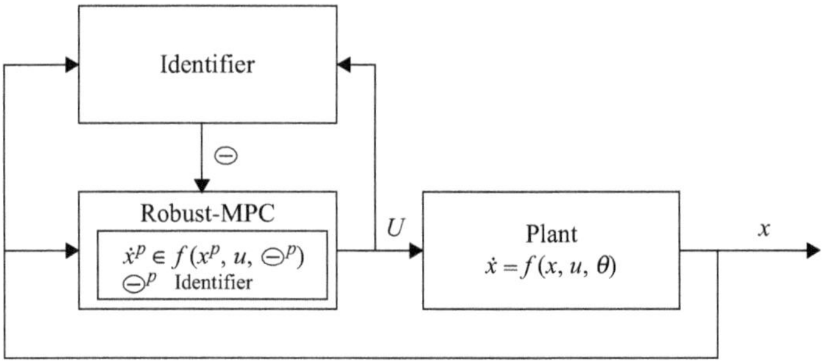

Figure 6.2 Adaptive robust MPC structure

A key result of both of Chapters 6 and 7 is that the internal model of the adaptive mechanism can be approximated suboptimally, to absolutely any degree of suboptimality. This implies that the adaptive methodology itself (in particular the form in Chapter 7) is amenable to real-time calculations, in the sense that its computational requirements can be restricted to within any desired margin of that of the underlying robust controller.

6.3 Minimally conservative approach

6.3.1 Problem description

The problem of interest is to achieve robust regulation, by means of state feedback, of the system state to some compact target set $\Sigma_x^o \in \mathbb{R}^n$. Optimality of the resulting trajectories are measured with respect to the accumulation of some instantaneous penalty (i.e., stage cost) $L(x, u) \geq 0$, which may or may not have physical significance. Furthermore, the state and input trajectories are required to obey pointwise constraints $(x, u) \in \mathbb{X} \times \mathbb{U} \subseteq \mathbb{R}^n \times \mathbb{R}^m$.

It is assumed that the system dynamics are not fully known, with uncertainty stemming from both unmodeled static nonlinearities and additional exogenous inputs. As such, the dynamics are assumed to be of the general form

$$\dot{x} = f(x, u, \theta, d(t)) \tag{6.8}$$

where f is a locally Lipschitz vector function of state $x \in \mathbb{R}^n$, control input $u \in \mathbb{R}^m$, disturbance input $d \in \mathbb{R}^d$, and constant parameters $\theta \in \mathbb{R}^p$. The entries of θ may represent physically meaningful model parameters (whose values are not exactly known *a priori*), or alternatively they could be parameters associated with any (finite) set of universal basis functions used to approximate unknown nonlinearities. The disturbance $d(t)$ represents the combined effects of actual exogenous inputs, neglected system states, or static nonlinearities lying outside the span of θ (such as the truncation error resulting from using a finite basis).

Assumption 6.3.1. *$\theta \in \Theta^o$, where Θ^o is a known compact subset of $\mathbb{R}^p$.* ■

Assumption 6.3.2. *$d(\cdot) \in \mathcal{D}_\infty$, where $\mathcal{D}_\infty$ is the set of all right-continuous $\mathcal{L}^\infty$-bounded functions $d:\mathbb{R}\to\mathcal{D}$; that is, composed of continuous subarcs $d_{[a,b)}$, and satisfying $d(\tau) \in \mathcal{D}$, $\forall \tau \in \mathbb{R}$, with $\mathcal{D} \subset \mathbb{R}^d$ a compact vector space.* ■

Unlike much of the robust or adaptive MPC literature, we do not necessarily assume exact knowledge of the system equilibrium manifold, or its stabilizing equilibrium control map. Instead, we make the following (weaker) set of assumptions.

Assumption 6.3.3. *Letting $\Sigma_u^o \subseteq \mathbb{U}$ be a chosen compact set, assume that $L: \mathbb{X} \times \mathbb{U} \to \mathbb{R}_{\geq 0}$ is continuous, $L(\Sigma_x^o, \Sigma_u^o) \equiv 0$, and $L(x,u) \geq \underline{\gamma}_L(\|(x,u)\|_{\Sigma_x^o\times\Sigma_u^o})$, $\underline{\gamma}_L \in \mathcal{K}_\infty$. As well, assume that Assumption 4.2.1 holds, with (4.3) interpreted as*

$$\min_{(u,\theta,d)\in\mathbb{U}\times\Theta^o\times\mathcal{D}} \left(\frac{L(x,u)}{\|f(x,u,\theta,d)\|}\right) \geq \frac{c_2}{\|x\|_{\Sigma_x^o}} \quad \forall x \in \mathbb{X} \setminus B(\Sigma_x^o, c_1) \tag{6.9}$$

■

Definition 6.3.4. *For each $\Theta \subseteq \Theta^o$, let $\Sigma_x(\Theta) \subseteq \Sigma_x^o$ denote the maximal (strongly) control-invariant subset for the differential inclusion $\dot{x} \in f(x,u,\Theta,\mathcal{D})$, using only controls $u \in \Sigma_u^o$.*

Assumption 6.3.5. *There exists a constant $N_\Sigma < \infty$, and a finite cover of Θ^o (not necessarily unique), denoted $\{\Theta\}^\Sigma$, such that*

i. *the collection $\{\mathring{\Theta}\}^\Sigma$ is an open cover for the interior $\mathring{\Theta}^o$.*
ii. *$\Theta \in \{\Theta\}^\Sigma$ implies $\Sigma_x(\Theta) \neq \emptyset$.*
iii. *$\{\Theta\}^\Sigma$ contains at most N_Σ elements.* ■

The most important requirement of Assumption 6.3.3 is that, since the exact location (in $\mathbb{R}^n \times \mathbb{R}^m$) of the equilibrium[5] manifold is not known *a priori*, $L(x,u)$ must be identically zero on the entire region of equilibrium candidates $\Sigma_x^o \times \Sigma_u^o$. One example of how to construct such a function would be to define $L(x,u) = \rho(x,u)\overline{L}(x,u)$, where $\overline{L}(x,u)$ is an arbitrary penalty satisfying $(x,u) \notin \Sigma_x^o \times \Sigma_u^o \implies \overline{L}(x,u) > 0$, and $\rho(x,u)$ is a smoothed indicator function of the form

$$\rho(x,u) = \begin{cases} 0 & (x,u) \in \Sigma_x^o \times \Sigma_u^o \\ \dfrac{\|(x,u)\|_{\Sigma_x^o\times\Sigma_u^o}}{\varepsilon_\rho} & 0 < \|(x,u)\|_{\Sigma_x^o\times\Sigma_u^o} < \varepsilon_\rho \\ 1 & \|(x,u)\|_{\Sigma_x^o\times\Sigma_u^o} \geq \varepsilon_\rho \end{cases} \tag{6.10}$$

The restriction that $L(x,u)$ is strictly positive definite with respect to $\Sigma_x^o \times \Sigma_u^o$ is made for convenience, and could be relaxed to positive semi-definite using an approach

[5] We use the word "equilibrium" loosely in the sense of control-invariant subsets of the target Σ_x^o, which need not be actual equilibrium *points* in the traditional sense.

similar to that of Reference 71 as long as $L(x, u)$ satisfies an appropriate detectability assumption (i.e., as long as it is guaranteed that all trajectories remaining in $\{x \mid \exists u\ s.t.\ L(x,u) = 0\}$ must asymptotically approach $\Sigma_x^o \times \Sigma_u^o$).

The first implication of Assumption 6.3.5 is that for any $\theta \in \Theta^o$, the target Σ_x^o contains a stabilizable "equilibrium" $\Sigma(\theta)$ such that the regulation problem is well-posed. Second, the openness of the covering in Assumption 6.3.5 implies a type of "local-ISS" property of these equilibria with respect to perturbations in small neighborhoods Θ of θ. This property ensures that the target is stabilizable given "sufficiently close" identification of the unknown θ, such that the adaptive controller design is tractable.

6.4 Adaptive robust controller design framework

6.4.1 *Adaptation of parametric uncertainty sets*

Unlike standard approaches to adaptive control, this work does not involve explicitly generating a parameter estimator $\hat{\theta}$ for the unknown θ. Instead, the parametric uncertainty set Θ^o is adapted to gradually eliminate sets which do not contain θ. To this end, we define the infimal uncertainty set

$$\mathcal{Z}(\Theta, x_{[a,b]}, u_{[a,b]}) \triangleq \{\theta \in \Theta \mid \dot{x}(\tau) \in f(x(\tau), u(\tau), \theta, \mathcal{D}), \quad \forall \tau \in [a,b]\} \tag{6.11}$$

By definition, $\mathcal{Z}$ represents the best-case performance that could be achieved by any identifier, given a set of data generated by (6.8), and a prior uncertainty bound Θ. Since exact online calculation of (6.11) is generally impractical, we assume that the set $\mathcal{Z}$ is approximated online using an arbitrary estimator Ψ. This estimator must be chosen to satisfy the following conditions.

Criterion 6.4.1. *$\Psi(\cdot,\cdot,\cdot)$ is designed such that for $a \le b \le c$, and for any $\Theta \subseteq \Theta^o$,*

1. $\mathcal{Z} \subseteq \Psi$
2. $\Psi(\Theta, \cdot, \cdot) \subseteq \Theta$, *and closed*
3. $\Psi(\Theta_1, x_{[a,b]}, u_{[a,b]}) \subseteq \Psi(\Theta_2, x_{[a,b]}, u_{[a,b]})$, *for* $\Theta_1 \subseteq \Theta_2 \subseteq \Theta^o$
4. $\Psi(\Theta, x_{[a,b]}, u_{[a,b]}) \supseteq \Psi(\Theta, x_{[a,c]}, u_{[a,c]})$
5. $\Psi(\Theta, x_{[a,c]}, u_{[a,c]}) \equiv \Psi(\Psi(\Theta, x_{[a,b]}, u_{[a,b]}), x_{[b,c]}, u_{[b,c]})$

The set Ψ represents an approximation of $\mathcal{Z}$ in two ways. First, both Θ^o and Ψ can be restricted *a priori* to any class of finitely parameterized sets, such as linear polytopes, quadratic balls, and so on. Second, contrary to the actual definition of (6.11), Ψ can be computed by *removing* values from Θ^o as they are determined to violate the differential inclusion model. As such, the search for infeasible values can be terminated at any time without violating Criterion 6.4.1.

The closed loop dynamics of (6.8) then take the form

$$\dot{x} = f(x, \kappa_{mpc}(x, \Theta(t)), \theta, d(t)), \quad x(t_0) = x_0 \tag{6.12a}$$

$$\Theta(t) = \Psi(\Theta^o, x_{[t_0,t]}, u_{[t_0,t]}) \tag{6.12b}$$

where $\kappa_{mpc}(x, \Theta)$ represents the MPC feedback policy, detailed in Section 6.4.2. In practice, the (set-valued) controller state Θ could be generated using an update law $\dot{\Theta}$ designed to gradually contract the set (satisfying Criterion 6.4.1). However, the given statement of (6.12b) is more general, as it allows for $\Theta(t)$ to evolve discontinuously in time, as may happen, for example, when the sign of a parameter can suddenly be conclusively determined.

6.4.2 Feedback-MPC framework

In the context of min–max robust MPC, it is well known that feedback-MPC, because of its ability to account for the effects of future feedback decisions on disturbance attenuation, provides significantly less conservative performance than standard open-loop MPC implementations. In the following, the same principle is extended to incorporate the effects of future parameter adaptation.

In typical feedback-MPC fashion, the receding horizon control law in (6.12) is defined by minimizing over *feedback policies* $\kappa : \mathbb{R}_{\geq 0} \times \mathbb{R}^n \times \overline{\text{cov}}\,\{\Theta^o\} \to \mathbb{R}^m$ as

$$u = \kappa_{mpc}(x, \Theta) \triangleq \kappa^*(0, x, \Theta) \tag{6.13a}$$

$$\kappa^* \triangleq \arg\min_{\kappa(\cdot,\cdot,\cdot)} J(x, \Theta, \kappa) \tag{6.13b}$$

where $J(x, \Theta, \kappa)$ is the (worst-case) cost associated with the optimal control problem:

$$J(x, \Theta, \kappa) \triangleq \max_{\substack{\theta \in \Theta \\ d(\cdot) \in \mathcal{D}_\infty}} \int_0^T L(x^p, u^p) d\tau + W(x_f^p, \hat{\Theta}_f) \tag{6.14a}$$

$$\text{s.t.} \quad \forall \tau \in [0, T]$$

$$\frac{d}{d\tau} x^p = f(x^p, u^p, \theta, d), \quad x^p(0) = x \tag{6.14b}$$

$$\hat{\Theta}(\tau) = \Psi_p(\Theta(t), x^p_{[0,\tau]}, u^p_{[0,\tau]}) \tag{6.14c}$$

$$x^p(\tau) \in \mathbb{X} \tag{6.14d}$$

$$u^p(\tau) \triangleq \kappa(\tau, x^p(\tau), \hat{\Theta}(\tau)) \in \mathbb{U} \tag{6.14e}$$

$$x_f^p \triangleq x^p(T) \in \mathbb{X}_f(\hat{\Theta}_f) \tag{6.14f}$$

$$\hat{\Theta}_f \triangleq \Psi_f(\Theta(t), x^p_{[0,T]}, u^p_{[0,T]}) \tag{6.14g}$$

Throughout the remainder, we denote the optimal cost $J^*(x, \Theta) \triangleq J(x, \Theta, \kappa^*)$, and furthermore we drop the explicit constraints (6.14d)–(6.14f) by assuming the definitions of L and W have been extended as follows:

$$L(x, u) = \begin{cases} L(x, u) < \infty & (x, u) \in \mathbb{X} \times \mathbb{U} \\ +\infty & \text{otherwise} \end{cases} \tag{6.15a}$$

$$W(x, \Theta) = \begin{cases} W(x, \Theta) < \infty & x \in \mathbb{X}_f(\Theta) \\ +\infty & \text{otherwise} \end{cases} \tag{6.15b}$$

The parameter identifiers Ψ_p and Ψ_f in (6.14) represent internal model approximations of the actual identifier Ψ, and must satisfy both Criterion 6.4.1 and the following criterion.

Criterion 6.4.2. *For identical arguments,* $\mathcal{Z} \subseteq \Psi \subseteq \Psi_f \subseteq \Psi_p$.

Remark 6.4.3. *We distinguish between different identifiers to emphasize that, depending on the frequency at which calculations are called, differing levels of accuracy can be applied to the identification calculations. The ordering in Criterion 6.4.2 is required for stability, and implies that identifiers existing within faster timescales provide more conservative approximations of the uncertainty set.*

There are two important characteristics which distinguish (6.14) from a standard (non-adaptive) feedback-MPC approach. First, the future evolution of $\hat{\Theta}$ in (6.14c) is fed back into both (6.14b) and (6.14e). The benefits of this feedback are analogous to those of adding state-feedback into the MPC calculation; the resulting cone of possible trajectories $x^p(\cdot)$ is narrowed by accounting for the effects of future adaptation on disturbance attenuation, resulting in less conservative worst-case predictions.

The second distinction is that both W and $\mathbb{X}_f$ are parameterized as functions of $\hat{\Theta}_f$, which reduces the conservatism of the terminal cost. Since the terminal penalty W has the interpretation of the "worst-case cost-to-go," it stands to reason that W should decrease with decreased parametric uncertainty. In addition, the domain $\mathbb{X}_f$ would be expected to enlarge with decreased parametric uncertainty, which in some situations could mean that a stabilizing CLF-pair $(W(x, \Theta), \mathbb{X}_f(\Theta))$ can be constructed even when no such CLF exists for the original uncertainty Θ^o. This effect is discussed in greater depth in Section 6.5.2.

6.4.3 Generalized terminal conditions

To guide the selection of $W(x_f, \hat{\Theta}_f)$ and $\mathbb{X}_f(\hat{\Theta}_f)$ in (6.14), it is important to outline (sufficient) conditions under which (6.12)–(6.14) can guarantee stabilization to the target Σ_x^o. The statement given here is extended from the set of such conditions for robust MPC from Reference 126 that was outlined in Sections 3.1 and 6.1.1.

For reasons that are explained later in Section 6.5.2, it is useful to present these conditions in a more general context in which $W(\cdot, \Theta)$ is allowed to be $\mathcal{LS}$-continuous with respect to x, as may occur if W is generated by a switching mechanism. This adds little additional complexity to the analysis, since (6.14) is already discontinuous due to constraints.

Criterion 6.4.4. *The set-valued terminal constraint function* $\mathbb{X}_f : \overline{\text{cov}}\{\Theta^o\} \to \overline{\text{cov}}\{\mathbb{X}\}$ *and terminal penalty function* $W : \mathbb{R}^n \times \overline{\text{cov}}\{\Theta^o\} \to [0, +\infty]$ *are such that for each* $\Theta \in \overline{\text{cov}}\{\Theta^o\}$, *there exists* $k_f(\cdot, \Theta) : \mathbb{X}_f \to \mathbb{U}$ *satisfying*

1. $\mathbb{X}_f(\Theta) \neq \emptyset$ *implies that* $\Sigma_x^o \cap \mathbb{X}_f(\Theta) \neq \emptyset$, *and* $\mathbb{X}_f(\Theta) \subseteq \mathbb{X}$ *is closed*
2. $W(\cdot, \Theta)$ *is* $\mathcal{LS}$*-continuous with respect to* $x \in \mathbb{R}^n$
3. $k_f(x, \Theta) \in \mathbb{U}$, *for all* $x \in \mathbb{X}_f(\Theta)$

4. $\mathbb{X}_f(\Theta)$ *and* $\Sigma_x(\Theta) \subseteq \{\Sigma_x^o \cap \mathbb{X}_f(\Theta)\}$ *are both strongly positively invariant with respect to the differential inclusion* $\dot{x} \in f(x, k_f(x,\Theta), \Theta, \mathcal{D})$
5. $\forall x \in \mathbb{X}_f(\Theta)$, *and denoting* $\mathcal{F} \triangleq f(x, k_f(x,\Theta), \Theta, \mathcal{D})$,

$$\max_{f\in\mathcal{F}}\left(L(x, k_f(x,\Theta)) + \liminf_{\substack{v\to f\\ \delta\downarrow 0}}\left(\frac{W(x+\delta v,\ \Theta) - W(x,\Theta)}{\delta}\right)\right) \le 0$$

■

Although condition Criterion 6.4.4 point 5 is expressed in a slightly non-standard form, it embodies the standard interpretation that W must be decreasing by at least an amount $-L(x, k_f(x,\Theta))$ along all vector fields in the closed-loop differential inclusion $\mathcal{F}$; i.e., $W(x,\Theta)$ is a robust-CLF (in an appropriate non-smooth sense) on the domain $\mathbb{X}_f(\Theta)$. Lyapunov stability involving $\mathcal{LS}$-continuous functions is thoroughly studied in Reference 40, and provides a meaningful sense in which W can be considered a "robust-CLF" despite its discontinuous nature.

It is important to note that for the purposes of Criterion 6.4.4, $W(x,\Theta)$ and $\mathbb{X}_f(\Theta)$ are parameterized by the set Θ, but the criterion imposes no restrictions on their functional dependence with respect to the Θ argument. This Θ-dependence is required to satisfy the following criteria:

Criterion 6.4.5. *For any* $\Theta_1, \Theta_2 \in \overline{cov}\{\Theta^o\}$ *such that* $\Theta_1 \subseteq \Theta_2$,

1. $\mathbb{X}_f(\Theta_2) \subseteq \mathbb{X}_f(\Theta_1)$
2. $W(x,\Theta_1) \le W(x,\Theta_2), \quad \forall x \in \mathbb{X}_f(\Theta_2)$ ■

Designing W and $\mathbb{X}_f$ as functions of Θ satisfying Criteria 6.4.4 and 6.4.5 may appear prohibitively complex; however, the task is greatly simplified by noting that neither criterion imposes any notion of continuity of W or $\mathbb{X}_f$ with respect to Θ. A constructive design approach exploiting this fact is presented in Section 6.5.2.

6.4.4 Closed-loop stability

Theorem 6.4.6 (Main result). *Given system* (6.8), *target* Σ_x^o, *and penalty L satisfying Assumptions* 6.3.1–6.3.3*rd,* 6.3.5, *assume the functions* Ψ, Ψ_p, Ψ_f, W *and* $\mathbb{X}_f$ *are designed to satisfy Criteria* 6.4.1, 6.4.2, 6.4.4 *and* 6.4.5. *Furthermore, let* $\mathcal{X}_0 \triangleq \mathcal{X}_0(\Theta^o) \subseteq \mathbb{X}$ *denote the set of initial states, with uncertainty* $\Theta(t_0) = \Theta^o$, *for which* (6.14) *has a solution. Then under* (6.12), Σ_x^o *is feasibly asymptotically stabilized from any* $x_0 \in \mathcal{X}_0$. ■

Remark 6.4.7. *As indicated by Assumption* 6.3.5, *the existence of an invariant target set* $\Sigma_x^o(\Theta^o)$, *robust to the full parametric uncertainty* Θ^o, *is not required for Theorem* 6.4.6 *to hold. The identifier* $\hat{\Theta}_f$ *must be contained in a sufficiently small neighborhood of (the worst-case)* θ *such that nontrivial* $\mathbb{X}_f(\hat{\Theta}_f)$ *and* $W(\cdot,\hat{\Theta}_f)$ *exist, for* (6.14) *to be solvable. While this imposes a minimum performance requirement on* Ψ_f, *it enlarges the domain* $\mathcal{X}_0$ *for which the problem is solvable.*

6.5 Computation and performance issues

6.5.1 *Excitation of the closed-loop trajectories*

Contrary to much of the adaptive control literature, including adaptive-MPC approaches such as Reference 125, the result of Theorem 6.4.6 does not depend on any auxiliary excitation signal, nor does it require any assumptions regarding the persistency or quality of excitation in the closed-loop behavior.

Instead, any benefits to the identification which result from injecting excitation into the input signal are predicted by (6.14c) and (6.14g), and thereby are automatically accounted for in the posed optimization. In the particular case where $\Psi_p \equiv \Psi_f \equiv \Psi$, then the controller generated by (6.14) will automatically inject the exact type and amount of excitation which optimizes the cost $J^*(x, \Theta)$; that is, the closed-loop behavior (6.12) could be considered "optimally-exciting." Unlike most *a priori* excitation signal design methods, excess actuation is not wasted in trying to identify parameters which have little impact on the closed-loop performance (as measured by J^*).

As Ψ_p and Ψ_f deviate from Ψ, the convergence result of Theorem 6.4.6 remains valid. However, the non-smoothness of $J^*(x, \Theta)$ (with respect to both x and Θ) makes it difficult to quantify the impact of these deviations on the closed-loop behavior. Qualitatively, small changes to Ψ_p or Ψ_f yielding increasingly conservative identification would generally result in the optimal control solution injecting additional excitation to compensate for the de-sensitized identifier. However, if the changes to Ψ_p or Ψ_f are sufficiently large such that the injection of additional excitation is insufficient to prevent a discontinuous increase in J^*, then it is possible that the optimal solution may suddenly involve *less* excitation than previously, to instead reduce actuation energy. Clearly this behaviour is the result of nonconvexities in the optimal control problem (6.13), which is inherently a nonconvex problem even in the absence of the adaptive mechanisms proposed here.

6.5.2 *A practical design approach for W and $\mathbb{X}_f$*

Proposition 6.5.1. *Let $\{(W^i, \mathbb{X}_f^i)\}$ denote a finitely-indexed collection of terminal function candidates, with indices $i \in \mathcal{I}$, where each pair $(W^i, \mathbb{X}_f^i)$ satisfies Criteria* 6.4.4 *and* 6.4.5. *Then*

$$W(x, \Theta) \triangleq \min_{i \in \mathcal{I}}\{W^i(x, \Theta)\}, \quad \mathbb{X}_f(\Theta) \triangleq \bigcup_{i \in \mathcal{I}}\{\mathbb{X}_f^i(\Theta)\} \tag{6.16}$$

satisfy Criteria 6.4.4 *and* 6.4.5. ■

Using Proposition 6.5.1, it is clear that one approach to constructing $W(\cdot, \cdot)$ and $\mathbb{X}_f(\cdot)$ is to use a collection of pairs of the form

$$\left(W^i(x, \Theta), \mathbb{X}_f^i(\Theta)\right) = \begin{cases} \left(W^i(x), \mathbb{X}_f^i\right) & \Theta \subseteq \Theta^i \\ (+\infty, \emptyset) & \text{otherwise} \end{cases}$$

This collection can be obtained as follows:

1. **Generate a finite collection $\{\Theta^i\}$ of sets covering Θ^o.**
   - The elements of the collection can, and should, be overlapping, nested, and ranging in size.
   - Categorize $\{\Theta^i\}$ in a hierarchical (i.e., "tree") structure such that
     i. level 1 (i.e., the top) consists of Θ^o. (Assuming $\Theta^o \in \{\Theta^i\}$ is w.l.o.g., since $W(\cdot, \Theta^o) \equiv +\infty$ and $\mathbb{X}_f(\Theta^o) = \emptyset$ satisfy Criteria 6.4.4 and 6.4.5),
     ii. every set in the l'th vertical level is nested inside one or more "parents" on level $l-1$,
     iii. at every level, the "horizontal peers" constitute a cover[6] of Θ^o.
2. **For every set $\Theta^j \in \{\Theta^i\}$, calculate a robust CLF $W^j(\cdot) \equiv W^j(\cdot, \Theta^j)$, and approximate its domain of attraction $\mathbb{X}_f^j \equiv \mathbb{X}_f^j(\Theta^j)$.**
   - Generally, $W^j(\cdot, \Theta^j)$ is selected first, after which $\mathbb{X}_f(\Theta^j)$ is approximated as either a maximal level set of $W^j(\cdot, \Theta^j)$, or as some other controlled-invariant set.
   - Since the elements of $\{\Theta^i\}$ need not be unique, one could actually define multiple $(W^i, \mathbb{X}_f^i)$ pairs associated with the same Θ^j.
   - While not an easy task, this is a very standard robust-control calculation. As such, there is a wealth of tools in the robust control and viability literatures (see, e.g., Reference 17) to tackle this problem.
3. **Calculate $W(\cdot, \Theta)$ and $\mathbb{X}_f(\Theta)$ online:**
   i. Given Θ, identify all sets that are active: $\mathcal{I}^* = \mathcal{I}^*(\Theta) \triangleq \{j \mid \Theta \subseteq \Theta^j\}$. Using the hierarchy, test only immediate children of active parents.
   ii. Given x, search over the active indices to identify $\mathcal{I}_f^* = \mathcal{I}_f^*(x, \mathcal{I}^*) \triangleq \{j \in \mathcal{I}^* \mid x \in \mathbb{X}_f^j\}$. Define $W(x, \Theta) \triangleq \min_{j \in \mathcal{I}_f^*} W^j(x)$ by testing indices in $\mathcal{I}_f^*$, setting $W(x, \Theta) = +\infty$ if $\mathcal{I}_f^* = \emptyset$.

Remark 6.5.2. *Although the above steps assume that Θ^j is selected before $\mathbb{X}_f^j$, an alternative approach would be to design the candidates $W^j(\cdot)$ on the basis of a collection of* parameter values $\hat{\theta}^j$. *Briefly, this could be constructed as follows:*

1. *Generate a grid of values $\{\theta^i\}$ distributed across Θ^o.*
2. *Design $W^j(\cdot)$ based on a certainty-equivalence model for $\hat{\theta}^j$ (e.g., by linearization). Specify $\mathbb{X}_f^j$ (likely as a level set of W^j), and then approximate the maximal neighborhood Θ^j of $\hat{\theta}^j$ such that Criterion* 6.4.4 *holds.*
3. *For the same (θ^j, W^j) pair, multiple $(W^j, \mathbb{X}_f^j)$ candidates can be defined corresponding to different Θ^j.*

6.6 Robustness issues

One could argue that if the disturbance model $\mathcal{D}$ in (6.8) encompasses *all* possible sources of model uncertainty, then the issue of robustness is completely addressed

[6] Specifically, the interiors of all peers must together constitute an open cover.

by the min–max formulation of (6.14). In practice this is not realistic, since it is generally desirable to explicitly consider significant disturbances only, or to exclude $\mathcal{D}$ entirely if Θ represents the dominant uncertainty. The lack of nominal robustness to model error in constrained NMPC is a well-documented problem, as discussed in Reference 70. In particular, References 69 and 119 establish nominal robustness (for "accurate-model," discrete-time MPC) in part by implementing the constraint $x \in \mathbb{X}$ as a succession of strictly nested sets. We present here a modification to this approach, that is, relevant to the current adaptive framework.

In addition to ensuring robustness of the controller itself, using methods similar to those mentioned above, it is equally important to ensure that the adaptive mechanism Ψ, including its internal models Ψ_f and Ψ_p, exhibits at least some level of nominal robustness to unmodeled disturbances. By Criterion 6.4.1.4, the online estimation *must* evolve in a nested fashion and therefore the true θ must never be inadvertently excluded from the estimated uncertainty set. Therefore, just as $\mathcal{Z}$ in (6.11) defined a best-case bound around which the identifiers in the previous sections could be designed, we present here a modification of (6.11) which quantifies the type of conservatism required in the identification calculations for the identifiers to possess nominal robustness.

For any $\gamma, \varepsilon \geq 0$, and with $\tau_a \triangleq \tau - a$, we define the following modification of (6.11):

$$\mathcal{Z}^{\varepsilon,\gamma}(\Theta, x_{[a,b]}, u_{[a,b]}) \triangleq \{\theta \in \Theta \mid B(\dot{x}, \varepsilon + \gamma\tau_a) \cap f(B(x, \gamma\tau_a), u, \theta, \mathcal{D}) \neq \emptyset, \quad \forall \tau\}. \tag{6.17}$$

Equation (6.17) provides a conservative outer approximation of (6.11) such that $\mathcal{Z} \subseteq \mathcal{Z}^{\varepsilon,\gamma}$. The definition in (6.17) accounts for two different types of conservatism that can be introduced into the identification calculations. First, the parameter $\varepsilon > 0$ represents a minimum tolerance for the distance between actual derivative information from trajectory $x_{[a,b]}$ and the model (6.8) when determining if a parameter value can be excluded from the uncertainty set. For situations where the trajectory $x_{[a,b]}$ is itself a *prediction*, as is the case for the internal models Ψ_f and Ψ_p, the parameter $\gamma > 0$ represents increasingly relaxed tolerances applied along the length of the trajectory. Throughout the following we denote $\mathcal{Z}^{\varepsilon} \equiv \mathcal{Z}^{\varepsilon,0}$, with analogous notations for Ψ, Ψ_f, and Ψ_p.

The following technical property of definition (6.17) is useful toward establishing the desired robustness claim.

Claim 6.6.1. *For any* $a<b<c$, $\gamma \geq 0$, *and* $\varepsilon \geq \varepsilon' \geq 0$, *let* $x'_{[a,c]}$ *be an arbitrary, continuous perturbation of* $x_{[a,b]}$ *satisfying*

i. $$\|x'(\tau) - x(\tau)\| \leq \begin{cases} \gamma(\tau - a) & \tau \in [a, b] \\ \gamma(b - a) & \tau \in [b, c] \end{cases}$$

ii. $\|\dot{x}'(\tau) - \dot{x}(\tau)\| \leq \begin{cases} \varepsilon - \varepsilon' + \gamma(\tau - a) & \tau \in [a,b] \\ \gamma(b-a) & \tau \in [b,c] \end{cases}$

Then, $\mathcal{Z}^{\varepsilon,\gamma}$ satisfies

$$\mathcal{Z}^{\varepsilon,\gamma}\left(\mathcal{Z}^{\varepsilon'}(\Theta, x'_{[a,b]}, u_{[a,b]}), x'_{[b,c]}, u_{[b,c]}\right) \subseteq \mathcal{Z}^{\varepsilon,\gamma}(\Theta, x_{[a,c]}, u_{[a,c]}). \tag{6.18}$$

■

Based on (6.17), we are now able to detail sufficient conditions under which the stability claim of Theorem 6.4.6 holds in the presence of small, unmodeled disturbances. For convenience, the following proposition is restricted to the situation where the only discontinuities in $W(x,\Theta)$ and $\mathbb{X}_f(\Theta)$ are those generated by a switching mechanism (as per Proposition 6.5.1) between a set of candidates $\{W^i(x,\Theta), \mathbb{X}_f^i(\Theta)\}$ that are individually *continuous* on $x \in \mathbb{X}_f^i(\Theta)$ (i.e., a strengthening of Criterion 6.4.4 point 2). With additional complexity, the proposition can be extended to general $\mathcal{LS}$-continuous penalties $W(x,\Theta)$.

Proposition 6.6.2. *Assume that the following modifications are made to the design in Section* 6.4*:*

i. *$W(x,\Theta)$ and $\mathbb{X}_f(\Theta)$ are constructed as per Proposition* 6.5.1*, but with Criterion* 6.4.4 *point* 2 *strengthened to require the individual $W^i(x,\Theta)$ to be continuous w.r.t $x \in \mathbb{X}_f^i(\Theta)$.*
ii. *For some design parameter $\varepsilon_x > 0$,* (6.15) *and* (6.16) *are redefined as:*

$$\tilde{L}(\tau,x,u) = \begin{cases} L(x,u) & (x,u) \in \overleftarrow{B}(\mathbb{X}, \varepsilon_x \frac{\tau}{T}) \times \mathbb{U} \\ +\infty & \text{otherwise} \end{cases}$$

$$\tilde{W}^i(x,\Theta) = \begin{cases} W^i(x) & x \in \overleftarrow{B}(\mathbb{X}_f^i(\Theta), \varepsilon_x) \\ +\infty & \text{otherwise} \end{cases}$$

iii. *The individual sets $\mathbb{X}_f^i$ are specified such that there exists $\varepsilon_f > 0$, for which Criterion* 6.4.4 *point* 4 *holds for every inner approximation $\overleftarrow{B}(\mathbb{X}_f^i(\Theta), \varepsilon_x')$, $\varepsilon_x' \in [0, \varepsilon_x]$, where positive invariance is with respect to all flows generated by the differential inclusion $\dot{x} \in B(f(x, k_f^i(x,\Theta), \Theta, \mathcal{D}), \varepsilon_f)$*
iv. *Using design parameters $\varepsilon > \varepsilon' > 0$ and $\gamma > 0$, the identifiers are modified as follows:*
   - *Ψ in* (6.12b) *is replaced by $\Psi^{\varepsilon'} \equiv \Psi^{\varepsilon',0}$*
   - *Ψ_p and Ψ_f in* (6.14) *are replaced by $\Psi_p^{\varepsilon,\gamma}$ and $\Psi_f^{\varepsilon,\gamma}$, respectively*

   where the new identifiers are assumed to satisfy Criteria 6.4.1 *and* 6.4.2*, and a relation of the form* (6.18).

Then for any compact subset $\bar{\mathcal{X}}_0 \subseteq \mathcal{X}_0(\Theta^o)$, there exists $c^* = c^*(\gamma, \varepsilon_x, \varepsilon_f, \varepsilon, \varepsilon', \bar{\mathcal{X}}_0) > 0$ such that, for all $x_0 \in \bar{\mathcal{X}}_0$ and for all disturbances $\|d_2\| \leq c \leq c^*$, the target Σ_x^o and the actual dynamics

$$\dot{x} = f(x, \kappa_{mpc}(x, \Theta(t)), \theta, d(t)) + d_2(t), \quad x(t_0) = x_0 \tag{6.19a}$$

$$\Theta(t) = \Psi^{\varepsilon'}(\Theta^o, x_{[t_0,t]}, u_{[t_0,t]}) \tag{6.19b}$$

are ISS; i.e., there exists $\alpha_d \in \mathcal{K}$ such that $x(t)$ asymptotically converges to $B(\Sigma_x^o, \alpha_d(c))$. ■

6.7 Example problem

To demonstrate the versatility of our approach, we consider the following nonlinear system:

$$\begin{aligned} \dot{x}_1 &= -x_1 + |2\sin(x_1 + \pi\theta_1) + 1.5\theta_2 - x_1 + x_2|x_1 + d_1(t) \\ \dot{x}_2 &= 10\,\theta_{4a}\theta_{4b}x_1(u + \theta_3) + d_2(t) \end{aligned}$$

The uncertainty $\mathcal{D}$ is given by $|d_1|, |d_2| \leq 0.1$, and Θ^o by $\theta_1, \theta_2, \theta_3 \in [-1, 1]$, and $\theta_{4a} \in \{-1, +1\}$, $\theta_{4b} \in [0.5, 1]$. The control objective is to achieve regulation of x_1 to the set $x_1 \in [-0.2, 0.2]$, subject to the constraints $\mathbb{X} \triangleq \{|x_1| \leq M_1 \text{ and } |x_2| \leq M_2\}$, $\mathbb{U} \triangleq \{|u| \leq M_u\}$, with $M_1, M_2 \in (0, +\infty]$ and $M_u \in (1, +\infty]$ any given constants. The dynamics exhibit several challenging properties: (i) state constraints, (ii) nonlinear parameterization of θ_1 and θ_2, (iii) potential open-loop instability with finite escape, (iv) uncontrollable linearization, (v) unknown sign of control gain, and (vi) exogenous disturbances. This system is not stabilizable by any non-adaptive approach (MPC or otherwise), and furthermore fits very few, if any, existing frameworks for adaptive control.

One key property of the dynamics (which is arguably necessary for the regulation objective to be well-posed) is that for any *known* $\theta \in \Theta$ the target is stabilizable and nominally robust. This follows by observing that the surface

$$s \triangleq 2\sin(x_1 + \pi\theta_1) + 1.5\theta_2 - x_1 + x_2 = 0$$

defines a sliding mode for the system, with a robustness margin $|s| \leq 0.5$ for $|x_1| \geq 0.2$. This motivates the design choices:

$$\begin{aligned} \mathbb{X}_f(\Theta) &\triangleq \{x \in \mathbb{X} \,|\, -M_2 \leq \underline{\Gamma}(x_1, \Theta) \leq x_2 \leq \overline{\Gamma}(x_1, \Theta) \leq M_2\} \\ \underline{\Gamma} &\triangleq x_1 - 1.5\underline{\theta}_2 - 2\sin(x_1 + \pi\theta_1^{avg}) - 2\pi(\underline{\theta}_1 - \theta_1^{avg}) + 0.5 \\ \overline{\Gamma} &\triangleq x_1 - 1.5\overline{\theta}_2 - 2\sin(x_1 + \pi\theta_1^{avg}) - 2\pi(\overline{\theta}_1 - \theta_1^{avg}) - 0.5 \end{aligned}$$

where $\overline{\theta}_i$, $\underline{\theta}_i$ denote upper and lower bounds corresponding to $\Theta \subseteq \Theta^o$, and $\theta^{avg} \triangleq \frac{\overline{\theta} + \underline{\theta}}{2}$. The set $\mathbb{X}_f(\Theta)$ satisfies Criterion 6.4.5 and is nonempty for any Θ such that $\overline{\theta}_2 - \underline{\theta}_2 + \pi(\overline{\theta}_1 - \underline{\theta}_1) \leq 0.5$, that defines minimum thresholds for the performance of Ψ_f and the amount of excitation in solutions to (6.14).

It can be shown that $|s| \leq 0.5 \overset{\forall\theta\in\Theta^o}{\Longrightarrow} |x_1 - x_2| \leq 4$, and that $\mathbb{X}_f(\Theta)$ is control-invariant using $u \in [-1, 1]$, as long as the sign θ_{4a} is known. This motivates the definitions $\Sigma_u^o \triangleq [-1, 1]$, $\Sigma_1 = [-0.2, 0.2]$, $\Sigma_{12} = [-4, 4]$, and $\Sigma_x^o \triangleq \{x \mid (x_1, x_1 - x_2) \in \Sigma_1 \times \Sigma_{12}\}$, plus the modification of $\mathbb{X}_f(\Theta)$ above to contain the explicit requirement $\Theta_{4a} = \{-1, +1\} \implies \mathbb{X}_f(\Theta) = \emptyset$. Then on $x \in \mathbb{X}_f(\Theta)$, the cost functions $W(x, \Theta) \triangleq \frac{1}{2}\|x_1\|_{\Sigma_1}^2$ and $L(x, u) \triangleq \frac{1}{2}\left(\|x_1\|_{\Sigma_1}^2 + \|x_1 - x_2\|_{\Sigma_{12}}^2 + \|u\|_{\Sigma_u^o}^2\right)$ satisfy all the claims of Criterion 6.4.4, since $W \equiv L \equiv 0$ on $x \in \mathbb{X}_f \cap \Sigma_x^o$, and on $x \in \mathbb{X}_f \setminus \Sigma_x^o$ one has:

$$\dot{W} \leq \|x_1\|_{\Sigma_1}\left(-\frac{1}{2}|x_1| + 0.1\right) \leq -\frac{1}{2}\|x_1\|_{\Sigma_1}^2 \leq -L(x, u).$$

6.8 Conclusions

In this chapter, we have demonstrated the methodology for adaptive MPC proposed in Section 6.2, in which the adverse effects of parameter identification error are explicitly minimized using a robust MPC approach. As a result, it is possible to address both state and input constraints within the adaptive framework. Another key advantage of this approach is that the effects of future parameter estimation can be incorporated into the optimization problem, raising the potential to significantly reduce the conservativeness of the solutions, especially with respect to design of the terminal penalty. While the results presented here are conceptual, in that they are generally intractable to compute due to the underlying min–max feedback-MPC framework, this chapter provides insight into the maximum performance that could be attained by incorporating adaptation into a robust-MPC framework.

6.9 Proofs for Chapter 6

6.9.1 *Proof of Theorem 6.4.6*

This proof will follow the so-called "direct method" of establishing stability by directly proving strict decrease of $J^*(x(t), \Theta(t))$, for all $x \notin \Sigma_x^o$. Stability analysis involving $\mathcal{LS}$-continuous Lyapunov functions (e.g., Reference 40, Theorem 4.5.5) typically involves the proximal subgradient $\partial_p J^*$ (a generalization of ∇J), which is a somewhat ambiguous quantity in the context here given (6.12b). Instead, this proof exploits an alternative framework involving subderivates (generalized Dini-derivatives), which is equivalent by Reference 40, Proposition 4.5.3. Together, the following two conditions can be shown sufficient to ensure decrease of J^*, where $\mathcal{F} \triangleq f(x, \kappa_{mpc}(x, \Theta(t)), \Theta(t), \mathcal{D})$

i. $$\max_{f\in\mathcal{F}} \overrightarrow{D} J^*(x, \Theta) \triangleq \max_{f\in\mathcal{F}} \liminf_{\substack{v\to f\\ \delta\downarrow 0}} \frac{J^*(x + \delta v, \Theta(t+\delta)) - J^*(x, \Theta(t))}{\delta} < 0$$

ii. $$\min_{f\in\mathcal{F}} \overleftarrow{D} J^*(x, \Theta) \triangleq \min_{f\in\mathcal{F}} \limsup_{\substack{v\to f\\ \delta\downarrow 0}} \frac{J^*(x - \delta v, \Theta(t-\delta)) - J^*(x, \Theta(t))}{\delta} > 0$$

that is, J^* is decreasing on both open future and past neighborhoods of t, for all $t \in \mathbb{R}$, where $\overrightarrow{D}J^*,\ \overleftarrow{D}J^* \in [-\infty, +\infty]$.

To prove condition (i), let x^p, L^p, W^p, $\hat{\Theta}^p$ correspond to any worst-case minimizing solution of $J^*(x(t), \Theta(t))$, defined on $\tau \in [0, T]$. Additional notations which will be used: $T_\delta \triangleq T + \delta$, $\hat{\Theta}^p_T \triangleq \hat{\Theta}_f(T)$, $\hat{\Theta}^p_{T_\delta} \triangleq \hat{\Theta}_f(T_\delta)$; that is, both sets represent solutions of the terminal identifier Ψ_f, evaluated along $x^p_{[0,T]}$ and $x^p_{[0,T_\delta]}$, respectively. Likewise, for an arbitrary argument $\mathbb{S} \in \{\hat{\Theta}^p_T,\ \hat{\Theta}^p_{T_\delta}\}$, we define $W^p_T(\mathbb{S}) \triangleq W(x^p(T), \mathbb{S})$ and $W^p_{T_\delta}(\mathbb{S}) \triangleq W(x^p(T_\delta), \mathbb{S})$.

With the above notations, it can be seen that if the minimizing solution $x^p_{[0,T]}$ were extended to $\tau \in [0, T_\delta]$ by implementing the feedback $u^p(\tau) = k_f(x^p(\tau), \hat{\theta}^p_T)$ on $\tau \in [T, T_\delta]$ (i.e., with $\hat{\theta}^p_T$ fixed), then Criterion 6.4.4 point 5 guarantees the inequality

$$\lim_{\delta \downarrow 0} \frac{1}{\delta}\Big(\delta L(x^p_T, k_f(x^p_T, \hat{\Theta}^p_T)) + W^p_{T_\delta}(\hat{\Theta}^p_T) - W^p_T(\hat{\Theta}^p_T)\Big) \leq 0.$$

Using this fact, the relationship (i) follows from:

$$\begin{aligned}
\max_{f\in\mathcal{F}} \overrightarrow{D}J^*(x,\Theta) &= \max_{f\in\mathcal{F}} \liminf_{\substack{v\to f\\ \delta\downarrow 0}} \frac{1}{\delta}\left[J^*(x+\delta v, \Theta(t+\delta)) - \int_0^T L^p d\tau - W^p_T(\hat{\Theta}^p_T)\right]\\
&\leq \max_{f\in\mathcal{F}} \liminf_{\substack{v\to f\\ \delta\downarrow 0}} \frac{1}{\delta}\left[J^*(x+\delta v, \Theta(t+\delta)) - \int_0^\delta L^p d\tau - \int_\delta^T L^p d\tau - W^p_T(\hat{\Theta}^p_T)\right.\\
&\qquad \left. - \Big(\delta L(x^p_T, k_f(x^p_T, \hat{\Theta}^p_T)) + W^p_{T_\delta}(\hat{\Theta}^p_T) - W^p_T(\hat{\Theta}^p_T)\Big)\right]\\
&\leq \max_{f\in\mathcal{F}} \liminf_{\substack{v\to f\\ \delta\downarrow 0}} \frac{1}{\delta}\left[J^*(x+\delta v, \Theta(t+\delta)) - \int_\delta^T L^p d\tau - \int_T^{T_\delta} L^p d\tau - W^p_{T_\delta}(\hat{\Theta}^p_T) - \delta L^p|_\delta\right]\\
&\leq \max_{f\in\mathcal{F}} \lim_{\delta\downarrow 0} \frac{1}{\delta}\left[J^*(x^p(\delta), \hat{\Theta}^p(\delta)) - \int_\delta^{T_\delta} L^p d\tau - W^p_{T_\delta}(\hat{\Theta}^p_{T_\delta})\right] - \delta L^p|_\delta\\
&\leq -L(x, \kappa_{mpc}(x,\Theta))
\end{aligned}$$

The final inequalities are achieved by recognizing:

- The $\int L^p d\tau + W^p$ term is a (potentially) suboptimal cost on the interval $[\delta, T_\delta]$, starting from the point $(x^p(\delta), \hat{\Theta}_p(\delta))$.
- The relation $\hat{\Theta}^p_{T_\delta} \subseteq \hat{\Theta}^p_T$ holds by Criterion 6.4.1 point 4, which implies by Criterion 6.4.5 point 2 that $W^p_{T_\delta}(\hat{\Theta}^p_{T_\delta}) \leq W^p_{T_\delta}(\hat{\Theta}^p_T)$.
- By Criterion 6.4.2, $\Theta(t+\delta) \triangleq \Psi(\Theta(t), x_{[0,\delta]}, u_{[0,\delta]}) \subseteq \Psi_p(\Theta(t), x_{[0,\delta]}, u_{[0,\delta]})$, along any locus connecting x and $x + \delta v$.
- The $\liminf_v$ applies over all sequences $\{v_k\} \to f$, of which the particular sequence $\{v(\delta_k) = \frac{x^p(\delta_k) - x}{\delta}\}$ is a member.
- There exists an arbitrary perturbation of the sequence $\{v(\delta_k)\}$ satisfying $\Psi_p(\Theta(t), x_{[0,\delta]}) = \hat{\Theta}^p(\delta)$. The $\liminf_v$ includes the limiting cost $J^*(x^p(\delta), \hat{\Theta}^p(\delta))$ of any such perturbation of $\{v(\delta_k)\}$.

- The cost $J^*(x^p(\delta), \hat{\Theta}^p(\delta))$ is optimal on $[\delta, T_\delta]$, and passes through the same point $(x^p(\delta), \hat{\Theta}_p(\delta))$ as the trajectory defining the L^p and W^p expressions. Thus, the bracketed expression is non-positive.

For the purposes of condition (ii), let x^v denote a solution to the prediction model (6.14b) for initial condition $x^v(-\delta) = x - \delta v$. Condition (ii) then follows from:

$$\min_{f\in\mathcal{F}} \overleftarrow{D} J^*(x,\Theta) = \min_{f\in\mathcal{F}} \limsup_{\substack{v\to f\\ \delta\downarrow 0}} \frac{1}{\delta}\left[\int_{-\delta}^{T-\delta} L^v d\tau + W^v_{T-\delta}(\hat{\Theta}^v_{T-\delta}) - J^*(x,\Theta)\right]$$

$$\geq \min_{f\in\mathcal{F}} \limsup_{\substack{v\to f\\ \delta\downarrow 0}} \frac{1}{\delta}\Bigg[\delta L^v|_{-\delta} + \int_{0}^{T-\delta} L^v d\tau + W^v_{T-\delta}(\hat{\Theta}^v_{T-\delta}) - J^*(x,\Theta)$$

$$+ \left(\delta L(x^v_{T-\delta}, k_f(x^v_{T-\delta}, \hat{\Theta}^v_{T-\delta})) + W^v_T(\hat{\Theta}^v_{T-\delta}) - W^v_{T-\delta}(\hat{\Theta}^v_{T-\delta})\right)\Bigg]$$

$$\geq \min_{f\in\mathcal{F}} \limsup_{\substack{v\to f\\ \delta\downarrow 0}} \frac{1}{\delta}\left[\delta L^v|_{-\delta} + \int_{0}^{T} L^v d\tau + W^v_T(\hat{\Theta}^v_{T-\delta}) - J^*(x,\Theta)\right]$$

$$\geq \min_{f\in\mathcal{F}} \lim_{\delta\downarrow 0} \frac{1}{\delta}\left[\delta L^p|_{-\delta} + \int_{0}^{T} L^p d\tau + W^p_T(\hat{\Theta}^p_T) - J^*(x,\Theta)\right]$$

$$\geq L(x, \kappa_{mpc}(x,\Theta))$$

The above derivation made use of the fact that the reverse subderivate $\overleftarrow{D} W$ satisfies

$$\min_{f\in\mathcal{F}} \limsup_{\substack{v\to f\\ \delta\downarrow 0}} \left(-L(x-\delta v,\, k_f(x-\delta v,\, \Theta)) + \left(\frac{W(x-\delta v,\, \Theta) - W(x,\Theta)}{\delta}\right)\right) \geq 0$$

which follows from a combination of Criterion 6.4.4 point 5 and the $\mathcal{LS}$-continuity of W.

Using the above inequalities for $\overleftarrow{D} J^*(x,\Theta)$ and $\overrightarrow{D} J^*(x,\Theta)$ together with Assumption 6.3.3, it follows that $J^*(t)$ is strictly decreasing on $x \notin \Sigma^o_x$ and non-increasing on $x \in \Sigma^o_x$. It follows that $\lim_{t\to\infty}(x,\Theta)$ must converge to an invariant subset of $\Sigma^0_x \times \overline{\text{cov}}\{\Theta^o\}$. Assumption 6.3.1 guarantees that such an invariant subset exists, since it implies $\exists\varepsilon^* > 0$ such $\Sigma_x(B(\theta^*, \varepsilon^*)) \neq \emptyset$, with θ^* the actual unknown parameter in (6.8). Continued solvability of (6.14) as $(x(t), \Theta(t))$ evolve follows by: (1) $x(\tau) \notin \mathcal{X}_0(\Theta(\tau)) \Rightarrow J^*(\tau) = +\infty$ and (2) if $x(t) \in \mathcal{X}_0(\Theta(t))$ and $x(t') \notin \mathcal{X}_0(\Theta(t'))$, then $(t'-t) \downarrow 0$ contradicts either condition (i) at time t, or (ii) at time t'.

6.9.2 Proof of Proposition 6.5.1

The fact that Criterion 6.4.5 holds is a direct property of the union and min operations for the closed sets $\mathbb{X}^i_f$, and the fact that the Θ-dependence of individual $(W^i, \mathbb{X}^i_f)$ satisfies Criterion 6.4.5. For the purposes of Criterion 6.4.4, the Θ argument is a constant, and is omitted from notation. Properties Criterion 6.4.4 point 1 and 2 follow directly by (6.16), the closure of $\mathbb{X}_f^i$, and (2.2). Define

$$\mathcal{I}_f(x) = \{i \in \mathcal{I} \mid x \in \mathbb{X}_f^i \quad \text{and} \quad W(x) = W^i(x)\}$$

Denoting $\mathcal{F}^i \triangleq f(x, k_f^i(x), \Theta, \mathcal{D})$, the following inequality holds for every $i \in \mathcal{I}_f(x)$:

$$\max_{f^i \in \mathcal{F}^i} \liminf_{\substack{v \to f^i \\ \delta \downarrow 0}} \frac{W(x + \delta v) - W(x)}{\delta} \leq \max_{f^i \in \mathcal{F}^i} \liminf_{\substack{v \to f^i \\ \delta \downarrow 0}} \frac{W^i(x + \delta v) - W(x)}{\delta} \leq -L(x, k_f^i(x))$$

It then follows that $u = k_f(x) \triangleq k_f^{i(x)}(x)$ satisfies Criterion 6.4.4 point 5 for any arbitrary selection rule $i(x) \in \mathcal{I}_f(x)$ (from which Criterion 6.4.4 point 3 is obvious). Condition Criterion 6.4.4 point 4 follows from continuity of the $x(\cdot)$ flows, and observing that by (6.15), Criterion 6.4.4 point 5 would be violated at any point of departure from $\mathbb{X}_f$.

6.9.3 Proof of Claim 6.6.1

By contradiction, let θ^* be a value contained in the left-hand side of (6.18), but not in the right-hand side. Then by (6.17), there exists $\tau \in [a, c]$ (i.e., $\tau_a \equiv (\tau - a) \in [0,\ c - a]$) such that

$$f(B(x, \gamma\tau_a), u, \theta^*, \mathcal{D}) \cap B(\dot{x}, \varepsilon + \gamma\tau_a) = \emptyset \tag{6.20}$$

Using the bounds indicated in the claim, the following inclusions hold when $\tau \in [a, b]$:

$$f(x', u, \theta^*, \mathcal{D}) \subseteq f(B(x, \gamma\tau_a), u, \theta^*, \mathcal{D}) \tag{6.21a}$$

$$B(\dot{x}', \varepsilon') \subseteq B(\dot{x}, \varepsilon + \gamma\tau_a) \tag{6.21b}$$

Combining (6.21) and (6.20) yields

$$f(x', u, \theta^*, \mathcal{D}) \cap B(\dot{x}', \varepsilon') = \emptyset \quad \Rightarrow \quad \theta^* \notin \mathcal{Z}^{\varepsilon'}(\Theta, x'_{[a,\tau]}, u_{[a,\tau]}) \tag{6.22}$$

which violates the initial assumption that θ^* is in the LHS of (6.18). Meanwhile, for $\tau \in [b, c]$ the inclusions

$$f(B(x', \gamma\tau_b), u, \theta^*, \mathcal{D}) \subseteq f(B(x, \gamma\tau_a), u, \theta^*, \mathcal{D}) \tag{6.23a}$$

$$B(\dot{x}', \varepsilon + \gamma\tau_b) \subseteq B(\dot{x}, \varepsilon + \gamma\tau_a) \tag{6.23b}$$

yield the same contradictory conclusion:

$$f(B(x', \gamma\tau_b), u, \theta^*, \mathcal{D}) \cap B(\dot{x}', \varepsilon + \gamma\tau_b) = \emptyset \tag{6.24a}$$

$$\Rightarrow \theta^* \notin \mathcal{Z}^{\varepsilon,\gamma}\left(\mathcal{Z}^{\varepsilon'}(\Theta, x'_{[a,b]}, u_{[a,b]}), x'_{[b,\tau]}, u_{[b,\tau]}\right) \tag{6.24b}$$

It therefore follows that the containment indicated in (6.18) necessarily holds.

6.9.4 Proof of Proposition 6.6.2

It can be shown that Assumption 6.3.3, together with the compactness of Σ_x, is sufficient for an analog of Claim 4.2.2 to hold (i.e., with J^*_∞ interpreted in a min–max sense). In other words, the cost $J^*(x, \Theta)$ satisfies

$$\alpha_l(\|x\|_{\Sigma_x^o}, \Theta) \leq J^*(x, \Theta) \leq \alpha_h(\|x\|_{\Sigma_x^o}, \Theta)$$

for some functions α_l, α_h which are class-$\mathcal{K}_\infty$ w.r.t. x, and whose parameterization in Θ satisfies $\alpha_i(x, \Theta_1) \leq \alpha_i(x, \Theta_2)$, $\Theta_1 \subseteq \Theta_2$. We then define the compact set $\bar{\mathcal{X}}_0^\uparrow \triangleq \{x \mid \min_{\Theta \in \overline{\text{cov}}\{\Theta^o\}} J^*(x, \Theta) < \max_{x_0 \in \bar{\mathcal{X}}_0} \alpha_h(\|x_0\|_{\Sigma_x^o}, \Theta^0)\}$.

By a simple extension of Reference 90, Theorem 4.19, the ISS property follows if it can be shown that there exists $\alpha_c \in \mathcal{K}$ such that $J^*(x, \Theta)$ satisfies

$$x \in \bar{\mathcal{X}}_0^\uparrow \setminus B(\Sigma_x^o, \alpha_c(c)) \Rightarrow \begin{cases} \max_{f \in \mathcal{F}_c} \overrightarrow{D} J^*(x, \Theta) < 0 \\ \min_{f \in \mathcal{F}_c} \overleftarrow{D} J^*(x, \Theta) > 0 \end{cases} \tag{6.25}$$

where $\mathcal{F}_c \triangleq B(f(x, \kappa_{mpc}(x, \Theta(t)), \Theta(t), \mathcal{D}), c)$. To see this, it is clear that J decreases until $x(t)$ enters $B(\Sigma_x^o, \alpha_c(c))$. While this set is not necessarily invariant, it is contained within an invariant, compact level set $\Omega(c, \Theta) \triangleq \{x \mid J^*(x, \Theta) \leq \alpha_h(\alpha_c(c), \Theta)\}$. By Criterion 6.4.1 point 4, the evolution of $\Theta(t)$ in (6.19b) must approach some constant interior bound Θ^∞, and thus $\lim_{t\to\infty} x(t) \in \Omega(c, \Theta^\infty)$. Defining $\alpha_d(c) \triangleq \max_{x \in \Omega(c, \Theta^\infty)} \|x\|_{\Sigma_x^o}$ completes the proposition, if c^* is sufficiently small such that $B(\Sigma_x^o, \alpha_d(c^*)) \subseteq \bar{\mathcal{X}}_0^\uparrow$.

Next, we only prove decrease in the forward direction, since the reverse direction follows analogously, as it did in the proof of Theorem 6.4.6. Using similar procedure and notation as the Theorem 6.4.6 proof, $x^p_{[0,T]}$ denotes any worst-case prediction at (t, x, Θ), extended to $[T, T_\delta]$ via k_f, that is assumed to satisfy the specifications of Proposition 6.6.2. Following the proof of Theorem 6.4.6,

$$\begin{aligned}
&\max_{f \in \mathcal{F}_{c^*}} \overrightarrow{D} J^*(x, \Theta) \\
&\quad \leq \max_{f \in \mathcal{F}} \liminf_{\substack{v \to f \\ \delta \downarrow 0}} \frac{1}{\delta} \left[J^*(x + \delta v, \Theta(t+\delta)) - \int_\delta^{T_\delta} L^p d\tau - W^p_{T_\delta}(\hat{\Theta}^p_T) \right] - L^p|_\delta \\
&\quad \leq \max_{f \in \mathcal{F}} \liminf_{\substack{v \to f \\ \delta \downarrow 0}} \frac{1}{\delta} \left[J^*(x + \delta v, \Theta(t+\delta)) - \int_\delta^{T_\delta} L^v d\tau - W^v_{T_\delta}(\hat{\Theta}^v_{T_\delta}) \right] - L^p|_\delta \\
&\qquad + \frac{1}{\delta} \left[\int_\delta^{T_\delta} L^v d\tau + W^v_{T_\delta}(\hat{\Theta}^v_{T_\delta}) - \int_\delta^{T_\delta} L^p d\tau - W^p_{T_\delta}(\hat{\Theta}^p_T) \right]
\end{aligned} \tag{6.26}$$

where L^v, W^v denote costs associated with a trajectory $x^v_{[0,T_\delta]}$ satisfying the following:

- initial conditions $x^v(0) = x$, $\Theta^v(0) = \Theta$
- generated by the same worst-case $\hat{\theta}$ and $d(\cdot)$ as $x^p_{[0,T_\delta]}$

- dynamics of form (6.19) on $\tau \in [0, \delta]$, and of form (6.14b),(6.14c) on $\tau \in [\delta, T_\delta]$, with the trajectory passing through $x^v(\delta) = x + \delta v$, $\Theta^v_p(\delta) = \Theta(t+\delta)$
- the $\min_\kappa$ in (6.14) is constrained such that $\kappa^v(\tau, x^v, \Theta^v) = \kappa^p(\tau, x^p, \Theta^p)$; that is, $u^v_{[0,T_\delta]} \equiv u^p_{[0,T_\delta]} \equiv u_{[0,T_\delta]}$

Let K_f denote a Lipschitz constant of (6.8) with respect to x, over the compact domain $\bar{\mathcal{X}}_0^\uparrow \times \Theta^o \times \mathcal{D}$. Then, using the comparison lemma [90, Lemma 3.4] one can derive the bounds

$$\tau \in [0, \delta]: \quad \begin{cases} \|x^v - x^p\| \le \frac{c}{K_f}(e^{K_f\tau} - 1) \\ \|\dot{x}^v - \dot{x}^p\| \le c\, e^{K_f\tau} \end{cases} \tag{6.27a}$$

$$\tau \in [\delta, T_\delta]: \quad \begin{cases} \|x^v - x^p\| \le \frac{c}{K_f}(e^{K_f\delta} - 1)\, e^{K_f(\tau-\delta)} \\ \|\dot{x}^v - \dot{x}^p\| \le c(e^{K_f\delta} - 1)\, e^{K_f(\tau-\delta)} \end{cases} \tag{6.27b}$$

As $\delta \downarrow 0$, the above inequalities satisfy the conditions of Claim 6.6.1 as long as $c^* < \min\{\gamma, (\varepsilon - \varepsilon'), \gamma e^{K_f T}, \frac{\gamma}{K_f} e^{K_f T}\}$, thus yielding

$$\hat{\Theta}^v_f = \Psi^{\varepsilon,\gamma}_f(\Psi^{\varepsilon'}(\Theta, x^v_{[0,\delta]}, u_{[0,\delta]}), x^v_{[\delta,T_\delta]}, u_{[\delta,T_\delta]}) \subseteq \Psi^{\varepsilon,\gamma}_f(\Theta, x^p_{[0,T_\delta]}, u_{[0,T_\delta]}) = \hat{\Theta}^p_f$$

as well as the analog $\hat{\Theta}^v_p(\tau) \subseteq \hat{\Theta}^p_p(\tau)$, $\forall \tau \in [0, T_\delta]$.

Since $x^p_{[0,T]}$ is a feasible solution of the original problem from (t, x, Θ) with $\tau \in [0, T]$, it follows for the new problem posed at time $t + \delta$ that x^v is feasible with respect to the appropriate inner approximations of $\mathbb{X}$ and $\mathbb{X}^{i^*}_f(\hat{\Theta}^p_T) \subseteq \mathbb{X}_f(\hat{\Theta}^v_{T_\delta})$ (where i^* denotes an active terminal set for x^p_f) if

$$\|x^v - x^p\| \le \begin{cases} \delta \dfrac{\varepsilon_x}{T} & \tau \in [\delta, T] \\ \delta\varepsilon_f & \tau \in [T, T_\delta] \end{cases}$$

which holds by (6.27) as long as $c^* < \min\{\varepsilon_f, \frac{\varepsilon_x}{T}\}e^{-K_f T}$. Using arguments from the proof of Theorem 6.4.6, the first term in (6.26) can be eliminated, leaving:

$$\begin{aligned}
&\max_{f \in \mathcal{F}_c} \overrightarrow{D} J^*(x, \Theta) \\
&\quad \le \max_{f \in \mathcal{F}} \liminf_{\substack{v \to f \\ \delta \downarrow 0}} \frac{1}{\delta}\left[\int_\delta^{T_\delta} L^v d\tau + W^v_{T_\delta}(\hat{\Theta}^v_{T_\delta}) - \int_\delta^{T_\delta} L^p d\tau - W^p_{T_\delta}(\hat{\Theta}^p_T)\right] - L^p|_\delta \\
&\quad \le \max_{f \in \mathcal{F}} \liminf_{\substack{v \to f \\ \delta \downarrow 0}} \frac{1}{\delta}\left(\int_\delta^{T_\delta} K_L\|x^v - x^p\| d\tau + K_W\|x^v(T) - x^p(T)\| - L^p|_\delta\right) \\
&\quad \le \lim_{\delta \downarrow 0}\left(\frac{c(e^{K_f\delta} - 1)}{K_f\delta}[K_W + TK_L]e^{K_f T} - L^p|_\delta\right) \\
&\quad \le -L(x, k_{MPC}(x, \Theta)) + c(K_W + TK_L)e^{K_f T} \\
&\quad < 0 \qquad \forall x \in \bar{\mathcal{X}}_0^\uparrow \backslash B(\Sigma^o_x, \alpha_c(c))
\end{aligned}$$

with $\alpha_c \in \mathcal{K}$ given by

$$\alpha_c(c) \triangleq \underline{\gamma}_L^{-1}\left(c(K_W + TK_L)\, e^{K_f T}\right)$$

where K_W is a Lipschitz constant of $W^{i^*}(x, \Theta)$ over the compact domain $\bar{\mathcal{X}}_0^{\uparrow} \cap \mathbb{X}_f^{i^*}(\Theta)$, maximal over all $\Theta \in \overline{\text{cov}}\,\{\Theta^o\}$. Likewise, K_L is a Lipschitz constant of $L(x, u)$ with respect to x, maximal over $u \in \mathbb{U}$.

This proves the forward case in (6.25), with the reverse case following similarly. As argued previously, this is sufficient to yield the ISS property of (6.19) with respect to $\|d_2\| \leq c \leq c^*$, which completes the proof.

Chapter 7
Computational aspects of robust adaptive MPC

7.1 Problem description

In contrast to the previously discussed general dynamics (6.8), for the purposes of this chapter the system of interest is assumed to be of the following parameter-affine form

$$\dot{x} = f(x,u) + g(x,u)\theta + d(t) \tag{7.1}$$

where $\theta \in \mathbb{R}^p$ represents unknown parameters, which are assumed to lie within an initially known, compact ball around the origin $\theta \in \Theta_0 = B(0, M_\theta)$. As before, the system is assumed to be subject to pointwise constraints $x \in \mathbb{X} \subseteq \mathbb{R}^n$ and $u \in \mathbb{U} \subseteq \mathbb{R}^m$, and the mappings $f : \mathbb{X} \times \mathbb{U} \to \mathbb{R}^n$, $g : \mathbb{X} \times \mathbb{U} \to \mathbb{R}^n \times \mathbb{R}^m$ are assumed to be both locally Lipschitz in x and continuous in u. The disturbance $d(t)$ is assumed to satisfy a known pointwise bound $\|d(t)\| \leq M_d < \infty$ for all $t \in \mathbb{R}$. Although the results of this chapter technically hold for any $M_d < \infty$, in practice it is assumed that $M_d \ll M_\theta$.

The control objective is to feasibly stabilize x to a given compact set Σ_x^o, that is not necessarily robustly invariant with respect to the full uncertainty Θ_0. Following the development in Section 6.3.1, it is assumed that a set Σ_u^o and instantaneous cost $L(x,u)$ are selected such that Assumptions 6.3.3 and 6.3.5 are satisfied, where Definition 6.3.4 is interpreted in the context of (7.1).

The robust control design proposed in Section 7.2 depends on the knowledge of appropriate Lipschitz bounds for the x-dependence of the dynamics $f(x,u)$ and $g(x,u)$, and for the penalty functions $L(x,u)$ and $W(x,\Theta)$. To this end, we assume the following:

Assumption 7.1.1. *A set of functions* $\mathcal{L}_j^o : \overline{cov}\{\mathbb{X}\} \times \mathbb{U} \to \mathbb{R}_{\geq 0}$, $j \in \{f, g, L\}$ *and* $\mathcal{L}_W^o : \overline{cov}\{\mathbb{X}\} \times \overline{cov}\{\Theta_o\} \to \mathbb{R}_{\geq 0}$ *are known which satisfy*

$$\mathcal{L}_f^o(\Omega_x, u) \geq \min\left\{\mathcal{L}_f \mid \sup_{x_1, x_2 \in \Omega_x} \left(\|f(x_1,u) - f(x_2,u)\| - \mathcal{L}_f\|x_1 - x_2\|\right) \leq 0\right\} \tag{7.2a}$$

$$\mathcal{L}_g^o(\Omega_x, u) \geq \min\left\{\mathcal{L}_g \mid \sup_{x_1, x_2 \in \Omega_x} \left(\|g(x_1,u) - g(x_2,u)\| - \mathcal{L}_g\|x_1 - x_2\|\right) \leq 0\right\} \tag{7.2b}$$

$$\mathcal{L}_L^o(\Omega_x, u) \geq \min\left\{\mathcal{L}_L \;\Big|\; \sup_{x_1, x_2 \in \Omega_x} \big(L(x_1, u) - L(x_2, u) - \mathcal{L}_L\|x_1 - x_2\|\big) \leq 0\right\} \tag{7.2c}$$

$$\mathcal{L}_W^o(\Omega_x, \Theta) \geq \min\left\{\mathcal{L}_W \;\Big|\; \sup_{x_1, x_2 \in \Omega_x} \big(W(x_1, \Theta) - W(x_2, \Theta) - \mathcal{L}_W\|x_1 - x_2\|\big) \leq 0\right\} \tag{7.2d}$$

where (7.2b) *is interpreted in the sense of an induced norm. Furthermore, the* $\mathcal{L}^o$ *are such that for any sets* $\Omega_x^1 \subseteq \Omega_x^2 \subseteq \mathbb{X}$ *and any* $u \in \mathbb{U}$, $\Theta \subseteq \Theta_0$, *it follows that* $\mathcal{L}_j^o(\Omega_x^1, u) \leq \mathcal{L}_j^o(\Omega_x^2, u), j \in \{f, g, L\}$, *and likewise* $\mathcal{L}_W^o(\Omega_x^1, \Theta) \leq \mathcal{L}_W^o(\Omega_x^2, \Theta)$. ■

We emphasize that the Lipschitz bounds in Assumption 7.1.1 are with respect to the x-dependence only, where the functions $\mathcal{L}_f^o$, $\mathcal{L}_g^o$, $\mathcal{L}_L^o$ can be parameterized in terms of u. The vector fields f, g and the function L need only to be continuous in u, and the functions $\mathcal{L}^o$ themselves need only to be piecewise continuous with respect to either of their arguments. Similarly, Assumption 7.1.1 deals only with the x-dependence of $W(x, \Theta)$, where the Θ-dependence is characterized in Section 7.2.3.

Remark 7.1.2. *Definition of the functions* $\mathcal{L}_f^o$, $\mathcal{L}_g^o$, $\mathcal{L}_L^o$, $\mathcal{L}_w^o$ *can involve varying degrees of offline or online calculation. One could specify* $\mathcal{L}_j^o$, $j \in \{f, g, L\}$, *to be constants, determined offline to be maximal over compact* $\mathbb{X}$ *and* $\mathbb{U}$. *Alternatively, one could perform the indicated search online, thus providing the tightest possible bound. A practical compromise could be to perform a partial search offline, for example, by defining the* $\mathcal{L}_j^o$ *to be of the form* $\mathcal{L}_j^o(\Omega_x, u) \triangleq \widehat{\mathcal{L}_j^o}(\max_{x' \in \Omega_x}\|x'\|, \|u\|)$, *where the surface* $\widehat{\mathcal{L}_j^o} : \mathbb{R}_{\geq 0} \times \mathbb{R}_{\geq 0} \to \mathbb{R}_{\geq 0}$ *is calculated offline by searching over level sets of* $\|x\| \times \|u\|$ *in* $\mathbb{R}^n \times \mathbb{R}^m$. *Then the online search is limited to the evaluation of* $\max_{x' \in \Omega_x} \|x'\|$.

Remark 7.1.3. *One important technique for reducing the conservativeness of the robust-control design in this chapter would be to use weighted norms of the form* $\|x\|^2 = x^T \Gamma_x x$, *where* $\Gamma_x = \Gamma_x^T > 0$ *can be selected to provide the tightest possible Lipschitz bounds. Similarly, the conservativeness of the adaptive mechanism in the subsequent sections could be adjusted by incorporating a weighting matrix* $\Gamma_\theta = \Gamma_\theta^T > 0$ *into the norm used to define the ball* Θ_0. *This is equivalent to performing a transformation of the form* $\bar{\theta} = \Gamma_\theta^{-\frac{1}{2}}\theta$, $\bar{g}(x, u) = g(x, u)\Gamma_\theta\Gamma_\theta^{-\frac{1}{2}}$. *For convenience of presentation, explicit tuning matrices of this type are omitted throughout the chapter under the pretense that they could be incorporated via such transformations.*

7.2 Adaptive robust design framework

7.2.1 Method for closed-loop adaptive control

The adaptive controller presented in Section 6.3 was premised upon the idea of adapting a *set-valued* description of the parameter uncertainty, rather than the more typical adaptation of nominal parameter estimates. In this chapter, the *set* being adapted is restricted to be a ball around a nominal estimate $\hat{\theta}$, whose radius is based upon a (scalar) error bound $z_\theta \geq \|\tilde{\theta}\|$, where $\tilde{\theta} \triangleq \theta - \hat{\theta}$ is the (unknown) identification error. A defining characteristic of the result presented in this chapter is the propagation of this uncertainty ball in a manner that satisfies the essential characteristics required of the general identifiers of Chapter 6.3.

The results presented throughout Sections 7.2 and 7.3 assume that an output of the form $y = \dot{x} + e(t)$ is continuously available, by either measurement or calculation, in which $e(t)$ represents an arbitrary error with known bound $\|e(t)\| \leq M_e < \infty, \forall t \in \mathbb{R}$. A discussion on removing this assumption by the incorporation of an asymptotically convergent observer is contained in Section 7.4.

The adaptive controller is implemented according to Algorithm 7.2.1 below, that updates successively the uncertainty set Θ in a discrete fashion. It is important to note that, although a continuous-time identifier $\dot{\psi}$ is calculated online, the underlying robust control calculation is based upon the fixed value $\hat{\theta}$, which is only updated at the reset step. Similarly, the uncertainty radius z_θ is only contracted during a reset, despite the fact that the contraction ratio $\varepsilon_z(t)$ is continuously recalculated online. The feedbacks κ_{mpc} and κ_ε appearing in the algorithm will be specified in Section 7.2.2.

The behavior of Algorithm 7.2.1 is technically that of a hybrid system, and thus all states would be most properly described as evolving in a hybrid time domain as in Chapter 4. However, since successive resets of the algorithm are necessarily separated by intervals of nonzero (real) time, we simplify the notation by omitting the explicit dependence upon an event-time coordinate. Instead, we denote pre-reset and post-reset values at time t_i using t_i^- and t_i^+, respectively.[1]

Algorithm 7.2.1. *Specify design constants* $T_{\max} > 0,\ c_\lambda > 0,$ *and* $0 < \underline{\varepsilon_z} \ll 1$. *Starting from time* t_0, *the controller is implemented in the following iterative fashion:*

1. ***Initialization:*** $z_\theta = M_\theta, \hat{\theta} = 0$.
2. ***While*** *the following condition holds (where* $\underline{\lambda}\{\cdot\}$ *denotes the least eigenvalue):*

$$\left(\Psi_2 < \frac{1}{2}\varepsilon_z^*(2 - \varepsilon_z^*)z_\theta^2 + (M_d + M_e)\Psi_1\right) \textit{ AND } \left(\underline{\lambda}\{\Phi_2\} < \frac{(M_d + M_e)\Phi_1 + c_\lambda}{(1 - \varepsilon_z^*)z_\theta}\right) \tag{7.3a}$$

Implement *the following dynamic feedback (recalling* $y \triangleq \dot{x} + e(t)$*) over a maximal interval* $t \in [t_i, t_{i+1})$*:*

$$u \triangleq \kappa_{mpc}(x, \hat{\theta}, z_\theta, \phi, \Phi, \psi, \Psi) \tag{7.3b}$$

$$\varepsilon_z^* \triangleq \kappa_\varepsilon(x, \hat{\theta}, z_\theta, \phi, \Phi, \psi, \Psi) \geq \underline{\varepsilon_z} \tag{7.3c}$$

[1] For the states which evolve differentially, values at t_i^- and t_i^+ are respectively equivalent to standard left and right limits $\lim_{\delta \uparrow 0} t_i + \delta$ and $\lim_{\delta \downarrow 0} t_i + \delta$, respectively. However, since $\varepsilon_z^*(t)$ is defined algebraically, its trajectory $\varepsilon_z^*(\cdot)$ need not be continuous in any meaningful sense, and thus $\varepsilon_z^*(t_i^-) \neq \lim_{\tau \uparrow t_i} \varepsilon_z^*(\tau)$ is possible. This implies that the same evaluation of ε_z^* must be applied for both the test condition and reset mappings in (7.4).

$$\dot{\psi} = g^T(x,u)(y - f(x,u) - g(x,u)\psi) \qquad \psi(t_i^+) = \hat{\theta} \tag{7.3d}$$

$$\dot{\Psi}_1 = \|y - f(x,u) - g(x,u)\psi\| \qquad \Psi_1(t_i^+) = 0 \tag{7.3e}$$

$$\dot{\Psi}_2 = \|y - f(x,u) - g(x,u)\psi\|^2 \qquad \Psi_2(t_i^+) = 0 \tag{7.3f}$$

$$\dot{\phi} = g^T(x,u)(y - f(x,u) - g(x,u)\hat{\theta}) \qquad \phi(t_i^+) = 0 \tag{7.3g}$$

$$\dot{\Phi}_1 = \|g(x,u)\| \qquad \Phi_1(t_i^+) = 0 \tag{7.3h}$$

$$\dot{\Phi}_2 = g^T(x,u)g(x,u) \qquad \Phi_2(t_i^+) = 0 \tag{7.3i}$$

3. ***When*** *the following condition is (first) satisfied at an arbitrary time* $t = t_{i+1}$,

$$\left(\Psi_2 \geq \frac{1}{2}\varepsilon_z^*(2-\varepsilon_z^*)z_\theta^2 + (M_d + M_e)\Psi_1\right) \; \textbf{\textit{OR}} \; \left(\underline{\lambda}\{\Phi_2\} \geq \frac{(M_d + M_e)\Phi_1 + c_\lambda}{(1-\varepsilon_z^*)z_\theta}\right) \tag{7.4a}$$

Then *select the new parameter estimate*

$$\hat{\theta}^o \triangleq \begin{cases} \psi & \Psi_2 \geq \frac{1}{2}\varepsilon_z^*(2-\varepsilon_z^*)z_\theta^2 + (M_d + M_e)\Psi_1 \\ \hat{\theta} + \Phi_2^{-1}\phi & \textit{otherwise} \end{cases} \tag{7.4b}$$

and perform the resets (denoting $z_\theta^+ \equiv z_\theta(t_{i+1}^+)$, *etc.):*

$$z_\theta^+ = \left(1 - \varepsilon_z^*\left(t_{i+1}^-\right)\right) z_\theta\left(t_{i+1}^-\right) \tag{7.4c}$$

$$\hat{\theta}^+ = \Upsilon_\theta\left(\hat{\theta}^o, \hat{\theta}\left(t_{i+1}^-\right), z_\theta\left(t_{i+1}^-\right), \varepsilon_z^*\left(t_{i+1}^-\right)\right) \tag{7.4d}$$

$$\Upsilon_\theta\left(\hat{\theta}^o, \hat{\theta}, z_\theta, \varepsilon_z^*\right) \triangleq \hat{\theta} + (\hat{\theta}^o - \hat{\theta})\frac{\sqrt{\varepsilon_z^*(2-\varepsilon_z^*)}\; z_\theta}{\|\hat{\theta}^o - \hat{\theta}\|}\,\mathrm{sat}_0^1\left(\frac{\|\hat{\theta}^o - \hat{\theta}\|}{\sqrt{\varepsilon_z^*(2-\varepsilon_z^*)}\; z_\theta}\right) \tag{7.4e}$$

where $\mathrm{sat}_0^1(\cdot)$ *denotes saturation to the scalar interval* $[0, 1] \subset \mathbb{R}$.

4. ***Iterate*** *back to Step 2, incrementing* $i := i + 1$. ■

The above algorithm differs from many approaches for online estimation and adaptive control, such as the results in References 92, 125, 151, and 152, that are based upon continuous re-solving of least-squares estimation problems over moving horizons of past history. In contrast, Algorithm 7.2.1 actually contains two independent identification mechanism, as represented by the groups (ψ, Ψ_1, Ψ_2) and (ϕ, Φ_1, Φ_2). Both of these mechanisms are driven entirely by *current* measurement signals, and are not as computationally expensive as moving horizon estimation approaches.

The underlying principle behind Algorithm 7.2.1 is that the uncertainty description $\theta \in B(\hat{\theta}, z_\theta)$ remains fixed until such time as one of the identifiers can generate $\hat{\theta}^o$ satisfying $\theta \in B(\hat{\theta}^o, (1-\varepsilon_z^*)z_\theta)$, for some contraction $\varepsilon_z^* \in (0, 1)$. The asymptotic identifier ψ in (7.3d) is driven by error observed between the plant and model, with dynamics

$$\dot{\tilde{\psi}} = -g(x,u)^T\left(g(x,u)\tilde{\psi} + e(t) + d(t)\right), \qquad \tilde{\psi} \triangleq \theta - \psi. \tag{7.5}$$

Together, the accumulators (7.3e) and (7.3f) measure the worst-case progress of the identifier in the presence of noise, and trigger a reset if $\|\tilde{\psi}\| \leq (1-\varepsilon_z^*) z_\theta$ can be guaranteed. Although the identifier of (7.3d)–(7.3f) could technically be omitted, its inclusion improves transient performance of the adaptation by providing quick response when significant plant-model mismatch is observed.

The second estimator (7.3g)–(7.3i) accumulates the effects of any plant-model mismatch based upon the *constant* value $\hat{\theta}$. Only once the regressor matrix Ψ_2 achieves sufficient magnitude in all of its eigen-directions is the least-squares estimation problem then solved. This means that rather than *assuming* a persistency of excitation condition holds over horizons of pre-specified length, the algorithm selects the interval length according to the observed excitation in the system. The advantage of this estimator, versus that in (7.3d)–(7.3f), is that it continues to trigger contractions of the uncertainty bound z_θ even in the absence of observed plant-model mismatch (e.g., when $\hat{\theta}$ has converged to θ, but z_θ remains large). In order to prevent "double accounting" of the excitation in the system, it is necessary that all accumulators be reset to zero irrespective of the specific identifier that triggered the update. Admittedly, this means that some useful excitation can be lost during the reset of Ψ_2. However, this approach does avoid the failure to register excitation due to a moving window that has been selected too short, as in References 92, 151, and 152.

Although the test condition (7.4a) ensures that $\theta \in B(\hat{\theta}^o, (1-\varepsilon_z) z_\theta)$ immediately prior to reset, there is no guarantee that this ball is contained within the previous $B(\hat{\theta}, z_\theta)$. It is expected from Criterion 6.4.1 that the underlying robust controller will require an uncertainty description Θ whose reset behavior satisfies both $\theta \in \Theta(t_i^+)$ and $\Theta(t_i^+) \subseteq \Theta(t_i^-)$. The following proposition demonstrates how the update (7.4d) can be used to generate such a set.

Proposition 7.2.2. *Define the set* $\Theta(t) \equiv \Theta(\hat{\theta}(t), z_\theta(t), \varepsilon_z^*(t)) \triangleq B(\hat{\theta}, \zeta_\theta)$, *where* $\zeta_\theta \equiv \zeta_\theta(z_\theta, \varepsilon_z^*) \triangleq \left(\sqrt{\frac{2-\varepsilon_z^*}{\varepsilon_z^*}}\right) z_\theta$, *and assume that* (7.3c) *is specified as being any constant* $\varepsilon^*(t) \equiv \varepsilon_z \in [\underline{\varepsilon}_z, 1]$, $\forall t \geq t_0$. *Then the reset behavior of* Θ *under Algorithm* 7.2.1 *satisfies*

a. $\Theta(t_i^+) \subseteq \Theta(t_i^-)$.
b. $\theta \in B(\hat{\theta}(t_i^-), z_\theta(t_i^-)) \subseteq \Theta(t_i^-)$ *implies* $\theta \in B(\hat{\theta}(t_i^+), z_\theta(t_i^+)) \subseteq \Theta(t_i^+)$.

which by induction implies

c. $\Theta(\tau_2) \subseteq \Theta(\tau_1)$, *for all* $t_0 \leq \tau_1 \leq \tau_2$.
d. $\theta \in B(\hat{\theta}(\tau_1), z_\theta(\tau_1)) \subseteq \Theta(\tau_1)$ *implies* $\theta \in B(\hat{\theta}(\tau_2), z_\theta(\tau_2)) \subseteq \Theta(\tau_2)$, $\forall t_0 \leq \tau_1 \leq \tau_2$. ∎

The intuition which underlies the claim of Proposition 7.2.2 is demonstrated in Figure 7.1, for the worst-case scenario in which the unknown θ lies on the boundary $\theta \in \partial B(\hat{\theta}, z_\theta)$. It can be seen from the figure that saturating the step-size

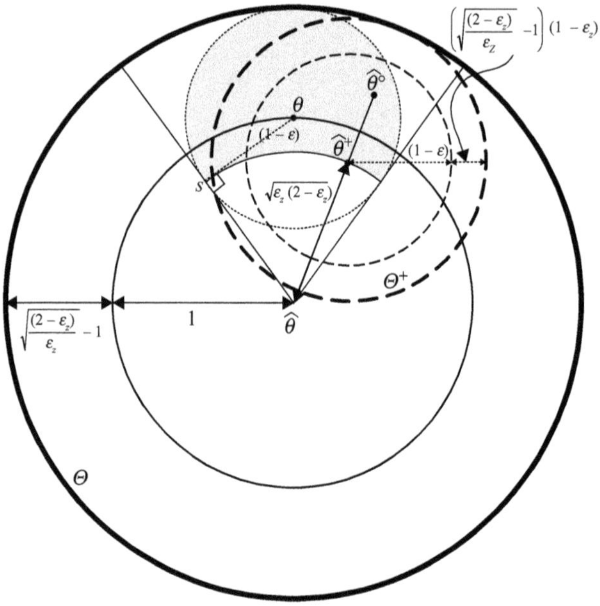

Figure 7.1 *Nested evolution of the uncertainty set* Θ *during a reset in Algorithm 7.2.1. Bold outer circles denote* Θ*, for pre-reset (solid) and post-reset (dash) conditions. Dimensions normalized with respect to* $z_\theta(t_i^-)$

$\|\hat{\theta}^+ - \hat{\theta}^-\| \le \sqrt{\varepsilon_z(2-\varepsilon_z)}\, z_\theta$ preserves the nested property of Θ, while at the same time ensuring that for any $\hat{\theta} \in B(\theta, z_\theta)$, the saturation does not violate $\hat{\theta}^+ \in B(\theta, z_\theta)$.

The set Θ is clearly a conservative outer bound for the actual uncertainty $B(\hat{\theta}, z_\theta)$, since the additional margin $(1-\varepsilon_z)\,z_\theta$ is required solely to ensure that the successive Θ are nested. However, we note that as $\varepsilon_z \to 1$, then Θ approaches the actual uncertainty $B(\hat{\theta}, z_\theta)$, and Algorithm 7.2.1 reduces to a (non-adaptive) implementation of the underlying robust controller.

7.2.2 Finite-horizon robust MPC design

To achieve robustness of the closed-loop system in the presence of constraints, the parametric uncertainty can be explicitly accounted for using tools from the robust-MPC literature. In Chapter 6.3, this was achieved by selecting a min–max feedback MPC as the underlying robust controller. It would be trivial to show that the feedback-MPC of Chapter 6.3 could be replaced with a simpler, more conservative controller based on open-loop model predictions, as long as the calculations are still performed within a min–max framework. However, it is our goal here to demonstrate that adaptation can be incorporated into even the most simple of robust control designs which do not make use of min–max calculations. In the following, the idea of using simple Lipschitz-based error bounds to generate an uncertainty cone around a nominal state prediction is extended[2] from Reference 119.

The resulting feedbacks κ_{mpc} and κ_ε in (7.3b) and (7.3c) are of the form:

$$\kappa_{mpc}(x, \hat{\theta}, z_\theta, \phi, \Phi, \psi, \Psi) = u^{p*}(0) \tag{7.6a}$$

$$\kappa_\varepsilon(x, \hat{\theta}, z_\theta, \phi, \Phi, \psi, \Psi) = \varepsilon_z^* \tag{7.6b}$$

$$(u^{p*}_{[0,T^*]}, \; \varepsilon_z^*, T^*) \triangleq \arg\min_{\substack{u^p_{[0,T]} \\ \varepsilon_z \in [\underline{\varepsilon_z}, 1] \\ T \in [0, T_{\max}]}} \; J(x, u^p_{[0,T]}, T, \hat{\theta}^J, z^J_\theta, \varepsilon_z, \Phi^J) \tag{7.6c}$$

$$(\hat{\theta}^J, z^J_\theta, \Phi^J_1, \Phi^J_2) \triangleq \begin{cases} (\psi, \, (1-\varepsilon_z) z_\theta, \, 0, 0) & \Psi_2 \geq \frac{1}{2}\varepsilon_z (2-\varepsilon_z) \, z_\theta^2 + (M_d + M_e)\Psi_1 \\ (\hat{\theta} + \Phi_2^{-1}\phi, \, (1-\varepsilon_z) \, z_\theta, \, 0, 0) & \underline{\lambda}\{\Phi_2\} \geq \dfrac{(M_d + M_e)\Phi_1 + c_\lambda}{(1-\varepsilon_z) \, z_\theta} \\ (\hat{\theta}, z_\theta, \Phi_1, \Phi_2) & \text{otherwise} \end{cases}$$

where the function $\zeta_\theta(z_\theta, \varepsilon_z)$ is defined in Proposition 7.2.2. The underlying robust-control calculation used to evaluate J (detailed below) is based upon the expanded uncertainty set Θ of radius ζ_θ defined in Proposition 7.2.2, rather than the actual uncertainty z_θ. The optimization of ε_z dictates the conservatism of the approximation of z_θ by ζ_θ. In doing so, it follows that the overall cost can never exceed that of a non-adaptive implementation (i.e., realizable by selecting $\varepsilon_z = 1$). Note that the testing of the reset conditions in (7.6) prior to solving the robust-MPC problem improves performance but is not essential, and can be omitted (i.e., by setting $(\hat{\theta}^J, z^J_\theta, \Phi^J) \equiv (\hat{\theta}, z_\theta, \Phi)$) if calculating Φ_2^{-1} within the optimization is not practical.

The cost J is defined as follows:

$$J(x, u^p_{[0,T]}, T, \hat{\theta}^J, z^J_\theta, \varepsilon_z, \Phi^J) = \int_0^T L'(x^p, u^p, z^p_x) \, d\tau + W'(x^p(T), z^p_x(T), \zeta^p_\theta(T), \Theta) \tag{7.7a}$$

$$\text{s.t. for } \tau \in [0, T]: \quad \dot{x}^p = f(x^p, u^p) + g(x^p, u^p)\hat{\theta}^J, \quad x^p(0) = x \tag{7.7b}$$

$$\dot{z}^p_x = \left(\mathcal{L}^p_f + \mathcal{L}^p_g M_\Theta\right) z^p_x + \|g(x^p, u^p)\| \zeta^J_\theta + M_d, z^p_x(0) = 0 \tag{7.7c}$$

$$X^p(\tau) \subseteq \mathbb{X}, \quad u(\tau) \in \mathbb{U} \tag{7.7d}$$

$$X^p(T) \subseteq \mathbb{X}'_f(\zeta^p_\theta(T), \Theta) \tag{7.7e}$$

[2] The manner in which we explicitly incorporate the error bound into the cost functional (7.7a) by defining L' and W' relaxes the very restrictive assumptions in Reference 119 regarding strict decrease of $\dot{W}$, and allows us to quantify the robustness beyond the simple nominal robustness proven in Reference 119.

where
$$L'(x^p, u^p, z_x^p) \triangleq L(x^p, u^p) + \mathcal{L}_L^p z_x^p \tag{7.7f}$$
$$W'(x^p, z_x^p, \zeta_\theta^p, \Theta) \triangleq \max_{\Theta_T^p \in \Omega(\zeta_\theta^p, \Theta)} \{W(x^p, \Theta_T^p) + \mathcal{L}_W^p(X^p, \Theta_T^p) z_x^p\} \tag{7.7g}$$
$$\mathbb{X}_f'(\zeta_\theta^p, \Theta) \triangleq \bigcap_{\Theta_T^p \in \Omega(\zeta_\theta^p, \Theta)} \mathbb{X}_f(\Theta_T^p) \tag{7.7h}$$
$$X^p(\tau) \triangleq B(x^p(\tau), z_x^p(\tau)) \tag{7.7i}$$
$$\zeta_\theta^J \triangleq \zeta_\theta(z_\theta^J,\ \varepsilon_z) \tag{7.7j}$$
$$\Theta \triangleq B(\hat{\theta}^J, \zeta_\theta^J) \tag{7.7k}$$
$$\Omega(\zeta_\theta^p, \Theta) \triangleq \left\{ \Theta_T^p \mid \exists \bar{\theta} \text{ s.t. } \Theta_T^p \triangleq B(\bar{\theta}, \zeta_\theta^p(T)) \subseteq \Theta \right\} \tag{7.7l}$$
$$\mathcal{L}_j^p(\tau) \triangleq \mathcal{L}_j^o(X^p(\tau), u^p(\tau)), \ \ j \in \{f, g, L\} \tag{7.7m}$$
$$M_\Theta \triangleq \text{sat}_0^{M_\theta}(\|\hat{\theta}^J\| + \zeta_\theta^J) \tag{7.7n}$$

We note that ζ_θ^J in (7.7j) is a constant with respect to prediction time τ in problem (7.7). In contrast, $\zeta_\theta^p(T)$ is the final value in a predicted trajectory $\zeta_\theta^p(\tau)$, whose computation[3] is discussed in Section 7.3. The interpretation of (7.7g), (7.7k) is that W' is defined according to the worst-case ball of radius $\zeta_\theta^p(T)$ existing in Θ.

The state $z_x^p(\tau)$ provides an upper bound on the deviation $\max_{\theta^* \in \Theta} \|x^p(\tau) - x^*(\tau)\|$, where x^p denotes the nominal estimate of (7.7b), while x^* denotes the worst-case solution for the actual system (7.1). Therefore, $X_{[0,T]}^p$ has the form of a quadratic cone, centered around the nominal trajectory $x_{[0,T]}^p$, with radius defined by $z_{x\,[0,T]}^p$. Furthermore, (7.7f) and (7.7g) are defined such that evaluation of L' and W' along the nominal trajectory x^p provides an upper bound on the worst-case values L and W for the cone X^p.

Using only a nominal model prediction (i.e., (7.7b,c)), it is not possible to predict the future behavior of the parameter estimator $\hat{\theta}$, as was possible in the min–max framework of Chapter 6.3. However, based upon the excitation of the nominal prediction $x_{[0,T]}^p$, one can generate from $X_{[0,T]}^p$ a lower bound on the future contraction of the uncertainty-radius z_θ (or equivalently ζ_θ, since they contract together, and ε_z is presumed constant with respect to the prediction time-coordinate). Even without knowledge of future values of $\hat{\theta}$, the fact that z_θ contracts faster along some trajectories is potentially useful to the optimization. This is why (7.7g,h) are defined according to the worst-case future behavior of $\hat{\theta}$, given the predicted error bound $\zeta_\theta^p(T)$. By Proposition 7.2.2, this worst-case search can be restricted to balls contained within the current uncertainty set $\Theta = B(\hat{\theta}^J, \zeta_\theta^J)$.

[3] Note that the indicated dependence of (7.7) on the arguments ε_z and Φ is due solely to the dependence of the estimator $\zeta_\theta^p(T)$.

Remark 7.2.3. *The implication of (7.7c, f) is that the functions $\mathcal{L}_j^o$, $j \in \{f, g, L\}$ are continuously re-calculated along $\tau \in [0, T]$ as $X^p(\tau)$ expands. If this is not feasible, the variables $\mathcal{L}_j^p$ can be treated as constants with respect to $\tau \in [0, T]$, and their values can be added as minimization variables in* (7.6c)*, subject to the additional constraints (assuming the $\mathcal{L}_j^o$ are appropriately re-defined to accept a set-valued argument Ω_u in place of u)*

$$\mathcal{L}_j^p \geq \mathcal{L}_j^o\left(\bigcup_{\tau\in[0,T]} X^p(\tau), \bigcup_{\tau\in[0,T]} \{u(\tau)\}\right), \quad j = \{f, g, L\}. \tag{7.7o}$$

Remark 7.2.4. *The greatest source of conservatism in the calculation of* (7.7) *is in the generation of the error bound $z_x^p \geq \|x^p - x\|$ in* (7.7c)*, which exhibits unstable linear growth. At the expense of increased online computation, the results in the remainder of this chapter could be re-developed with* (7.7c) *replaced by the* **significantly** *less conservative bound (where $f^p \equiv f(x^p, u)$, $f^s \equiv f(s, u)$, etc.)*

$$z_x^p = \sqrt{2V_x}, \quad z_x^p(0) = V_x(0) = 0 \tag{7.8a}$$

$$\dot{V}_x = \max_{s\in\partial B(x^p, z_x^p)} \left\{(x^p - s)^T(f^p - f^s + g^p\zeta_\theta^J) + \|(x^p - s)^T(g^p - g^s)\|M_\Theta + \|x^p - s\|M_d\right\} \tag{7.8b}$$

This results in the construction of the tightest quadratic cone centered around the nominal estimate x^p. The substantial benefit of using (7.8) *over* (7.7c) *is to account for stabilizing terms appearing in $f(x, u)$ that can partially counteract the effects of the uncertainty.*

7.2.3 Stability of the underlying robust MPC

The identifier $\zeta_\theta^p \equiv \zeta_\theta(z_\theta^p, \varepsilon_z)$ in (7.7g, h, k) represents an internal model of the identifier state $\zeta_\theta(z_\theta, \varepsilon_z)$, which means that Algorithm 7.2.1, together with (7.6)–(7.7), defines an adaptive robust feedback structure of the form depicted in Figure 6.2. Noting that the underlying robust control calculation in (7.7) is non-standard, we first demonstrate the robust stability of the controller in the absence of this internal identification model.

The first step to establish robust stability requires a characterization of the requirements imposed on the terminal functions $W(x, \Theta)$ and $\mathbb{X}_f(\Theta)$, similar to those given previously in Section 6.4.3. Note that for the purpose of the following criteria, Θ is interpreted as an arbitrary set-valued parameter, that need not take the form of a ball.

Criterion 7.2.5. *The set-valued terminal constraint function $\mathbb{X}_f : \overline{cov}\{\Theta_0\} \to \overline{cov}\{\mathbb{X}\}$ and terminal penalty function $W : \mathbb{R}^n \times \overline{cov}\{\Theta_0\} \to [0, +\infty]$ are such that for each $\Theta \in \overline{cov}\{\Theta_0\}$, there exists $k_f(\cdot, \Theta) : \mathbb{X}_f \to \mathbb{U}$ satisfying*

1. *$\mathbb{X}_f(\Theta) \neq \emptyset$ implies that $\Sigma_x^o \cap \mathbb{X}_f(\Theta) \neq \emptyset$ and $\mathbb{X}_f(\Theta) \subseteq \mathbb{X}$ is closed,*
2. *$W(\cdot, \Theta)$ is locally Lipschitz w.r.t. $x \in \mathbb{X}_f(\Theta)$, satisfying* (7.2d)*,*
3. *$k_f(x, \Theta) \in \mathbb{U}$, $\forall x \in \mathbb{X}_f(\Theta)$,*

4. $\mathbb{X}_f(\Theta)$ *and* $\Sigma_x(\Theta) \subseteq \{\Sigma_x^o \cap \mathbb{X}_f(\Theta)\}$ *are both strongly positive invariant with respect to the differential inclusion* $\dot{x} \in \mathcal{F} \triangleq f(x, k_f(x,\Theta)) + g(x, k_f(x,\Theta))\Theta + \mathcal{D}$, *where* $\mathcal{D} \triangleq B(0, M_d)$.
5. *for all* $x \in \mathbb{X}_f(\Theta)$,

$$\max_{f' \in \mathcal{F}} \left(L(x, k_f(x,\Theta)) + \dot{W}(x,\Theta) \right) \leq 0$$

where $\dot{W}$ *denotes the forward derivative of* W *in the direction* f'. ■

As well, we assume that the Θ-dependence of the functions $W(x,\Theta)$ and $\mathbb{X}_f(\Theta)$ satisfies the following extension of Criterion 6.4.5.

Criterion 7.2.6. *For any* $\Theta_1, \Theta_2 \in \overline{cov}\{\Theta_0\}$ *such that* $\Theta_1 \subseteq \Theta_2$,

1. $\mathbb{X}_f(\Theta_2) \subseteq \mathbb{X}_f(\Theta_1)$
2. $W(x,\Theta_1) \leq W(x,\Theta_2)$, *for all* $x \in \mathbb{X}_f(\Theta_2)$
3. $\mathcal{L}_W(X^p,\Theta_1) \leq \mathcal{L}_W(X^p,\Theta_2)$, *for all* $X^p \subseteq \mathbb{X}$. ■

Remark 7.2.7. *The switching-based construction of* $W(x,\Theta_T^p)$ *and* $\mathbb{X}_f(\Theta_T^p)$ *in Section* 6.5.2 *can be incorporated into* (7.7) *by designing an indexed collection of candidates* $(W^i(x), \mathbb{X}_f^i, \mathcal{L}_W^o(X^p)^i, \Theta_0^i)$, *in which* $W^i(x)$ *is a robust CLF with respect to* $\theta \in \Theta_0^i \subseteq \Theta_0$, *with domain of attraction* $\mathbb{X}_f^i \subseteq \mathbb{X}$, *and Lipschitz bound* $\mathcal{L}_W^o(X^p)^i$. *Then* (7.7g) *and* (7.7h) *could be replaced with expressions of the form*

$$W' \triangleq \max_{\Theta_T^p \in \Omega} \ \min_{i \in \mathcal{I}_X(\Theta_T^p)} \{W'^i(x) + \mathcal{L}_W^o(X^p)^i\}, \quad \mathcal{I}_X \neq \emptyset \tag{7.9}$$

given definitions

$$\mathcal{I}_\Theta(\Theta_T^p) \triangleq \{i \mid \Theta_T^p \subseteq \Theta_0^i\}, \quad \mathcal{I}_X \triangleq \{i \in \mathcal{I}_\Theta \mid X^p(T) \subseteq \mathbb{X}_f'^i\} \tag{7.10}$$

As in Chapter 6.3, *this provides a constructive approach for ensuring that the* Θ-*dependence of* W' *and* $\mathbb{X}_f'$ *satisfies Criterion* 7.2.6. *However, to preserve clarity of presentation we do not pursue this extension here.* ■

Criteria 7.2.5 and 7.2.6 are sufficient to yield the following claim of robust stability, in the absence of an internal model for the identifier ζ_θ^p:

Theorem 7.2.8. (Robust stabilization) *Given system* (7.1), *target* Σ_x^o, *and penalty* L *satisfying the assumptions of Section* 7.1, *assume that functions* W *and* $\mathbb{X}_f$ *are designed to satisfy Criteria* 7.2.5 *and* 7.2.6. *Let* $\mathcal{X}_0 \equiv \mathcal{X}_0(\Theta_0)$ *denote the domain of attraction of a non-adaptive implementation of the robust controller* $u = \kappa_{mpc}(x, \hat{\theta}, z_\theta, \phi, \Phi, \psi, \Psi) \equiv \kappa_{mpc}(x, 0, M_\theta, 0, 0, 0, 0)$ *defined by* (7.6). *Furthermore, let* $\mathcal{X}_0^1 \triangleq \mathcal{X}_0^1(\Theta_0) \subseteq \mathbb{X}$ *denote the set of initial states, with uncertainty* $\Theta(t_0) = \Theta_0$, *for which* (7.6) *has a solution when* (7.7) *is calculated under the condition* $\zeta_\theta^p(T) \equiv \zeta_\theta^J$. *Then under the closed-loop control of Algorithm* 7.2.1, Σ_x^o *is feasibly asymptotically stabilized from any initial state* $x_0 \in \mathcal{X}_0^1$, *and furthermore* $\mathcal{X}_0^1 \supseteq \mathcal{X}_0$. ■

Theorem 7.2.8 proves robust stabilization for an adaptive control structure of the form depicted in Figure 6.1. Incorporation of an internal model of the identifier that yields a control structure of the form in Figure 6.2 is the subject of the next section.

7.3 Internal model of the identifier

Unlike the feedback-MPC results of Section 6.3, the trajectory x^p of (7.7b) and the integral component of the cost in (7.7a) are independent of the estimator $\zeta_\theta^p(\tau)$. This admittedly reduces the effectiveness of embedding an internal model of the identifier within the model predictions, since narrowing the uncertainty cone $X^p(\tau)$ by anticipating the effects of future adaptation was a major advantage highlighted in Section 6.3. However, the fact that $\zeta_\theta^p(T)$ appears in the terminal expressions of (7.7a) and (7.7e) implies that its estimation can improve the overall performance by reducing the conservatism of the terminal penalty (i.e., the robust-CLF estimate of the remaining cost-to-go). In particular, calculation of ζ_θ^p can counteract the effects of a large initial uncertainty Θ_0 that prevents the design of any suitable robust-CLF $W(x)$, or results in an unacceptably small domain $\mathbb{X}_f$.

For convenience, we define the function

$$G(x,u) \triangleq g(x,u)^T g(x,u), \tag{7.11}$$

about which we make the assumption.

Assumption 7.3.1. *A known function $\mathcal{L}_G^o : \overline{cov}\{\mathbb{X}\} \times \mathbb{U} \to \mathbb{R}_{\geq 0}$ satisfies the following Lipschitz-like bound for G:*

$$\mathcal{L}_G^o(\Omega_x, u) \geq \min \left\{ \mathcal{L}_G \;\middle|\; \sup_{\substack{x_1, x_2 \in \Omega_x \\ \|\mathbf{v}\|=1}} \left(\mathbf{v}^T \left[G(x_1,u) - G(x_2,u) \right] \mathbf{v} - \mathcal{L}_G \|x_1 - x_2\| \right) \leq 0 \right\} \tag{7.12}$$

where additionally $\Omega_x^1 \subseteq \Omega_x^2 \subseteq \mathbb{X}$ implies $\mathcal{L}_G^o(\Omega_x^1, u) \leq \mathcal{L}_G^o(\Omega_x^2, u)$. ■

In order to clarify discussions throughout the remaining sections,[4] we consider the following definition.

Definition 7.3.2. *Let A be an arbitrary symmetric matrix, with eigenvalues $\{\lambda^i\}$ and (ortho-normal) eigenvectors $\{\mathbf{v}^i\}$, and let $\|\mathbf{v}\| = 1$ be an arbitrary direction. Then the* ***magnitude of A in direction*** $\mathbf{v}$ *refers to the quantity $\mathbf{v}^T A \mathbf{v} = \sum_i \lambda^i \langle \mathbf{v}^i, \mathbf{v} \rangle$.* ■

While the above quantity could be viewed as defining a norm for the vector **v**, our perspective here relates to $\mathbf{v}^T A \mathbf{v}$ as quantifying the total eigen-contributions of A in the direction **v**.

[4] And for lack of better terminology.

The error bound $\zeta^p_{\theta\,[0,T]}$ can be determined from $x^p_{[0,T]}$, $u^p_{[0,T]}$, $z^p_{x\,[0,T]}$ using Algorithm 7.3.4 below, in a way that mimics the reset behavior of Algorithm 7.2.1. Note that Algorithm 7.2.1 can be implemented either simultaneously or sequentially with the generation of $x^p_{[0,T]}$, $u^p_{[0,T]}$, $z^p_{x\,[0,T]}$. For computational considerations, it may be desirable to limit the frequency (with respect to the prediction coordinate τ) with which the eigenvalues $\lambda\{\Phi\}$ are queried. This is why the procedure allows for specifying a schedule of times $\mathcal{I}_\lambda(t) \in [0, T]$ at which the eigenvalues are tested[5]. This schedule may contain any combination of points or intervals (i.e., of nonzero measure) in $[0, T]$, as long as it is defined relative to actual time in the following sense.

Assumption 7.3.3. *The schedule $\mathcal{I}_\lambda : \mathbb{R} \to [0, T]$ is selected such that $\tau \in \mathcal{I}_\lambda(t)$ implies $(\tau - \delta) \in \mathcal{I}_\lambda(t + \delta)$, for any $\tau \in [0, T]$, $\delta \in [0, \tau]$.* ■

In practice, such a $\mathcal{I}_\lambda(t)$ could be easily defined using modular arithmetic. Note that Assumption 7.3.3 need only apply in the forward direction $\delta \geq 0$. Qualitatively, this implies that the "density" of the sampling schedule may not increase along the prediction horizon.

Algorithm 7.3.4. *Given the trajectories $x^p_{[0,T]}$, $u^p_{[0,T]}$, $z^p_{x\,[0,T]}$:*

1. ***Initialize:*** $z^p_\theta = z^J_\theta(t)$, $\Phi^p_{10} = \Phi^J_1(t)$, $\Phi^p_{20} = \Phi^J_2(t)$
2. ***While** the following condition holds:*

$$\left(\tau \notin \mathcal{I}_\lambda\right) \textbf{\textit{OR}} \left(\underline{\lambda}\{\Phi^p_2\} < \frac{(M_d + M_e)\Phi^p_1 + c_\lambda}{(1 - \varepsilon_z)z^p_\theta}\right)$$

***Calculate** the following accumulation over a maximal interval of the form $\tau \in [\tau_i, \tau_{i+1})$, where $\tau_{i+1} \in (\tau_i, T]$:*

$$\dot{\Phi}^p_1 = \|g(x^p(\tau), u^p(\tau))\| + \mathcal{L}^o_g(X^p(\tau), u^p(\tau))z^p_x(\tau), \quad \Phi^p_1(\tau_i^+) = \Phi^p_{10} \tag{7.13a}$$

$$\dot{\Phi}^p_2 = G(x^p, u^p) - (\mathcal{L}^o_G(X^p, u^p)\, z^p_x)I, \quad \Phi^p_2(\tau_i^+) = \Phi^p_{20} \tag{7.13b}$$

3. ***When** the following condition is (first) satisfied, at some arbitrary $\tau = \tau_{i+1}$,*

$$\left(\tau \in \mathcal{I}_\lambda\right) \textbf{\textit{AND}} \left(\underline{\lambda}\{\Phi^p_2\} \geq \frac{(M_d + M_e)\Phi^p_1 + c_\lambda}{(1 - \varepsilon_z)z^p_\theta}\right) \tag{7.14}$$

***Then** perform the resets*

$$z^{p+}_\theta = (1 - \varepsilon_z)\, z^p_\theta(\tau^-_{i+1}) \qquad \Phi^{p\,+}_{10} = 0 \qquad \Phi^{p\,+}_{20} = 0 \tag{7.15}$$

4. ***If** $\tau < T$ **Then Iterate** to Step 2, **Else Exit*** ■

The most important property of (7.13) is that the lower bound provided by the eigenvalues of $\Phi^p_2(\tau)$, and the upper bound of $\Phi^p_1(\tau)$, are both path-independent

[5] A similar testing schedule could easily be incorporated into Algorithm 7.2.1, although the difference in timescales implies that assuming continuous testing of $\underline{\lambda}\{\Phi_2\}$ is more realistic for Algorithm 7.2.1 than for Algorithm 7.3.4.

bounds with respect to the antagonists $x^*_{[0,T]} \in X^p_{[0,T]}$. In other words, the magnitude of $\Phi_2^p(\tau)$, in any arbitrary direction, must accumulate slower than the corresponding accumulation in (7.3i) under the worst-case $\theta \in \Theta \equiv B(\hat{\theta}^J, z_\theta^J)$. This property is quantified in the following claim.

Claim 7.3.5. *At any instant t such that* $\left(\underline{\lambda}\{\Phi_2^p\} < \frac{(M_d+M_e)\Phi_1^p + c_\lambda}{(1-\varepsilon_z)z_\theta^p}\right)$, *the following is a consequence of* (7.13) *for any sufficiently small* $\delta > 0$*:*

$$\sup_{\substack{\|\mathbf{v}\|=1 \\ x^*_{[0,\delta]} \in \mathcal{X}^p_{[0,\delta]}}} \left(\mathbf{v}^T \left[\Phi_2^p(\delta) - \int_0^\delta G(x^*, u^p)\, d\tau - \Phi_2(t) \right] \mathbf{v} \right) \leq 0 \tag{7.16a}$$

$$\inf_{x^*_{[0,\delta]} \in \mathcal{X}^p_{[0,\delta]}} \left(\Phi_1^p(t) - \int_0^\delta \|g(x^*, u^p)\|\, d\tau - \Phi_1(\delta) \right) \geq 0 \tag{7.16b}$$

where the supremum is taken over arbitrary piecewise-continuous $x^*_{[0,\delta]} \in \mathcal{X}^p_{[0,\delta]}$. ■

Clearly, the likelihood of (7.13b) accumulating sufficient magnitude to trigger a reset worsens along the prediction horizon $\tau \in [0, T]$ as the error bound z_x^p grows. Practical considerations suggest that computational resources would be best employed by defining a testing schedule $\mathcal{I}_\lambda$ focussed on the early portions of the prediction horizon where the occurrence of a reset is the most likely.

Proposition 7.3.6. (Adaptive robust stabilization) *Let the identifier* $\zeta_\theta^p(T) \triangleq \zeta_\theta(z_\theta^p(T), \varepsilon_z)$ *in* (7.7) *be generated by Algorithm* 7.3.4. *Then the statement of Theorem* 7.2.8 *applies, for some new domain of attraction* $x_0 \in \mathcal{X}_0^2$ *satisfying* $\mathcal{X}_0^1 \subseteq \mathcal{X}_0^2 \subseteq \mathbb{X}$.

In the interest of improving the performance resulting from application of (7.13b), but without significantly increasing online computational complexity, Assumption 7.3.1 can be generalized to allow for the use of tighter Lipschitz-like bounds of the following form.

Assumption 7.3.7. *A known matrix-valued function* $\mathcal{M}_G^o : \overline{cov}\{\mathbb{X}\} \times \mathbb{U} \to \mathbb{R}^{p\times p}_{\succeq 0}$, *where* $\mathbb{R}^{p\times p}_{\succeq 0}$ *denotes the space of positive semi-definite symmetric matrices, satisfies the following:*

$$\mathcal{M}_G^o(\Omega_x, u) \geq \min \left\{ \mathcal{M}_G \in \mathbb{R}^{p\times p}_{\succeq 0} \;\middle|\; \sup_{\substack{x_1, x_2 \in \Omega_x \\ \|\mathbf{v}\|=1}} \left(\mathbf{v}^T \left[G(x_1, u) - G(x_2, u) - \mathcal{M}_G \|x_1 - x_2\| \right] \mathbf{v} \right) \leq 0 \right\} \tag{7.17}$$

where additionally, $\Omega_x^1 \subseteq \Omega_x^2 \subseteq \mathbb{X} \implies \sup_{\|\mathbf{v}\|=1} \mathbf{v}^T (\mathcal{M}_G^o(\Omega_x^1, u) - \mathcal{M}_G^o(\Omega_x^2, u))$ $\mathbf{v} \leq 0$. ■

Assumption 7.3.1 can be viewed as a special case of Assumption 7.3.7, in which the matrix function $\mathcal{M}_G^o$ happens to takes values within the class of positive scalar multiples of identity, and is hence parameterizable by $\mathcal{L}_G^o$. Similarly, one could moderate performance versus the complexity involved in defining $\mathcal{M}_G^o$ by searching over other subsets of $\mathbb{R}_{\geq 0}^{p\times p}$, such as the class of diagonal matrices $\mathcal{M}_G \in \mathbb{R}_{\geq 0}^{p\times p}$ parameterized by p scalar functions $\mathcal{L}_{G_i}$, $i = 1 \dots p$. This yields the following result.

Corollary 7.3.8. *Let the identifier $\zeta_\theta^p(T)$ in* (7.7) *be generated by Algorithm* 7.3.4, *with the update law in* (7.13b) *replaced by*

$$\dot{\Phi}_2^p \triangleq G(x^p, u^p) - \mathcal{M}_G^o(X^p, u^p) z_x^p \tag{7.18}$$

Then the statement of Theorem 7.2.8 applies, for some new domain of attraction $x_0 \in \mathcal{X}_0^3$ satisfying $\mathcal{X}_0^2 \subseteq \mathcal{X}_0^3 \subseteq \mathbb{X}$. ■

There is the potential to define more general update laws for (7.13), based upon knowledge of the eigenvectors of $G(x^p, u^p)$. This would provide a clear benefit in terms of performance, since it would then be possible to maintain the accumulator Φ_2^p in (7.13b) positive semi-definite, as the true Φ_2 in (7.3i) is guaranteed to be. However, this would result in a substantial increase in computational requirements, which in most cases could be put to more beneficial use elsewhere in the underlying robust controller calculation, such as implementing the modification in Remark 7.2.4.

7.4 Incorporating asymptotic filters

The results in Sections 7.2 and 7.3 were based on the assumption that an output signal of the form $y = \dot{x} + e(t)$ was available for identification purposes. If this is not the case, then it becomes necessary to construct a state predictor, whose prediction error drives the parameter identification. For example, one could use an approach similar to that in References 73 and 74 to decouple the dynamics of the observer and identification mechanisms. We will describe briefly the impact of this modification on the design presented in Sections 7.2 and 7.3.

The parameter estimator (ψ, Ψ) in (7.3d)–(7.3f) can be augmented with a state predictor of the form

$$\dot{\hat{x}}_\psi = f(x,u) + g(x,u)\psi + k_c(x - \hat{x}_\psi) + \frac{1}{k_c} c\dot{\psi}, \quad x_\psi(0) = x(0) \tag{7.19a}$$

with $k_c > 0$ a design constant, $\dot{\psi}$ an update law yet to be defined, and c the output of the filter

$$\dot{c} = k_c(g(x,u) - c), \quad c(0) = 0. \tag{7.19b}$$

Defining the error signal $\tilde{x}_\psi \triangleq x - \hat{x}_\psi$ and the quantity $\eta_\psi \triangleq \tilde{x} - \frac{1}{k_c} c\tilde{\psi}$, it can be shown that the dynamics of η_ψ satisfy

$$\dot{\eta}_\psi = -k_c \eta_\psi + d(t) \qquad \eta_\psi(0) = 0$$

and thus η_ψ is bounded $\|\eta_\psi\| \leq \frac{M_d}{k_c}$ for all $t \geq 0$. This means that $k_c\tilde{x} = c\tilde{\psi} + k_c\eta_\psi(t)$ can be viewed as a measurable replacement for the expression $(y - f(x,u) - g(x,u)\psi) \equiv g\tilde{\psi} + e(t) + d(t)$ used in (7.3d)–(7.3f). The new update laws then take the form

$$\dot{\psi} = k_c c^T \tilde{x}_\psi \qquad \dot{\Psi}_1 = \|k_c\tilde{x}_\psi\| \qquad \dot{\Psi}_2 = \|k_c\tilde{x}_\psi\|^2 \tag{7.19c}$$

and the test condition for Ψ in (7.4a) remains unchanged (interpreted with $M_e \equiv 0$).

A similar filter for the $(\phi,\ \Phi)$ identifier in (7.3g)–(7.3i) would take the form

$$\hat{x}_\phi = f(x,u) + g(x,u)\hat{\theta} + k_c(x - \hat{x}_\phi) \tag{7.20a}$$

$$\dot{\phi} = k_c\, c^T \tilde{x}_\phi \tag{7.20b}$$

$$\dot{\Phi}_1 = \|c\| \tag{7.20c}$$

$$\dot{\Phi}_2 = c^T c \tag{7.20d}$$

where c is given by (7.19b), and once again the corresponding test condition is unchanged. With the above definitions, it can be shown that all of the results in Section 7.2 hold; that is, satisfaction of condition (7.4a) is sufficient to ensure that $\|\hat{\theta}^o - \theta\| \leq (1 - \varepsilon_z^*)z_\theta$, and hence a contraction of z_θ is allowed.

Unfortunately, the approach in Section 7.3 for calculating the internal model z_θ^p becomes somewhat more challenging with (7.19) and (7.20). Although it is straightforward to predict the trajectory $c^p{}_{[0,T]}$ used in defining Φ_1^p and Φ_2^p, it becomes difficult to compensate for the effects of deviation away from the nominal trajectory, since analogous replacements for $\mathcal{L}_g^o$ and $\mathcal{L}_G^o$ would involve the propagation of the differential sensitivity equations for the filter c. Such an approach is generally not practical. A more tractable approach would be to recognize that $c \approx g(x,u)$ for sufficiently large k_c, although in practice, sensitivity to measurement noise will impose an upper limit on k_c. Nonetheless, this would motivate replacing (7.13) with

$$\dot{\Phi}_1^p = \|c(\tau)\| + k_\Phi \mathcal{L}_g^o(X^p(\tau), u^p(\tau))z_x^p(\tau), \quad \Phi_1^p(\tau_i^+) = \Phi_{10}^p \tag{7.21a}$$

$$\dot{\Phi}_2^p = c(\tau)^T c(\tau) - k_\Phi(\mathcal{L}_G^o(X^p, u^p)z_x^p)I, \qquad \Phi_2^p(\tau_i^+) = \Phi_{20}^p \tag{7.21b}$$

in which the constant $k_\Phi > 1$ is chosen "sufficiently large," dependent upon k_c.

7.5 Simulation example

In order to demonstrate the lack of robustness of nominal model MPC in the presence of state constraints, Grimm *et al.* [70] presented a state-constrained version of the classic "Artstein's Circles" problem first proposed in Reference 15. Although Artstein's problem is an interesting nonlinear control challenge for a variety of reasons,[6] the features most relevant toward the robustness results in Reference 70 are that the dynamics of the 2-state system are drift-free, while the affine control vector field steers

[6] The most notable being that the system does not admit any smooth Lyapunov function, and is thus not controllable by any continuous feedback [15]. The system contains a control-singularity at the origin.

the state toward the origin along directions of circular rotation (and hence the direction of rotation is determined by the sign of the input). The robustness issue addressed in Reference 70 corresponds to the question of robustly selecting the feasible direction of rotation when one path is blocked by a state constraint.

The example problem presented here is in some respects very different from Artstein's original problem,[7] yet it too meets the general description provided above, and exhibits the same lack of nominal robustness with respect to direction of rotation as studied in Reference 70.

7.5.1 System description

The system under consideration is given by the dynamics

$$\dot{x}_1 = -x_2 x_3 \qquad \dot{x}_2 = (x_1 - \theta_1)x_3 \qquad \dot{x}_3 = (1 + 0.5\theta_2)u \tag{7.22}$$

subject to the uncertainty $\|\theta\| \leq 0.2 \triangleq M_\theta$. The objective is to regulate to the target $\Sigma_x^o = \{x \in \mathbb{R}^3 \mid x_2 = 0\}$, subject to the constraints $\mathbb{X} = \{x \in \mathbb{R}^3 \mid |x_1| \leq 1$ and $|x_3| \leq 1\}$ and $\mathbb{U} = \{|u| \leq 1\}$. It can be seen that in the $x_1 - x_2$ plane, the flows of the system rotate around the (uncertain) point $(x_1, x_2) = (\theta_1, 0)$, with angular velocity x_3. As is shown in Figure 7.2, identification of the parameter θ_1 helps to expand the domain of feasible attraction by enabling the controller to select the correct direction of rotation to feasibly reach the x_1 axis. The true parameter values used in (7.22) are $\theta = (-0.18, 0.05)$.

The cost function is taken to be $L(x, u) = \frac{0.1}{1+M_\theta}|x_2| + 0.9\,|u|$, which can be seen to satisfy $L(x, u) \leq 1$ along any feasible trajectory of the uncertain system. It is assumed that an output signal $y = \dot{x} + e(t)$ is available, with accuracy $\|e\| \leq 0.01 \triangleq M_e$. Other parameters used in the controller are selected as $\underline{\varepsilon_z} = 0.05$, $c_\lambda = 0.005$, and $T_{\max} = 1$. For ease of computation, the optimal control problem (7.6) was minimized over the class of PWC control signals $u^p(\tau)$ supported by a uniform time-discretization of 0.1 seconds, with (7.6) only re-solved at the switching nodes.

7.5.2 Terminal penalty

Since it is known that the system flows along circular orbits, a terminal penalty satisfying Criterion 7.2.5 can be constructed by (analytically) calculating the minimum time to the target using control values $u \in \{-1, +1\}$. The calculation is relatively inexpensive, so it is computed both for the current uncertainty set Θ (based on the tightest square containing the ball $\Theta = B(\hat{\theta}, \zeta_\theta)$), as well as for each element in a covering consisting of 30 evenly distributed hypercubic subsets. Since the most dominant uncertainty is that of the θ_1 parameter, these subsets are chosen as hypercubes

[7] Our reason for modifying the dynamics from those proposed by Artstein was due primarily to the fact that linearly parameterized perturbations of the dynamics skew the circularity of the orbits, and do not lend themselves to any intuitive geometrical interpretation.

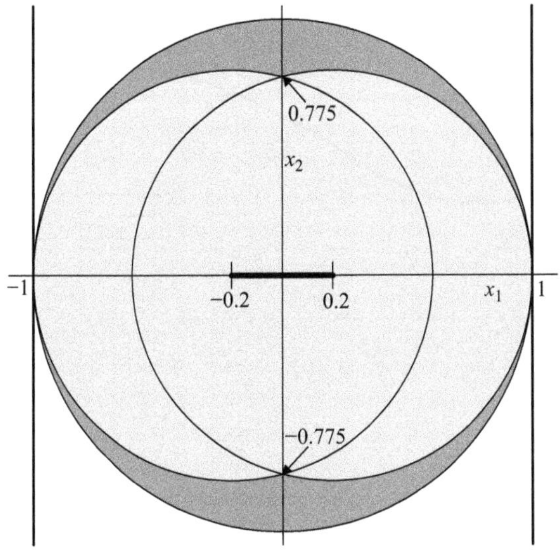

Figure 7.2 Stabilizable regions in the $x_1 - x_2$ plane. The lightly-shaded area represents the maximal region (projected onto $x_3 = 0$) stabilizable by a standard, non-adaptive robust-MPC controller. The darker area represents the limiting bound on the stabilizable region achievable by the proposed adaptive method, as the adaptation mechanism approaches perfect, instantaneous identification of θ

of size $\zeta_\theta^{\text{cov}} \times \zeta_\theta$ in the θ_1 and θ_2 coordinate directions respectively, where the dimension $\zeta_\theta^{\text{cov}} \triangleq \zeta_\theta^p(T) + \frac{\zeta_\theta - \zeta_\theta^p(T)}{30}$ is the smallest possible value for which an arbitrary ball $B(\bar{\theta}, \zeta_\theta^p(T)) \subseteq \Theta$ is guaranteed to be contained within at least one member of the subcover. Using the procedure outlined in Section 6.5.2, the terminal cost is then taken to be of the form $W(x_f, \Theta) = \min\{W^0(x_f), \max_{i=1\dots30}\{W^i(x_f)\}\}$, where W^0 represents the calculation involving the full uncertainty set (and assuming the convention $x_f \notin \mathbb{X}_f^j \implies W^j = +\infty$ as indicated in Section 6.5.2).

For each of the 31 candidate uncertainty sets Θ^j, the minimum-time to the x_1-axis is calculated by determining the amount of time (and radians traveled) spent in each of acceleration, coast, and deceleration phases (or decelerate-accelerate-decelerate if an overshoot is unavoidable), for both clockwise and counter clockwise rotation. The penalty W^j is then defined to be the lesser (feasible) of these two costs. In all cases, the worst-case value of θ_2 is given by $\hat{\theta}_2 + \zeta_\theta$, while the worst-case value of θ_1 is the limit value which results in the longer arc-length.

Satisfaction of the terminal state constraint $\mathbb{X}_f$ is determined in similar fashion by explicit testing for potential constraint violation along the minimum-time path. This simply involves testing whether the worst-case circular path intersects the constraint in question, and if so, whether the required stopping distance (in radians) is less than the worst-case distance to the constraint. In other words, the region $\mathbb{X}_f$ is defined

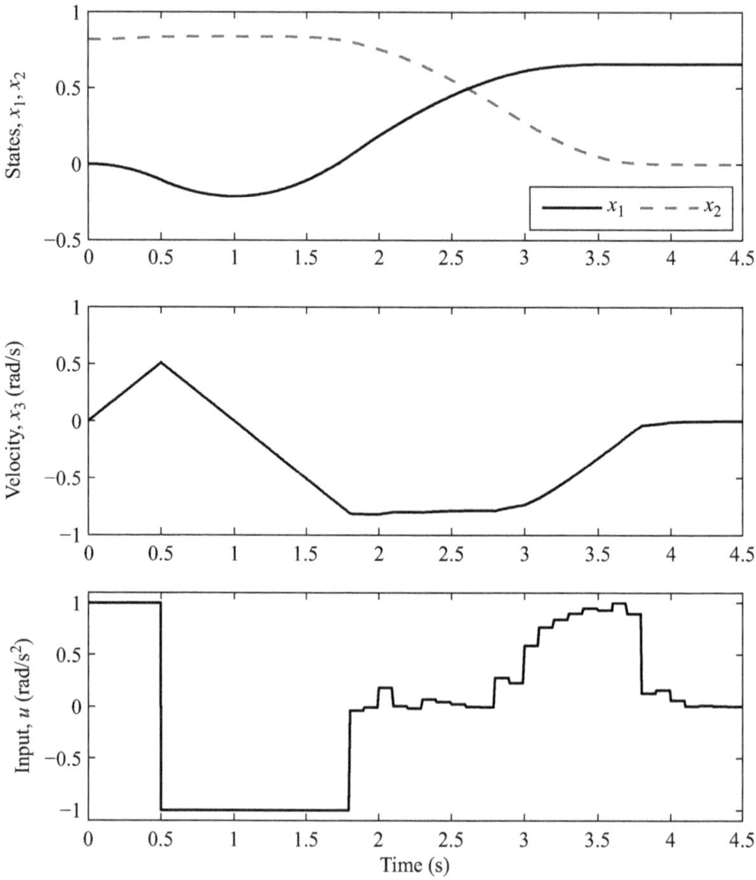

Figure 7.3 Closed-loop trajectories of the system states and applied control input for Example 7.5, for initial condition $(x_1, x_2, x_3) = (0, 0.82, 0)$

implicitly by declaring x_f to be in $\mathbb{X}_f$ if the minimum-time trajectory in at least one direction of rotation does not violate any constraints. The sequence of logical tests involved is rather lengthy, but involves a straightforward testing for each of the different manners in which a constraint could be violated. Since construction of the terminal penalty and constraint region are not the main contribution of this work, we omit further details.

7.5.3 Simulation results

The system was simulated from an initial state of $x_0 = (x_1, x_2, x_3)_0 = (0, 0.82, 0)$. Given the initial uncertainty bound $\|\theta\| \leq M_\theta = 0.2$, it is clear that rotation in at least one direction must result in feasible stabilization to the origin. However, without more accurate knowledge of θ_1 it is impossible to guarantee which direction is the

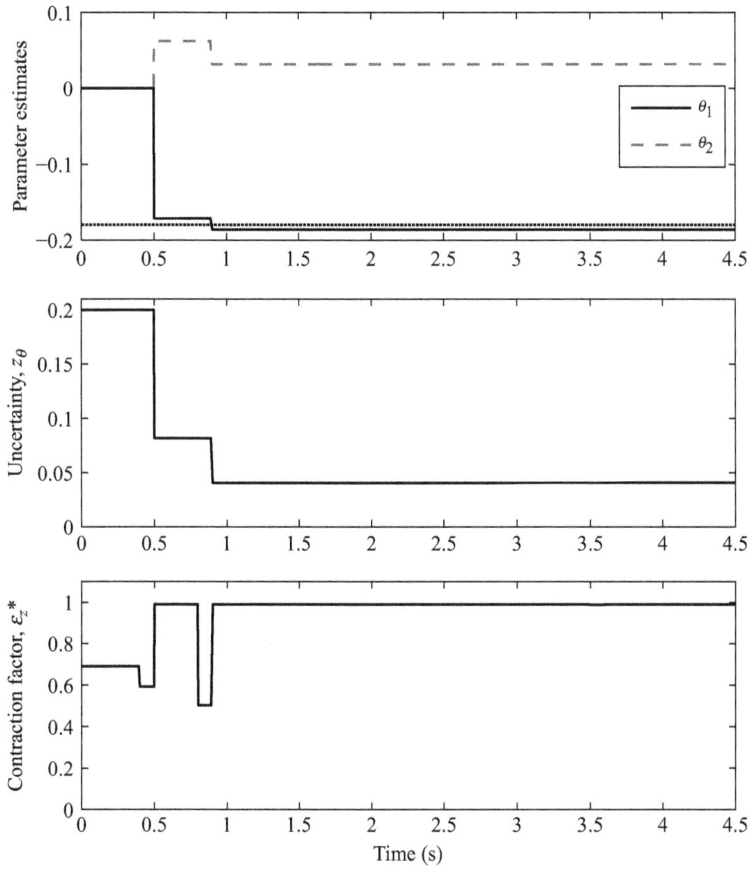

Figure 7.4 Closed-loop trajectories of the identifier states for Example 7.5, for initial condition $(x_1, x_2, x_3) = (0, 0.82, 0)$

feasible choice, and thus it can be seen from Figure 7.2 that these initial conditions are not stabilizable using a standard (non-adaptive) robust-MPC calculation (i.e., the necessary optimal control problem would not admit a feasible solution).

In contrast, the feasible closed-loop trajectories generated by the implementation of Algorithm 7.2.1 are presented in Figures 7.3 and 7.4. By choosing to initially perturb the system using fairly aggressive control action, the reduction in parametric uncertainty occurring at time $t = 0.5$ was predicted by the internal model of Algorithm 7.3.4 within the initial control problem solved at time $t = 0$. This allowed the controller to proceed with a feasible control solution, despite the fact that the initial (arbitrary) selection of counter clockwise rotation eventually proved to be the wrong choice.

Following the first reduction of z_θ at time $t = 0.5$, the new uncertainty set $\theta \in B(\hat{\theta}, z_\theta) = B([-0.17, 0.06], 0.082)$ enabled the controller to conclusively select clockwise rotation as the appropriate course of action. For this reason the optimal value of ε_z switched to $\varepsilon_z^* \approx 1$, the value which minimizes the diameter $z_x(\tau)$ of the uncertainty cone in (7.7c). However, by time $t = 0.9s$ it became possible to more precisely characterize $\theta \in B([-0.187, 0.032], 0.04)$, which briefly satisfied the current-time test condition in (7.6) for an optimal value $\varepsilon_z^* = 0.5$.

7.5.4 Discussion

A key property of the above example is that the control problem involves making a choice between various alternative outcomes (in this case the choice was essentially between stabilization to either the positive or negative x_1-axis, although in general the number of candidates need not be finitely indexed). In the presence of the full initial uncertainty $\theta \in B(\mathbf{0}, 0.2)$, the worst-case scenarios involve the extreme values $\theta_1 \in \{-0.2, 0.2\}$, and lead to the smaller stabilizable region indicated in Figure 7.2. In contrast, as the predicted uncertainty bound $z_\theta^p(T) \leq z_\theta$ approaches zero, the maximization in (7.7g) approaches a search for the single-worst value that a practically-known θ could take, which in this case would be $\theta = 0$. The reason the domain of attraction increases is because the worst-case value of θ_1 differs for the two candidate decisions; that is, choosing to stabilize to $\{x_1 < 0, x_2 = x_3 = 0\}$ involves a different worst-case θ_1 than the alternative target $\{x_1 > 0, x_2 = x_3 = 0\}$. Thus, by reducing the uncertainty bound it is guaranteed that at least one of these worst-case scenarios will improve, even if it can't be predicted in advance.

Admittedly, this type of behavior in the worst-case scenarios is not exhibited by all systems. In practice, these type of situations could occur whenever the controller must make some sort of decision, where the optimal choice is parameter-dependent. Examples could include such problems as obstacle-avoidance in the presence of some sort of uncertain bias (cross-wind, a preferential turning radius, etc.), or problems of balancing the load between parallel process equipment with different operating properties (reactor beds with different activity, heat exchangers with uncertain fouling effects, etc.).

For systems which do not exhibit the above type of behavior, that is, systems where there is a unique worst-case parameter value that applies to all candidate decisions, then the worst-case maximization in (7.7g) is unaffected by $z_\theta^p(T)$. In these cases, the proposed algorithm is not able to improve upon the domain of attraction of a non-adaptive robust controller (e.g., if the above simulation problem were modified such that the negative x_1-axis $\{x_1 < 0, x_2 = x_3 = 0\}$ were the only allowable target, then the proposed method would be unable to improve upon the domain of attraction of the non-adaptive controller). However, even without expanding the domain of attraction, there is still the potential for the internal model $z_\theta^p(\tau)$ to improve the *performance* of the closed-loop adaptive control. By designing the terminal penalty $W(x, \Theta)$ such that large values of $z_\theta^p(\tau)$ are explicitly penalized, the optimization will automatically favor trajectories that result in improved identification.[8] This enables the controller to preserve the benefits of an optimization-based approach to injecting excitation into the

[8] In fact, one could easily replace $L(x, u)$ with an appropriate $L(x, u, z_\theta^p)$ that likewise penalizes $z_\theta^p(\tau)$, which would have a similar effect.

system, as discussed in Chapter 6.3. By improving the closed-loop convergence rate of the identifier, the conservatism of the overall control action will be more quickly reduced as the closed-loop system evolves.

7.6 Summary

This chapter has demonstrated that by posing the adaptive control problem within the framework of a simplistic robust-MPC design, a numerically tractable adaptive feedback can be designed for a general class of constrained, parameter-affine nonlinear systems. Although the approach here is potentially quite conservative, it preserves many of the important properties of the more computationally challenging min–max feedback approach developed in Chapter 6. Embedding an internal model approximation of the identification mechanism within the controller predictions enables one to account for the impact of trajectory excitation within the optimal control problem, and to construct less-conservative terminal CLF-penalty functions based upon subsets of the original parametric uncertainty. The approach proposed here errs intentionally on the side of computational simplicity, just as Chapter 6 erred on the side of performance, irrespective of complexity. However, the most valuable implication of Chapters 6 and 7 is that the overall methodology presented in Section 6.2 is amenable to *any* robust-MPC design involving any desired tradeoff between computational complexity and conservatism of the solutions.

7.7 Proofs for Chapter 7

7.7.1 Proof of Proposition 7.2.2

Using the upper bound on the magnitude of the update in (7.4d), Claim (a) follows from a simple triangle inequality (see Figure 7.1):

$$\sup_{w\in\Theta^+} \|w-\hat{\theta}\| \leq \sup_{w\in\Theta^+} \|w-\hat{\theta}^+\| + \|\hat{\theta}^+ - \hat{\theta}\|$$

$$\leq (1-\varepsilon_z)\left(\sqrt{\frac{2-\varepsilon_z}{\varepsilon_z}}\, z_\theta\right) + \sqrt{\varepsilon_z(2-\varepsilon_z)}\, z_\theta = \left(\sqrt{\frac{2-\varepsilon_z}{\varepsilon_z}}\right) z_\theta$$

which proves $w \in \Theta^+ \implies w \in \Theta$, and hence $\Theta(t_i^+) \subseteq \Theta(t_i^-)$.

To address the second claim, we will first demonstrate that the reset condition (7.4a), together with the previous bound $\|\hat{\theta} - \theta\| \leq z_\theta$, together imply that $\|\hat{\theta}^o - \theta\| \leq (1-\varepsilon_z) z_\theta$. In other words, $\hat{\theta}^o \in B(\theta, (1-\varepsilon_z)z_\theta)$, the dotted circle in Figure 7.1.

Case 1: Reset based on Ψ

Using (7.5) we define the simple Lyapunov function (arguments of $g(x,u)$ omitted)

$$V_{\tilde{\psi}} = \frac{1}{2}\tilde{\psi}^T\tilde{\psi}, \quad \dot{V}_{\tilde{\psi}} = -\tilde{\psi}^T g^T (g\tilde{\psi} + e + d), \quad \forall t \in [t_{i-1}^+, t_i^-) \tag{7.23}$$

The inclusion $\psi(t_i^-) \in B(\theta, (1-\varepsilon_z)z_\theta)$ follows directly from

$$
\begin{aligned}
V_{\tilde{\psi}}(t_i^-) &= V_{\tilde{\psi}}(t_{i-1}^+) + \int_{t_{i-1}}^{t_i} \dot{V}\tilde{\psi}(\tau)\, d\tau \\
&\leq \frac{1}{2}z_\theta^2 - \int_{t_{i-1}}^{t_i} \tilde{\psi}^T g^T (g\tilde{\psi} + e + d)\, d\tau \\
&\leq \frac{1}{2}z_\theta^2 - \int_{t_{i-1}}^{t_i} (g\tilde{\psi} + e + d)^T (g\tilde{\psi} + e + d)\, d\tau \\
&\quad + \int_{t_{i-1}}^{t_i} (e + d)^T (g\tilde{\psi} + e + d)\, d\tau \\
&\leq \frac{1}{2}z_\theta^2 - \Psi_2 + (M_e + M_d)\Psi_1 \\
&\leq \frac{1}{2}z_\theta^2 - \left(\frac{1}{2}\varepsilon_z(2-\varepsilon_z)z_\theta^2 + (M_e + M_d)\Psi_1\right) + (M_e + M_d)\Psi_1 \\
&= \frac{1}{2}(1-\varepsilon_z)^2 z_\theta^2
\end{aligned}
$$

Case 2: Reset on Φ

Defining $\tilde{\theta} \triangleq \theta - \hat{\theta}$ and $\tilde{\theta}^o \triangleq \theta - \hat{\theta}^o$, we have that

$$
\begin{aligned}
V_{\tilde{\theta}}(t_i^-) &= \frac{1}{2}\,{\tilde{\theta}^o}^T\,\tilde{\theta}^o = \frac{1}{2}\|\theta - \hat{\theta} - \Phi_2^{-1}\phi\|^2 \\
&= \frac{1}{2}\left\|\tilde{\theta} - \left(\int_{t_{i-1}}^{t_i} g^T g\, d\tau\right)^{-1}\left[\int_{t_{i-1}}^{t_i} g^T g\tilde{\theta}\, d\tau + \int_{t_{i-1}}^{t_i} g^T(d(\tau) + e(\tau))\, d\tau\right]\right\|^2 \\
&\leq \frac{1}{2}\left|\frac{1}{\underline{\lambda}\{\Phi_2\}}\left(\int_{t_{i-1}}^{t_i} \|g\|\, d\tau\right)(M_d + M_e)\right|^2 \\
&\leq \frac{1}{2}\left|\left(\frac{(1-\varepsilon_z)z_\theta}{\left(\int_{t_{i-1}}^{t_i}\|g\|\, d\tau\right)(M_e + M_d) + c_\lambda}\right)\left(\int_{t_{i-1}}^{t_i} \|g\|\, d\tau\right)(M_e + M_d)\right|^2 \\
&\leq \frac{1}{2}(1-\varepsilon_z)^2 z_\theta^2
\end{aligned}
$$

This completes Case 2, and it can be concluded that in all cases, $\|\hat{\theta}^o - \theta\| \leq (1-\varepsilon_z)z_\theta$ in the pre-reset state t_i^-.

Having proven $\hat{\theta}^o \in B(\theta, (1-\varepsilon_z) z_\theta)$, it needs to be shown that the orthogonal projection in (7.4d) preserves $\hat{\theta}^+ \in B(\theta, (1-\varepsilon_z) z_\theta)$. For the worst-case situation depicted in Figure 7.1, values of $\hat{\theta}^o$ occurring in the shaded region are projected orthogonally onto the spherical shell $\partial B(\hat{\theta}, \sqrt{\varepsilon_z(2-\varepsilon_z)} z_\theta)$. The inclusion $\hat{\theta}^+ \in B(\theta, (1-\varepsilon_z) z_\theta)$ (i.e., that $\hat{\theta}^+$ remains in the dotted circle in Figure 7.1, for any $\hat{\theta}$ in the shaded region) then follows from the orthogonality of the intersection between $\partial B(\theta, (1-\varepsilon) z_\theta)$ and $\partial B(\hat{\theta}, \sqrt{\varepsilon_z(2-\varepsilon_z)} z_\theta)$ (i.e., angle $\angle \theta s \hat{\theta}$ in Figure 7.1). This orthogonality holds by

$$\left((1-\varepsilon_z) z_\theta\right)^2 + \left(\sqrt{\varepsilon_z(2-\varepsilon_z)}\, z_\theta\right)^2 = z_\theta^2. \tag{7.24}$$

Clearly, if the unknown θ lies in the interior of $B(\hat{\theta}, z_\theta)$ (i.e., the dotted circle in Figure 7.1 is translated downwards an arbitrary distance), then the inclusion $\hat{\theta}^+ \in B(\theta, (1-\varepsilon_z) z_\theta)$ will hold with additional conservatism. The implied bound $\|\hat{\theta}^+ - \theta\| \le (1-\varepsilon_z) z_\theta$ therefore completes the proof of claim (b).

Claims (c) and (d) follow directly by induction, given that for constant ε_z^*, $\Theta(\tau)$ is fixed over intervals of continuous evolution (i.e., during Step 2 of Algorithm 7.2.1), its reset behavior satisfies (a) and (b), and initially $\theta \in B(0, M_\theta) \equiv B(\hat{\theta}(t_0), z_\theta(t_0))$.

7.7.2 Proof of Theorem 7.2.8

The constraints (7.7d)–(7.7e) can be implicitly incorporated into the analysis by extending the interpretation of L' and W' in (7.7a) to involve the convention

$$L'(x^p, u^p, z_x^p) \equiv \begin{cases} (7.7\text{f}) & (X^p, u) \in \mathbb{X} \times \mathbb{U} \\ +\infty & \text{otherwise} \end{cases}$$

$$W'(x^p, z_x^p, \zeta_\theta^p, \Theta) \equiv \begin{cases} (7.7\text{g}) & X(T) \subseteq \mathbb{X}_f'(\zeta_\theta^p(T), \Theta) \\ +\infty & \text{otherwise} \end{cases}$$

Note that this has no impact on the Lipschitz constants $\mathcal{L}_L$, $\mathcal{L}_W$, which apply to L and W, but not L', W'.

It then follows that both stability and feasibility of the closed-loop dynamics are guaranteed if the optimal cost $J^*(x, \Theta, \Phi) \triangleq J(x, u^{p*}_{[0,T^*]}, T^*, \hat{\theta}^J, \zeta_\theta^J, \varepsilon_z^*, \Phi^J)$ resulting from (7.6) and (7.7) is strictly decreasing for $x \notin \Sigma_x^o$. To conclude the decrease of J^* over both flows and resets of the closed-loop behavior of Algorithm 7.2.1, the same two conditions as were used in Theorem 6.4.6 are sufficient:

i. $\max\limits_{f' \in \mathcal{F}} \overrightarrow{D} J^* \triangleq \max\limits_{f' \in \mathcal{F}} \liminf\limits_{\substack{v \to f' \\ \delta \downarrow 0}} \dfrac{J^*(x+\delta v, \Theta(t+\delta), \Phi(t+\delta)) - J^*(x, \Theta, \Phi)}{\delta} < 0$

ii. $\min\limits_{f' \in \mathcal{F}} \overleftarrow{D} J^* \triangleq \min\limits_{f' \in \mathcal{F}} \limsup\limits_{\substack{v \to f' \\ \delta \downarrow 0}} \dfrac{J^*(x-\delta v, \Theta(t-\delta), \Phi(t-\delta)) - J^*(x, \Theta, \Phi)}{\delta} > 0$

where $\mathcal{F} \triangleq f(x, \kappa_{mpc}(x)) + g(x, \kappa_{mpc}(x)) B(\hat{\theta}^J, \zeta_\theta^J) + B(0, M_d)$, with $\kappa_{mpc}(x)$ denoting the feedback (7.6a) (arguments omitted). Note that it is sufficient to take the worst case over only uncertainties in $\theta \in B(\hat{\theta}^J, \zeta_\theta^J) \subseteq B(\hat{\theta}, \zeta_\theta)$, since it follows from Proposition 7.2.2 that in the event that $(\hat{\theta}^J, \zeta_\theta^J) \neq (\hat{\theta}, \zeta_\theta)$ in (7.6), then $\theta \in B(\hat{\theta}^J, \zeta_\theta^J)$ is guaranteed.

To begin, we first dispose of the special case where $T = 0$. In this case, the requirement $x \in \mathbb{X}_f(\Theta)$ implies $J^*(x, \Theta, \Phi) \equiv W(x, \Theta)$, from which the combination of the Lipschitz continuity of W and Criterion 7.2.5 point 5 is sufficient to satisfy both conditions (i) and (ii). For the remainder, it is assumed that $0 < \delta < T$. We begin by proving condition (i) for $\overrightarrow{D} J^*$.

At an arbitrary time t, we let $(x^p, z_x^p, \zeta_\theta^p)$ denote solutions of (7.7), corresponding to a feasible solution $(u^p_{[0,T]}, \varepsilon_z, T)$ of the optimization problem (7.6) posed at time t. The trajectories therefore originate from initial conditions $(x^p, z_x^p, z_\theta^p)_{\tau=0} = (x(t), 0, z_\theta^J(t))$. Similarly, we denote by (x^v, z_x^v, z_θ^v) a set of solutions of (7.7) with initial conditions $(x^v, z_x^v, z_\theta^v)_{\tau=\delta} = (x(t) + \delta v, 0, z_\theta^J(t+\delta))$, but which correspond to the same choices of $u^v_{[\delta,T]} \equiv u^p_{[\delta,T]}$ and ε_z. Here $z_\theta^J(t+\delta)$ denotes the (potentially uncertain) value of z_θ^J that would result under the test in (7.6) at time $(t+\delta)$, using Algorithm 7.2.1, with open-loop implementation of $u_{[t,t+\delta]} = u^p_{[0,\delta]}$ while keeping ε_z^* fixed at $\varepsilon_z^*(\tau) \equiv \varepsilon_z^*(t)$ for $\tau \in [t, t+\delta]$. We similarly denote $\zeta_\theta^p(\tau) \triangleq \zeta_\theta(z_\theta^p(\tau), \varepsilon_z)$ and $\zeta_\theta^v(\tau) \triangleq \zeta_\theta(z_\theta^v(\tau), \varepsilon_z)$ for all $\tau \in [0, T]$, where $\varepsilon_z \equiv \varepsilon_z^*(t)$. Throughout the proof, the superscripts p or v on any expression will indicate association with either the x^p or x^v solutions.

At prediction time $\tau = 0$, it follows from (7.7b), (7.7c) that

$$\max_{f' \in \mathcal{F}} \|\dot{x}^p(0) - f'\| = \max_{\substack{\theta \in \Theta(t) \\ d \in \mathcal{D}}} \left\|g(x, u^p(0))\left(\hat{\theta}^J - \theta\right) + d\right\| \tag{7.25a}$$

$$\leq \left\|g(x, u^p(0))\right\| \zeta_\theta^J + M_d \tag{7.25b}$$

$$= \dot{z}_x^p(0) \tag{7.25c}$$

where the inequality results from Proposition 7.2.2.b. From definition (7.7i), the continuity of $x^p(\tau)$ and $z_x^p(\tau)$ ensures that there exists $v \equiv v(\delta)$ satisfying $\lim_{\delta \downarrow 0} v(\delta) = f'$ such that $(x + \delta v) \in X^p(\delta)$. Without loss of generality, we assume $x^v(\delta) \in X^p(\delta)$.

Since $z_x^v(\delta) = 0$, it follows that $\|x^p(\delta) - x^v(\delta)\| \leq z_x^p(\delta) - z_x^v(\delta)$. Defining the variable $e_z \triangleq z_x^p - z_x^v - \|x^p - x^v\|$, it follows that $X^v \subseteq X^p$ and $\mathcal{L}_j^v \leq \mathcal{L}_j^p, j \in \{f, g, L\}$ on the domain $e_z \in \mathbb{R}_{\geq 0}$. On this domain, the dynamics of e_z satisfy

$$\begin{aligned}
\dot{e}_z &= \dot{z}_x^p - \dot{z}_x^v - \|\dot{x}^p - \dot{x}^v\| \\
&= (\mathcal{L}_f^p + \mathcal{L}_g^p M_\Theta) z_x^p - (\mathcal{L}_f^v + \mathcal{L}_g^v M_\Theta) z_x^v \\
&\quad - \|f(x^p, u) - f(x^v, u) + (g(x^p, u) - g(x^v, u))\hat{\theta}^J\| \\
&\geq \left(\mathcal{L}_f^p + \mathcal{L}_g^p M_\Theta\right)(z_x^p - z_x^v) - \left(\mathcal{L}_f^p + \mathcal{L}_g^p \hat{\theta}^J\right)\|x^p - x^v\| \\
&\geq \left(\mathcal{L}_f^p + \mathcal{L}_g^p M_\Theta\right) e_z
\end{aligned}$$

from which the initial condition $e_z(\delta) \geq 0$ guarantees that $e_z(\tau) \geq 0$, $\forall \tau \in [\delta, T]$. This implies that $X^v(\tau) \subseteq X^p(\tau) \subseteq \mathbb{X}$ and $\mathcal{L}_j^v \leq \mathcal{L}_j^p$ for all $\tau \in [\delta, T]$, from which one obtains

$$\begin{aligned} L'(x^p, u) = L^p + \mathcal{L}_L^p z_x^p &\geq (L^v - \mathcal{L}^p \|x^p - x^v\|) + \mathcal{L}_L^p z_x^p \\ &\geq L^v + \mathcal{L}^p z_x^v \geq L'(x^v, u) \end{aligned} \tag{7.26}$$

for all $\tau \in [\delta, T]$. From the non-increase of z_θ, and the fact that ε_z is presumed constant, one has $\zeta_\theta^v(T) \equiv \zeta_\theta^J(z_\theta(t+\delta), \varepsilon_z(t)) \leq \zeta_\theta^J(t) \equiv \zeta_\theta^p(T)$. It then follows that $\Omega^v \triangleq \Omega(\zeta_\theta^v, \Theta(t+\delta)) \subseteq \Omega(\zeta_\theta^p, \Theta(t)) \triangleq \Omega^p$. This implies by (7.7h) that $X^v(T) \subseteq X^p(T) \subseteq \mathbb{X}_f'^p \subseteq \mathbb{X}_f'^v$, which yields

$$\begin{aligned} W'^p \triangleq \max_{\Theta_T^p \in \Omega^p} \left\{ W^p + \mathcal{L}_W^p z_x^p \right\} &\geq \max_{\Theta_T^p \in \Omega^p} \left\{ W(x^v, \Theta_T^p) - \mathcal{L}_W^p \|x^p - x^v\| + \mathcal{L}_W^p z_x^p \right\} \\ &\geq \max_{\Theta_T^p \in \Omega^p} \left\{ W(x^v, \Theta_T^p) + \mathcal{L}_W^p z_x^v \right\} \\ &\geq \max_{\Theta_T^v \in \Omega^v} \left\{ W^v + \mathcal{L}_W^v z_x^v \right\} \triangleq W'^v. \end{aligned} \tag{7.27}$$

This results in the desired conclusion

$$\begin{aligned} \max_{f' \in \mathcal{F}} \overrightarrow{D} J^* &\leq \max_{f' \in \mathcal{F}} \liminf_{\substack{v \to f' \\ \delta \downarrow 0}} \frac{1}{\delta} \left(\int_\delta^T (L'^v - L'^p)\, d\tau - \delta L^p|_0 + W'^v - W'^p \right) \\ &\leq -L(x, u) \end{aligned} \tag{7.28}$$

in which the first inequality holds by the suboptimality of the choices $(u_{[\delta,T]}^v, \varepsilon_z)$ used to generate the x^v solution at time $t + \delta$.

Condition (ii) can be proven in an analogous fashion to the above development. Because a similar analogy between (i) and (ii) was demonstrated in the proof of 6.4.6, our treatment here will be brief. This involves:

- Redefining the x^v trajectory to be defined over the interval $\tau \in [-\delta, T]$, satisfying initial conditions $(x^v, z_x^v, \zeta_\theta^v)_{\tau=-\delta} = (x(t) + \delta v, 0, \zeta_\theta^J(z_\theta(t-\delta), \varepsilon_z(t)))$, where the corresponding input $u_{[-\delta,T]}^v$ satisfies $u_{[-\delta,0]}^v \equiv u_{[t-\delta,t]}$ and $u_{[0,T]}^v \equiv u_{[0,T]}^p$. Without loss of generality, it is assumed $v(\delta)$ is such that $x^v(-\delta) \in \mathcal{X}$.
- Defining $e_z \triangleq z_x^v - z_x^p - \|x^v - x^p\|$, it can be shown $e_z(0) \geq 0$, and $\dot{e}_z \geq 0$ for all $\tau \in [0, T]$. This implies $X^p(\tau) \subseteq X^v(\tau)$ and $\mathcal{L}_j^p(\tau) \leq \mathcal{L}_j^v(\tau)$ for all $\tau \in [0, T]$.
- The analogs of (7.26) and (7.27) become

$$\begin{aligned} L'(x^v, u) &= L^v + \mathcal{L}_L^v z_x^v \geq L^p + \mathcal{L}^v z_x^p \geq L'(x^p, u) \\ W'^v &\geq \max_{\Theta_T^v \in \Omega^v} \left\{ W(x^p, \Theta_T^v) + \mathcal{L}_W^v z_x^p \right\} \geq \max_{\Theta_T^p \in \Omega^p} \left\{ W^p + \mathcal{L}_W^p z_x^p \right\} \triangleq W'^p. \end{aligned}$$

from which the desired result is obtained:

$$\begin{aligned} \min_{f' \in \mathcal{F}} \overleftarrow{D} J^* &\geq \min_{f' \in \mathcal{F}} \limsup_{\substack{v \to f' \\ \delta \downarrow 0}} \frac{1}{\delta} \left(\int_0^T L'^v - L'^p d\tau + \delta L^v|_{-\delta} + W'^v - W'^p \right) \\ &\geq L(x, u) \end{aligned}$$

This completes the proof of feasible stabilization from any point $x \in \mathcal{X}_0^1$. The containment $\mathcal{X}_0 \subseteq \mathcal{X}_o^1$ follows from the fact that suboptimally specifying $\varepsilon_z^* \equiv 1$ prevents the reset condition (7.4a) from ever being satisfied. This results in $\Theta(t) \equiv B(\hat{\theta}(t), z_\theta) \equiv \Theta_0$, and thus the control is equivalent to $\kappa_{mpc}(x, 0, M_\theta, 0, 0, 0, 0)$.

7.7.3 Proof of Claim 7.3.5

Since $\left(\underline{\lambda}\{\Phi_2^p(0)\} < \frac{(M_d+M_e)\Phi_1^p(0)+c_\lambda}{(1-\varepsilon_z)z_\theta^J}\right)$ we can assume $\delta > 0$ is small enough that no reset occurs in $\Phi^p_{2\,[0,\delta]}$. Then

$$\begin{aligned}
&\sup_{\substack{\|\mathbf{v}\|=1\\ x^*_{[0,\delta]}\in\mathcal{X}^p_{[0,\delta]}}} \left(\mathbf{v}^T\left[\Phi_2^p(\delta) - \int_0^\delta G|_{x^*}\,d\tau - \Phi_2(t)\right]\mathbf{v}\right)\\
&\quad= \sup_{\substack{\|\mathbf{v}\|=1\\ x^*_{[0,\delta]}\in\mathcal{X}^p_{[0,\delta]}}} \left(\mathbf{v}^T\left[\int_0^\delta \dot{\Phi}_2^p - G|_{x^*}\,d\tau\right]\mathbf{v}\right)\\
&\quad= \sup_{\substack{\|\mathbf{v}\|=1\\ x^*_{[0,\delta]}\in\mathcal{X}^p_{[0,\delta]}}} \left(\mathbf{v}^T\left[\int_0^\delta G|_{x^p} - \mathcal{L}_G(X^p,u)\,z_x^p\,I - G|_{x^*}\,d\tau\right]\mathbf{v}\right)\\
&\quad= \sup_{\substack{\|\mathbf{v}\|=1\\ x^*_{[0,\delta]}\in\mathcal{X}^p_{[0,\delta]}}} \left(\int_0^\delta \mathbf{v}^T\big[G|_{x^p} - G|_{x^*}\big]\mathbf{v} - \mathcal{L}_G(X^p,u)\,z_x^p\,d\tau\right)\\
&\quad\le \sup_{x^*_{[0,\delta]}\in\mathcal{X}^p_{[0,\delta]}} \left(\int_0^\delta \sup_{\|\mathbf{v}(\tau)\|=1} \mathbf{v}(\tau)^T\big[G|_{x^p} - G|_{x^*}\big]\mathbf{v}(\tau) - \mathcal{L}_G(X^p,u)z_x^p\,d\tau\right)\\
&\quad\le 0
\end{aligned} \tag{7.29}$$

as claimed. Note that it is not necessary to consider the scenario of "resets in $\Phi^*_{2\,[0,\delta]}$," since $x^*_{[0,\delta]}$ simply represents a point-wise maximizer of the indicated expression, and does not represent an actual trajectory of the system. The second inequality follows as (denoting $g^* \equiv g(x^*, u^p)$, $g^p \equiv g(x^p, u^p)$):

$$\begin{aligned}
&\inf_{x^*_{[0,\delta]}\in\mathcal{X}^p_{[0,\delta]}} \left(\Phi_1^p(t) - \int_0^\delta \|g^*\|\,d\tau - \Phi_1(\delta)\right)\\
&\quad\ge \inf_{x^*_{[0,\delta]}\in\mathcal{X}^p_{[0,\delta]}} \left(\int_0^\delta \|g^p\| - \|g^*\| + \mathcal{L}_g^o z_x^p\,d\tau\right)\\
&\quad\ge \inf_{x^*_{[0,\delta]}\in\mathcal{X}^p_{[0,\delta]}} \left(\int_0^\delta \|g^p\| - \|g^*\| + \|g^* - g^p\| - \mathcal{L}_g^o\|x^* - x^p\| + \mathcal{L}_g^o z_x^p\,d\tau\right)\\
&\quad\ge \inf_{x^*_{[0,\delta]}\in\mathcal{X}^p_{[0,\delta]}} \left(\int_0^\delta \|g^*\| - \|g^p + (g^* - g^p)\| + \mathcal{L}_g^o(z_x^p - \|x^* - x^p\|)\,d\tau\right)\\
&\quad\ge 0
\end{aligned}$$

7.7.4 Proof of Proposition 7.3.6

The identifier ζ_θ^p only factored into the proof of Theorem 7.2.8 in establishing (7.27). However, the previous assumption $z_\theta^p(T) \equiv z_\theta(t)$ was not explicitly used, other than to establish the relationship $z_\theta^v(T) \le z_\theta^p(T)$ upon which (7.27) depends. Therefore, to prove Proposition 7.3.6 it is sufficient to extend the proof of Theorem 7.2.8 by demonstrating that Algorithm 7.3.4, together with (7.13), results in $z_\theta^v(T) \le z_\theta^p(T)$.

We begin with the proof in the forward direction, to establish property (i) given in the proof of Theorem 7.2.8. Using notations from the proof of Theorem 7.2.8, we denote $\Phi^p_{1\,[0,T]}$ and $\Phi^p_{2\,[0,T]}$ to be solutions generated by Algorithm 7.3.4 from initial condition $(\Phi_1^p(0), \Phi_2^p(0)) = (\Phi_1(t), \Phi_2(t))$. Similarly, $\Phi^v_{1\,[0,T]}$ and $\Phi^v_{2\,[0,T]}$ denote solutions from $(\Phi_1^v(\delta), \Phi_2^v(\delta)) = (\Phi_1(t+\delta), \Phi_2(t+\delta))$, under the same conditions as described in the proof of Theorem 7.2.8.

Depending on whether any resets occur in $\Phi^p{}_{[0,\delta]}$ or $\Phi_{[t,t+\delta]}$, one of the following cases must hold for $\delta > 0$ sufficiently small.

Case 1, no resets: then $z_\theta^v(\delta) = z_\theta^p(\delta) = z_\theta^J$, and from Claim 7.3.5 it follows:

$$\sup_{\|\mathbf{v}\|=1} \mathbf{v}^T\big(\Phi^p(\delta) - \Phi^v(\delta)\big)\mathbf{v} \le \sup_{\substack{\|\mathbf{v}\|=1\\ x^*_{[0,\delta]}\in\mathcal{X}^p_{[0,\delta]}}} \mathbf{v}^T\left(\Phi^p(\delta) - \int_0^\delta G|_{x^*}\,d\tau - \Phi(t)\right)\mathbf{v} \le 0 \tag{7.30a}$$

$$\Phi_1^p(\delta) - \Phi_1^v(\delta) \ge \inf_{x^*_{[0,\delta]}\in\mathcal{X}^p_{[0,\delta]}} \left(\Phi_1^p(\delta) - \int_0^\delta \|g_{x^*}\|\,d\tau - \Phi_1(t)\right) \ge 0 \tag{7.30b}$$

Case 2, $\Phi_{[t,t+\delta]}$ contains reset, but not $\Phi^p_{[0,\delta]}$: then $z_\theta^v(\delta) < z_\theta^p(\delta)$
Case 3, both $\Phi_{[t,t+\delta]}$ and $\Phi^p_{[0,\delta]}$ contain reset: Taking $\delta > 0$ as sufficiently small, this implies that the resets are simultaneous. Then $z_\theta^v(\delta) = z_\theta^p(\delta) = (1-\varepsilon_z)z_\theta^J$, and (7.30) holds, by integrating forward from the time of the jump.
Case 4, $\Phi^p_{[0,\delta]}$ contains reset, but not $\Phi_{[t,t+\delta]}$: This would require that there exist a $\delta' \in (0,\delta)$ such that

$$\underline{\lambda}\{\Phi_2^p(\delta')\} \ge \frac{(M_e + M_d)\Phi_1^p(\delta') + c_\lambda}{(1-\varepsilon_z)z_\theta}$$

$$\ge \frac{(M_e + M_d)\Phi_1(t+\delta') + c_\lambda}{(1-\varepsilon_z)z_\theta} > \underline{\lambda}\{\Phi_2(t+\delta')\}.$$

However, the specific choice $\mathbf{v} = \underline{\mathbf{v}}(t+\delta')$ (i.e., associated with $\underline{\lambda}\{\Phi_2(t+\delta')\}$) violates (7.16) for time δ'. Therefore, this scenario cannot occur.

Proceeding by contradiction, we assume that $z_\theta^v(T) > z_\theta^p(T)$. It can then be seen from Algorithm 7.3.4 and the above cases that there must then exist arbitrary times $0 \le \tau_1 < \tau_2 < T$, such that: (i) $\tau_1, \tau_2 \in \mathcal{I}_\lambda(t)$, (ii) $z_\theta^p(\tau_1^+) = z_\theta^v(\tau_1^+)$, (iii) z_θ^p resets at τ_2 but z_θ^v does not, and (iv) no resets occur on $\tau \in (\tau_1, \tau_2)$.

Without loss of generality, we can take τ_1 as being infimal, which implies that either it is a switching time of z_θ^p, or $\tau_1 = 0$. In the following, we use the same notation as before to denote by time τ_2^- the pre-reset states that occur at time τ_2. Furthermore, we denote by $\underline{\mathbf{v}}^p(\tau)$ an eigen-direction corresponding to $\underline{\lambda}\{\Phi_2^p(\tau)\}$, with $\underline{\mathbf{v}}^v(\tau)$ defined analogously.

$$
\begin{aligned}
&\underline{\lambda}\{\Phi_2^p(\tau_2^-)\} - \underline{\lambda}\{\Phi_2^v(\tau_2^-)\} \\
&\quad = \underline{\mathbf{v}}^p(\tau_2^-)^T \Phi_2^p(\tau_2^-)\underline{\mathbf{v}}^p(\tau_2^-) - \underline{\mathbf{v}}^v(\tau_2^-)^T \Phi_2^v(\tau_2^-)\underline{\mathbf{v}}^v(\tau_2^-) \\
&\quad \le \underline{\mathbf{v}}^v(\tau_2^-)^T\big[\Phi_2^p(\tau_2^-) - \Phi_2^v(\tau_2^-)\big]\underline{\mathbf{v}}^v(\tau_2^-) \qquad \text{(hereinafter } \underline{\mathbf{v}}^v \equiv \underline{\mathbf{v}}^v(\tau_2^-)\text{)} \\
&\quad = \underline{\mathbf{v}}^{vT}\left[\int_{\tau_1}^{\tau_2} \left(G^p - \mathcal{L}_G^p z_x^p I - G^v + \mathcal{L}_G^v z_x^v I\right) d\tau\right]\underline{\mathbf{v}}^v + \underline{\mathbf{v}}^{vT}\big[\Phi_2^p(\tau_1) - \Phi_2^v(\tau_1)\big]\underline{\mathbf{v}}^v \\
&\quad \le \underline{\mathbf{v}}^{vT}\left[\int_{\tau_1}^{\tau_2} \left((G^v + \mathcal{L}_G^p\|x^p - x^v\|\, I) - \mathcal{L}_G^p z_x^p I - G^v + \mathcal{L}_G^v z_x^v I\right) d\tau\right]\underline{\mathbf{v}}^v \\
&\quad \le -\int_{\tau_1}^{\tau_2} \mathcal{L}_G^p(z_x^p - z_x^v - \|x^p - x^v\|)\, d\tau \equiv -\int_{\tau_1}^{\tau_2} \mathcal{L}_G^p e_z\, d\tau \le 0. \qquad (7.31a)
\end{aligned}
$$

Similarly,

$$
\begin{aligned}
\Phi_1^p(\tau_2^-) - \Phi_1^v(\tau_2^-) &\ge \int_{\tau_1}^{\tau_2} \left(\|g^p\| - \mathcal{L}_g^p z_x^p - \|g^v\| + \mathcal{L}_g^v z_x^v\right) d\tau \\
&\ge \int_{\tau_1}^{\tau_2} \left(\|g^p\|\|\|g^v - g^p\| - \|g^v\| + \mathcal{L}_g^p(z_x^p - z_x^v - \|x^p - x^v\|)\right) d\tau \\
&\ge \int_{\tau_1}^{\tau_2} \left(\|g^p\|\|\|g^v - g^p\| - \|g^v\| + \mathcal{L}_g^p e_z\right) d\tau \ge 0 \qquad (7.31b)
\end{aligned}
$$

However, if $\underline{\lambda}\{\Phi_2^p(\tau_2^-)\} \le \underline{\lambda}\{\Phi_2^v(\tau_2^-)\}$ and $\Phi_1^p(\tau_2^-) \ge \Phi_1^v(\tau_2^-)$, then by Algorithm 7.3.4 it is impossible that the assumed reset of z_θ^p at time τ_2 could precede a reset in z_θ^v, since $\varepsilon_z \equiv \varepsilon_z^p \equiv \varepsilon_z^v$. Hence no such τ_2 exists, implying $z_\theta^p(\tau) \ge z_\theta^v(\tau)$, for all $\tau \in [0, T]$. The result then follows from the proof of Theorem 7.2.8.

7.7.5 Proof of Corollary 7.3.8

The statement of Claim 7.3.5 can be seen to hold, since the final inequality in (7.29) takes the form

$$\sup_{x^*_{[0,\delta]} \in \mathcal{X}^p_{[0,\delta]}} \left(\int_0^\delta \sup_{\|\mathbf{v}(\tau)\|=1} \mathbf{v}(\tau)^T \left[G|_{x^p} - G|_{x^*} - \mathcal{M}^o_G(X^p, u) z^p_x \right] \mathbf{v}(\tau) \, d\tau \right) \leq 0 \quad (7.32)$$

This implies (7.30a) holds, and replacing $\mathcal{L}_G z_x I$ with $\mathcal{M}^o_G z_x$ throughout (7.31a) yields

$$\underline{\lambda}\{\Phi^p_2(\tau_2^-)\} - \underline{\lambda}\{\Phi^v_2(\tau_2^-)\} \leq - \int_{\tau_1}^{\tau_2} e_z(\tau) \left(\underline{\mathbf{v}}^{vT} \, \mathcal{M}^p_G(\tau) \, \underline{\mathbf{v}}^v \right) d\tau \leq 0 \qquad (7.33)$$

from which the result follows by the positive definiteness of $\mathcal{M}^o_G$.

Chapter 8

Finite-time parameter estimation in adaptive control

8.1 Introduction

There are two major approaches to online parameter identification of nonlinear systems. First is the identification of parameters as a part of state observer while the second deals with parameter identification as a part of controller. In the first approach, the observer is designed to provide state derivatives information and the parameters are estimated via estimation methods such as least squares method [131] and dynamic inversion [62]. The second trend of parameter identification is much more widespread, as it allows identification of systems with unstable dynamics. Algorithms in this area include parameter identification methods based on variable structure theory [162, 163] and those based on the notion of passivity [100].

In the conventional adaptive control algorithms, the focus is on the tracking of a given reference trajectory and in most cases parameter estimation errors are not guaranteed to converge to zero due to lack of excitation [80]. Parameter convergence is an important issue as it enhances the overall stability and robustness properties of the closed-loop adaptive systems [109]. Moreover, there are control problems whereby the reference trajectory is not known *a priori* but depends on the unknown parameters of the system dynamics. For example, in adaptive extremum-seeking control problems, the desired target is the operating setpoint that optimizes an uncertain cost function [73, 160].

Assuming the satisfaction of appropriate excitation conditions, asymptotic and exponential parameter convergence results are available for both linear and nonlinear systems. Some lower bounds which depends (nonlinearly) on the adaptation gain and the level of excitation in the system have been provided for some specific control and estimation algorithms [96, 118, 147]. However, it is not always easy to characterize the convergence rate.

The performance of any adaptive extremum-seeking control is dictated by the efficiency of its parameter adaptation procedure. This chapter presents a parameter estimation scheme that allows exact reconstruction of the unknown parameters in FT provided a given persistence of excitation (PE) condition is satisfied. The true parameter estimate is recovered at any time instant the excitation condition is satisfied. This condition requires the integral of a filtered regressor matrix to be invertible. The FT identification procedure assumes the state of the system $x(\cdot)$ is accessible for measurement but does not require the measurement or computation of the velocity state vector $\dot{x}(\cdot)$. The robustness of the estimation routine to bounded unknown disturbances or modeling errors is also examined. It is shown that the parameter estimation error can be rendered arbitrarily small for a sufficiently large filter gain.

A common approach to ensuring a PE condition in adaptive control is to introduce a perturbation signal as the reference input or to add it to the target setpoint or trajectory. The downside of this approach is that a constant PE deteriorates the desired tracking or regulation performance. Aside from the recent results on intelligent excitation signal design [10, 32], the standard approach has been to introduce such PE signal and remove it when the parameters are *assumed* to have converged. The fact that one has perfect knowledge of the convergence time in the proposed framework allows for a direct and immediate removal of the added PE signal. The results in this chapter have been published in Reference 5.

8.2 Problem description and assumptions

The system considered is the following nonlinear parameter affine system

$$\dot{x} = f(x,u) + g(x,u)\theta \tag{8.1}$$

where $x \in \mathbb{R}^{n_x}$ is the state and $u \in \mathbb{R}^{n_u}$ is the control input. The vector $\theta \in \mathbb{R}^{n_\theta}$ is the unknown parameter vector whose entries may represent physically meaningful unknown model parameters or could be associated with any finite set of universal basis functions. It is assumed that θ is uniquely identifiable and lie within an initially known compact set Θ^0. The n_x-dimensional vector $f(x,u)$ and the $(n_x \times n_\theta)$-dimensional matrix $g(x,u)$ are bounded and continuous in their arguments. System (8.1) encompasses the special class of linear systems,

$$\begin{aligned} f(x,u) &= A_0x + B_0u \\ g(x,u) &= [A_1x + B_1u,\ A_2x + B_2u,\ \ldots,\ A_{n\theta}x + B_{n\theta}u], \end{aligned}$$

where A_i and B_i for $i = 0 \ldots n_\theta$ are known matrices possibly time varying.

Assumption 8.2.1. *The following assumptions are made about system* (8.1).

1. *The state of the system $x(\cdot)$ is assumed to be accessible for measurement.*
2. *There is a known bounded control law $u = \alpha(\cdot)$ and a bounded parameter update law $\dot{\hat{\theta}}$ that achieves a primary control objective.* ■

The control objective can be to (robustly) stabilize the plant and/or to force the output to track a reference signal. Depending on the structure of the system (8.1), adaptive control design methods are available in the literature [98, 117].

For any given bounded control and parameter update law, the aim of this chapter is to provide the true estimates of the plant parameters in FT while preserving the properties of the controlled closed-loop system.

8.3 FT parameter identification

Let $\hat{x}$ denote the state predictor for (8.1), the dynamics of the state predictor is designed as

$$\dot{\hat{x}} = f(x,u) + g(x,u)\hat{\theta} + k_w(t)e + w\dot{\hat{\theta}}, \tag{8.2}$$

where $\hat{\theta}$ is a parameter estimate generated via any update law $\dot{\hat{\theta}}$, $k_w > 0$ is a design matrix, $e = x - \hat{x}$ is the prediction error, and w is the output of the filter

$$\dot{w} = g(x,u) - k_{ww}, \qquad w(t_0) = 0. \tag{8.3}$$

Denoting the parameter estimation error as $\tilde{\theta} = \theta - \hat{\theta}$, it follows from (8.1) and (8.2) that

$$\dot{e} = g(x,u)\tilde{\theta} - k_w\, e - w\dot{\hat{\theta}}. \tag{8.4}$$

The use of the filter matrix w in the above development provides direct information about parameter estimation error $\tilde{\theta}$ without requiring a knowledge of the velocity vector $\dot{x}$. This is achieved by defining the auxiliary variable

$$\eta = e - w\tilde{\theta} \tag{8.5}$$

with η, in view of (8.3, 8.4), generated from

$$\dot{\eta} = -k_w\eta, \qquad \eta(t_0) = e(t_0). \tag{8.6}$$

Based on the dynamics (8.2), (8.3), and (8.6), the main result is given by the following theorem.

Theorem 8.3.1. *Let $Q \in \mathbb{R}^{n_\theta \times n_\theta}$ and $C \in \mathbb{R}^{n_\theta}$ be generated from the following dynamics:*

$$\dot{Q} = w^T w, \qquad Q(t_0) = 0 \tag{8.7a}$$

$$\dot{C} = w^T(w\hat{\theta} + e - \eta), \qquad C(t_0) = 0 \tag{8.7b}$$

Suppose there exists a time t_c and a constant $c_1 > 0$ such that $Q(t_c)$ is invertible i.e.,

$$Q(t_c) = \int_{t_0}^{t_c} w^T(\tau)w(\tau)\,d\tau \succ c_1 I, \tag{8.8}$$

then

$$\theta = Q(t)^{-1}C(t) \qquad \textit{for all } t \geq t_c. \tag{8.9}$$

Proof: The result can be easily shown by noting that

$$Q(t)\theta = \int_{t_0}^{t} w^T(\tau)w(\tau)\big[\hat{\theta}(\tau) + \tilde{\theta}(\tau)\big]\,d\tau. \tag{8.10}$$

Using the fact that $w\tilde{\theta} = e - \eta$, it follows from (8.10) that

$$\theta = Q(t)^{-1} \int_{t_0}^{t} \dot{C}(\tau)\, d\tau = Q(t)^{-1} C(t) \tag{8.11}$$

and (8.11) holds for all $t \geq t_c$ since $Q(t) \succeq Q(t_c)$. ∎

The result in Theorem 8.3.1 is independent of the control u and parameter identifier $\dot{\hat{\theta}}$ structure used for the state prediction (8.2). Moreover, the result holds if a nominal estimate θ^0 of the unknown parameter (no parameter adaptation) is employed in the estimation routine. In this case, $\hat{\theta}$ is replaced with θ^0 and the last part of the state predictor (8.2) is dropped ($\dot{\hat{\theta}} = 0$).

Let

$$\theta^c \triangleq Q(t_c)^{-1}\, C(t_c) \tag{8.12}$$

The FT identifier (FTI) is given by

$$\hat{\theta}^c(t) = \begin{cases} \hat{\theta}(t), & \text{if } t < t_c \\ \theta^c, & \text{if } t \geq t_c. \end{cases} \tag{8.13}$$

The piecewise continuous function (8.13) can be approximated by a smooth approximation using the logistic functions

$$\psi_1 \triangleq \frac{\hat{\theta}(t)}{2}\big(1 - \tanh \nu_1(t - t_c)\big) = \frac{\hat{\theta}(t)}{1 + \exp^{2\nu_1(t-t_c)}} \tag{8.14a}$$

$$\psi_2 \triangleq \frac{\theta^c}{2}\big(1 + \tanh \nu_2(t - t_c)\big) = \frac{\theta^c}{1 + \exp^{-2\nu_2(t-t_c)}} \tag{8.14b}$$

$$\hat{\theta}^{\tilde{c}} = \psi_1 + \psi_2 \tag{8.14c}$$

where larger ν_1, ν_2 correspond to a sharper transition at $t = t_c$ and $\lim_{(\nu_1,\nu_2)\to\infty} \hat{\theta}^{\tilde{c}} = \hat{\theta}^c$. An example of such approximation is depicted in Figure 8.1 where the function

$$z(t) = \begin{cases} 6 + t^{0.3}, & \text{if } t < 5 \\ 4, & \text{otherwise} \end{cases}$$

is approximated by (8.14) with $\nu_1 = \nu_2 = 5$.

The invertibility condition (8.8) is equivalent to the standard PE condition required for parameter convergence in adaptive control. The condition (8.8) is satisfied if the regressor matrix g is PE. To show this, consider the filter dynamic (8.3), from which it follows that

$$w(t) = \int_{t_0}^{t} \exp^{-k_w(t-\tau)} g(\tau) d\tau = \frac{1}{s + k_w}[g(t)] \tag{8.15}$$

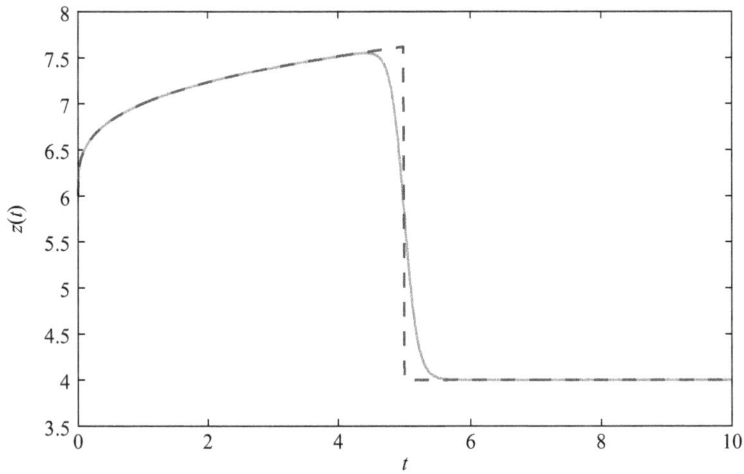

Figure 8.1 Approximation of a piecewise continuous function. The function $z(t)$ is given by the full line. Its approximation is given by the dotted line

Since $g(t)$ is PE by assumption and the transfer function $\frac{1}{s+k_w}$ is stable, minimum phase, and strictly proper, we know that $w(t)$ is PE [129]. Hence, there exists t_c and a c_1 for which (8.8) is satisfied. The superiority of the above design lies in the fact that the true parameter value can be computed at any time instant t_c the regressor matrix becomes positive definite and subsequently stop the parameter adaptation mechanism.

The procedure in Theorem 8.3.1 involves solving matrix valued ordinary differential equations (8.3, 8.7) and checking the invertibility of $Q(t)$ online. For computational considerations, the invertibility condition (8.8) can be efficiently tested by checking the determinant of $Q(t)$ online. Theoretically, the matrix is invertible at any time $\det(Q(t))$ becomes positive definite. The determinant of $Q(t)$ (which is a polynomial function) can be queried at pre-scheduled times or by propagating it online starting from a zero initial condition. One way of doing this is to include a scalar differential equation for the derivative of $\det(Q(t))$ as follows [68]:

$$\frac{d}{dt}\det(Q) = \text{Trace}\left(\text{Adjugate}(Q)\, w^T w\right), \qquad \det(Q(t_0)) = 0 \tag{8.16}$$

where Adjugate(Q), admittedly not a light numerical task, is also a polynomial function of the elements of Q.

8.3.1 Absence of PE

If the PE condition (8.8) is not satisfied, a given controller and the corresponding parameter estimation scheme preserve the system established closed-loop properties. When a bounded controller that is robust with respect to input $(\tilde{\theta}, \dot{\hat{\theta}})$ is known, it can be shown that the state prediction error e tends to zero as $t \to \infty$. An example of such robust controller is an ISS controller [98].

Theorem 8.3.2. *Suppose the design parameter k_w in (8.2) is replaced with $k_w(t) = k_{w_1} + k_{w_2}(t)$, $k_{w_1} > \frac{1}{4}I$, and $k_{w_2}(t) = \frac{1}{4}g\gamma g^T$. Then the state predictor (8.2) and the parameter update law*

$$\dot{\hat{\theta}} = \gamma(w^T + g^T)e \tag{8.17}$$

with $\gamma = \gamma^T > 0$ a design constant matrix, guarantee that

1. *$(e, \eta, \tilde{\theta}) \in \mathcal{L}_\infty$ and $(e, \eta) \to 0$ as $t \to \infty$*
2. *$\lim_{t\longrightarrow\infty} \hat{\theta}(t) = \bar{\theta}$, a constant.*

Proof:

1. Consider a Lyapunov function

$$V = \frac{1}{2}\left(e^Te + \tilde{\theta}^T\gamma^{-1}\tilde{\theta} + \eta^T\eta\right). \tag{8.18}$$

It follows from (8.4)–(8.6) and (8.17) that

$$\begin{aligned}\dot{V} &= -e^Tk_{w_1}e - e^T\left(\frac{1}{4}g\gamma g^T + w\gamma w^T + w\gamma g^T\right)e \\ &\quad - \tilde{\theta}^Tw^Tw\tilde{\theta} - \tilde{\theta}^Tw^T\eta - \eta^Tk_w(t)\eta \qquad (8.19)\\ &\le -e^Tk_{w_1}e - \lambda_{\min}(\gamma)\left\|w^Te + \frac{1}{2}g^Te\right\|^2 - \eta^Tk_{w_3}\eta - \left\|w\tilde{\theta} + \frac{1}{2}\eta\right\|^2 \qquad (8.20)\\ &\le -(e^Tk_{w_1}e + \eta^Tk_{w_3}\eta). \qquad (8.21)\end{aligned}$$

where $k_{w_3} = k_{w_1} - \frac{1}{4}$. This implies uniform boundedness of $(\eta, e, \tilde{\theta})$ as well as global asymptotic convergence of (η, e) to zero. Hence, it follows from (8.5) that $\lim_{t\longrightarrow\infty} w\tilde{\theta} = 0$.
2. This can be shown by noting from (8.17) that $\tilde{\theta}(t) = \tilde{\theta}(t_0) - \gamma\int_{t_0}^{t}(w^T + g^T)e\,d\sigma$. Since $g(\cdot)$ and e are bounded signals and $e \to 0$, the integral term exists and it is finite. ■

8.4 Robustness property

In this section, the robustness of the FTI to unknown bounded disturbances or modeling errors is demonstrated. Consider a perturbation of (8.1):

$$\dot{x} = f(x, u) + g(x, u)\theta + \vartheta(t, x, \theta) \tag{8.22}$$

where $\vartheta(\cdot)$ is a disturbance or modeling error term that satisfies $\|\vartheta(t)\| \leq M_\vartheta(t) < \infty$ If the PE condition (8.8) is satisfied and the disturbance term is known, the true unknown parameter vector is given by

$$\theta_\vartheta^c \triangleq \theta = Q(t)^{-1} \int_{t_0}^{t} w^T(\tau)[w(\tau)\hat{\theta}(\tau) + e(\tau) - \eta_\vartheta(\tau)]\, d\tau, \quad \text{for all } t \geq t_c, \tag{8.23}$$

with $e = x - \hat{x}$ and the signals $\hat{x}$, w, $\eta_\vartheta = e - w\theta$ generated from (8.2), (8.3), and

$$\dot{\eta_\vartheta} = -k_w \eta_\vartheta + \vartheta(\cdot), \qquad \eta_\vartheta(t_0) = e(t_0) \tag{8.24}$$

respectively.

Since $\vartheta(\cdot)$ is unknown, we provide a bound on the parameter identification error $\tilde{\theta^c} = \theta_\vartheta^c - \theta^c$ when (8.6) is used instead of (8.24).

Considering (8.9) and (8.23), it follows that

$$\tilde{\theta^c} = Q(t)^{-1} \int_{t_0}^{t} w^T(\tau)\,(-\eta_\vartheta(\tau) + \eta(\tau))\, d\tau \tag{8.25}$$

$$= -Q(t)^{-1} \int_{t_0}^{t} w^T(\tau)\tilde{\eta}(\tau)\, d\tau. \tag{8.26}$$

where $\tilde{\eta} = \eta_\vartheta - \eta$ is the output of

$$\dot{\tilde{\eta}} = -k_w \tilde{\eta} + \vartheta(\cdot), \qquad \tilde{\eta}(t_0) = 0. \tag{8.27}$$

Since $k_w \geq k_{w_1} > 0$, it follows that

$$\|\tilde{\eta}(t)\| \leq \frac{M_\vartheta}{k_{w_1}} \tag{8.28}$$

and hence

$$\|\tilde{\theta}^c(t)\| \leq \|Q(t)^{-1}\| \left\{ \frac{\bar{w} M_\vartheta (t - t_0)}{k_{w_1}} \right\}, \qquad \text{for all } t \geq t_c. \tag{8.29}$$

where $\bar{w} = \max_{\sigma \in [t_0, t]} \|w^T(\sigma)\|$.

This implies that the identification error can be rendered arbitrarily small by choosing a sufficiently large filter gain k_{w_1}. In addition, if the disturbance term ϑ and the system satisfies some given properties, then asymptotic convergence can be achieved as stated in the following theorem.

Theorem 8.4.1. *Suppose $\vartheta \in \mathcal{L}_p$, for $p = 1$ or 2 and $\lim_{t\to\infty} \lambda_{\min}(Q) = \infty$, then $\tilde{\theta^c} \to 0$ asymptotically with time.*

To prove this theorem, we need the following lemma.

Lemma 8.4.2. *[53]: Consider the system*

$$\dot{x}(t) = Ax(t) + u(t) \tag{8.30}$$

Suppose the equilibrium state $x_e = 0$ of the homogeneous equation is exponentially stable,

1. *if $u \in Ł_p$ for $1 < p < \infty$, then $x \in Ł_p$ and*
2. *if $u \in Ł_p$ for $p = 1$ or 2, then $x \to 0$ as $t \to \infty$.*

Proof of theorem 8.4.1. It follows from Lemma 8.4.2.2 that $\tilde{\eta} \to 0$ as $t \to \infty$ and therefore $\lim_{t\to\infty} \int_{t_0}^{t} w^T(\tau)\tilde{\eta}(\tau)\,d\tau$ is finite. So

$$\lim_{t\to\infty} \tilde{\theta^c} = \lim_{t\to\infty} \left\{ Q(t)^{-1} \int_{t_0}^{t} w^T(\tau)\tilde{\eta}(\tau)\,d\tau \right\} = 0. \tag{8.31}$$

■

8.5 Dither signal design

The problem of tracking a reference signal is usually considered in the study of parameter convergence and in most cases, the reference signal is required to provide sufficient excitation for the closed-loop system. To this end, the reference signal $y_r(t) \in \mathbb{R}^r$ is appended with a bounded excitation signal $d(t)$ as

$$y_{rd}(t) = y_r(t) + d(t) \tag{8.32}$$

where the auxiliary signal $d(t)$ is chosen as a linear combination of sinusoidal functions with $\hbar$ distinct frequencies:

$$d(t) := \sum_{k=1}^{\hbar} \underline{a}_k(t)\sin(\omega_k t) = \mathcal{A}(t)\zeta(t) \tag{8.33}$$

where

$$\mathcal{A}(t) = \begin{bmatrix} a_{11} & \cdots & a_{1\hbar} \\ \vdots & & \vdots \\ a_{r1} & \cdots & a_{r\hbar} \end{bmatrix}$$

is the signal amplitude matrix and

$$\zeta(t) = [\sin\omega_1 t \quad \ldots \quad \sin\omega_{\hbar} t]^T, \quad \omega_i \neq \omega_j \text{ for } i \neq j$$

is the corresponding sinusoidal function vector.

For this approach, it is sufficient to design the perturbation signal such that the regressor matrix g is PE. There are very few results on the design of persistently exciting (PE) input signals for nonlinear systems. By converting the closed-loop PE condition to a sufficient richness condition on the reference signal, attempts have been made to provide verifiable conditions for parameter convergence in some classes of nonlinear systems [4, 10, 109, 110].

8.5.1 Dither signal removal

Let $\mathcal{H} \leq (\hbar \times r)$ denotes the number of distinct elements in the dither amplitude matrix $\mathcal{A}(t)$ and let $a \in \mathbb{R}^{\mathcal{H}}$ be a vector of these distinct coefficients. The amplitude of the excitation signal is specified as

$$a(t) = \begin{cases} a, & \text{if } t < t_c \\ 0, & \text{otherwise} \end{cases} \tag{8.34}$$

or approximated by

$$a(t) \approx \frac{a}{1 + \exp^{2\nu(t-t_c)}} \tag{8.35}$$

where equality holds in the limit as $\nu \to \infty$.

8.6 Simulation examples

8.6.1 Example 1

We consider the following nonlinear system in parametric strict feedback form [110]:

$$\begin{aligned} \dot{x}_1 &= x_2 + \theta_1 x_1 \\ \dot{x}_2 &= x_3 + \theta_2 x_1 \\ \dot{x}_3 &= \theta_3 x_1^3 + \theta_4 x_2 + \theta_5 x_3 + (1 + x_1^2)u \\ y &= x_1, \end{aligned} \tag{8.36}$$

where $\theta^T = [\theta_1, \ldots, \theta_5]$ are unknown parameters. Using an adaptive backstepping design, the control and parameter update law presented in Reference 110 were used for the simulation. The pair stabilize the plant and ensure that the output y tracks a reference signal $y_r(t)$ asymptotically. For simulation purposes, parameter values are set to $\theta^T = [-1, -2, 1, 2, 3]$ as in Reference 110 and the reference signal is $y_r = 1$, which is sufficiently rich of order one. The simulation results for zero initial conditions are shown in Figure 8.2. Based on the convergence analysis procedure in Reference 110, all the parameter estimates cannot converge to their true values for this choice of constant reference. As confirmed in Figure 8.2, only θ_1 and θ_2 estimates are accurate. However, following the proposed estimation technique and implementing the FTI (8.14), we obtain the exact parameter estimates at $t = 17s$. This example demonstrates that, with the proposed estimation routine, it is possible to identify parameters using perturbation or reference signals that would otherwise not provide sufficient excitation for standard adaptation methods.

8.6.2 Example 2

To corroborate the superiority of the developed procedure, we demonstrate the robustness of the developed procedure by considering system (8.36) with added exogenous disturbances as follows:

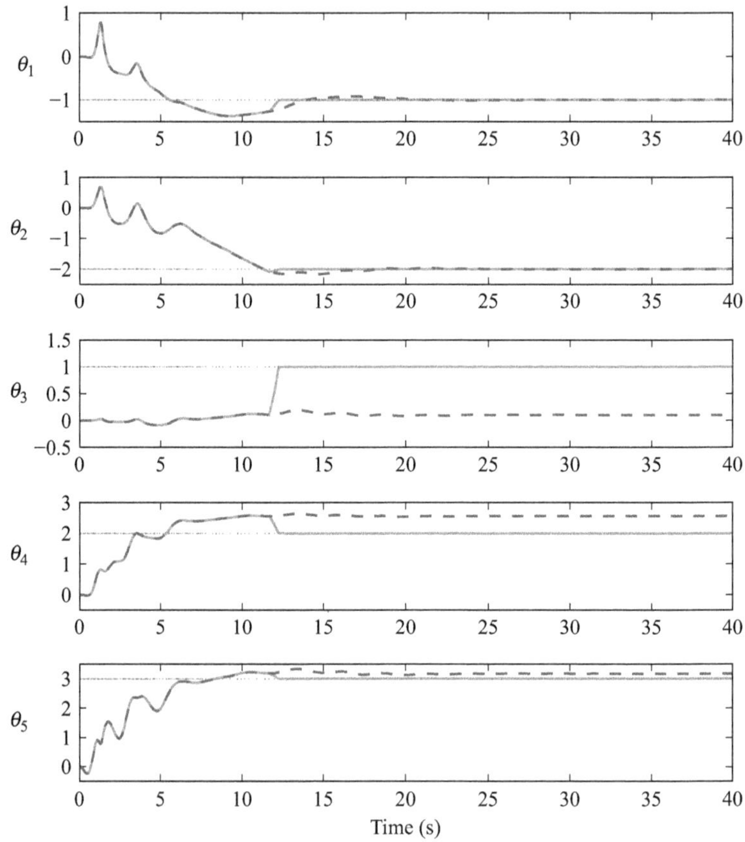

Figure 8.2 Trajectories of parameter estimates. Solid(–): FT estimates $\hat{\theta}^{\tilde{c}}$*; dashed(−−): standard estimates* $\hat{\theta}$ [110]*; dashdot(–.): actual value*

$$\begin{aligned}
\dot{x}_1 &= x_2 + \theta_1 x_1 + [1 \quad 0]\vartheta \\
\dot{x}_2 &= x_3 + \theta_2 x_1 + [1 \quad x_1]\vartheta \\
\dot{x}_3 &= \theta_3 x_1^3 + \theta_4 x_2 + \theta_5 x_3 + (1 + x_1^2)u + [0 \quad 1]\vartheta \\
y &= x_1,
\end{aligned} \tag{8.37}$$

where $\vartheta = [0.1\sin(2\pi t/5),\ 0.2\cos(\pi t)]^T$ and the tracking signal remains a constant $y_r = 1$.

The simulation result, Figure 8.3, shows convergence of the estimate vector to a small neighborhood of θ under FTI with filter gain $k_w = 1$ while no full parameter convergence is achieved with the standard identifier. The parameter estimation error

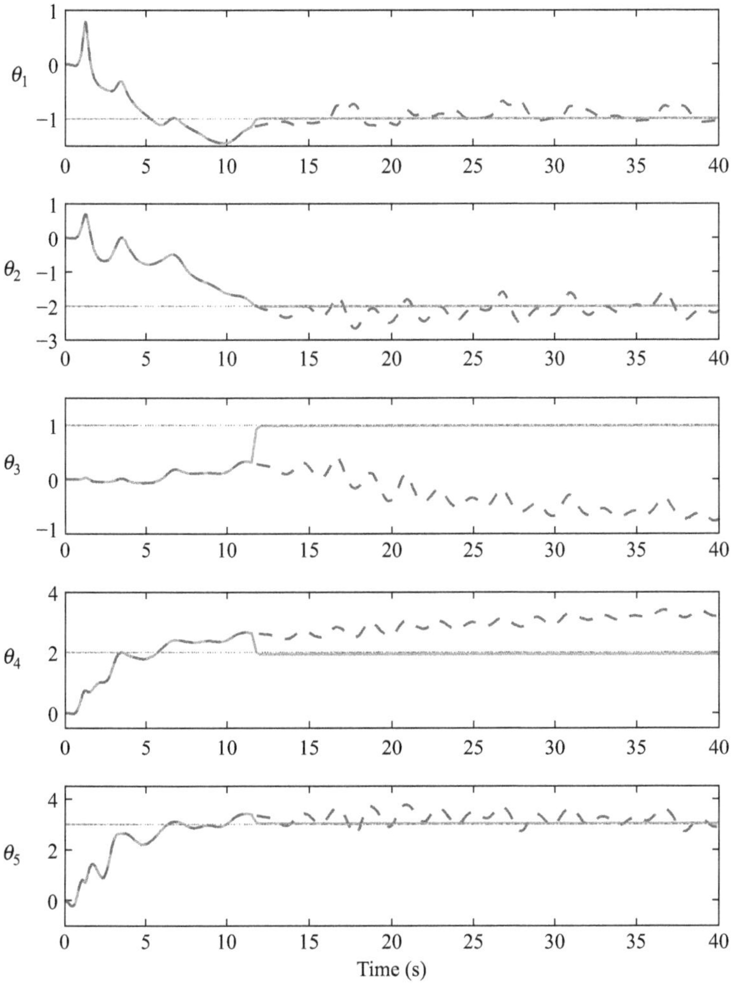

Figure 8.3 Trajectories of parameter estimates. Solid(–): FT estimates for the system with additive disturbance $\hat{\theta}^{\tilde{c}}$; *dashed(– –): standard estimates* $\hat{\theta}$ [110]; *dashdot(–.): actual value*

$\tilde{\theta}(t)$ is depicted in Figure 8.4 for different values of the filter gain k_w. The switching time for the simulation is selected as the time for which the condition number of Q becomes less than 20. It is noted that the time at which switching from standard adaptive estimate to FT estimate occurs increases as the filter gain increases. The convergence performance improves as k_w increases, however, no significant improvement is observed as the gain is increased beyond 0.5.

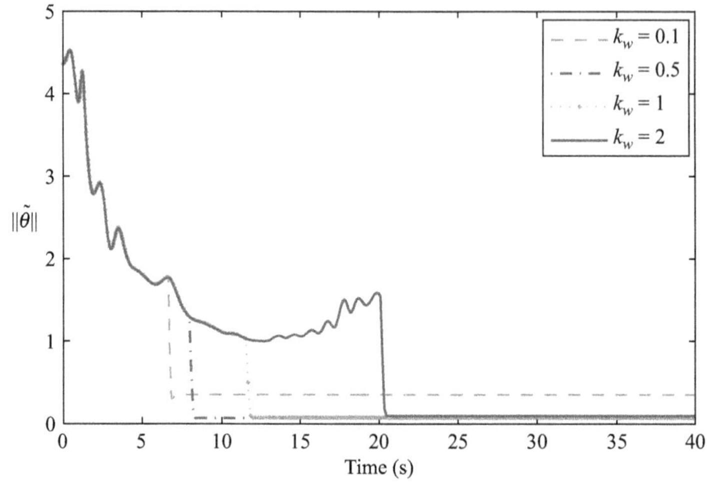

Figure 8.4 Parameter estimation error for different filter gains k_w

8.7 Summary

The work presented in this chapter transcends beyond characterizing the parameter convergence rate. A method is presented for computing the exact parameter value at a FT selected according to the observed excitation in the system. A smooth transition from a standard estimate to the FT estimate is proposed. In the presence of unknown bounded disturbances, the FTI converges to a neighborhood of the true value whose size is dictated by the choice of the filter gain. Moreover, the procedure preserves the system's established closed-loop properties whenever the required PE condition is not satisfied.

Chapter 9
Performance improvement in adaptive control

9.1 Introduction

The FT identification method developed in Chapter 8 has two distinguishing features. First, the true parameter estimate is obtained at any time instant the excitation condition is satisfied, and second, the procedure allows for a direct and immediate removal of any perturbation signal injected into the closed-loop system to aid in parameter estimation. However, the drawback of the FT identification algorithm is the requirement to check the invertibility of a matrix online and compute the inverse matrix when appropriate.

To avoid these concerns and enhance the applicability of the FT method in practical situations, the procedure is hereby exploited to develop a novel adaptive compensator that (almost) recovers the performance of the FTI. The compensator guarantees exponential convergence of the parameter estimation error at a rate dictated by the closed-loop system's excitation. It was shown how the adaptive compensator can be used to improve upon existing adaptive controllers. The modification proposed guarantees exponential stability of the parametric equilibrium provided the given PE condition is satisfied. Otherwise, the original system's closed-loop properties are preserved.

9.2 Adaptive compensation design

Consider the nonlinear system 8.1 satisfying Assumption 8.2.1 and the state predictor

$$\dot{\hat{x}} = f(x, u) + g(x, u)\theta^0 + k_w(x - \hat{x}) \tag{9.1}$$

where $k_w > 0$ and θ^0 is the nominal initial estimate of θ. If we define the auxiliary variable

$$\eta = x - \hat{x} - w(\theta - \theta^0) \tag{9.2}$$

and select the filter dynamic as

$$\dot{w} = g(x, u) - k_w w, \qquad w(t_0) = 0 \tag{9.3}$$

then η is generated by

$$\dot{\eta} = -k_w \eta, \quad \eta(t_0) = e(t_0). \tag{9.4}$$

Based on (9.1)–(9.4), our novel adaptive compensation result is given in the following theorem.

Theorem 9.2.1. *Let Q and C be generated from the following dynamics:*

$$\dot{Q} = w^T w, \qquad Q(t_0) = 0 \tag{9.5a}$$

$$\dot{C} = w^T(w\theta^0 + x - \hat{x} - \eta), \qquad C(t_0) = 0 \tag{9.5b}$$

and let t_c be the time such that $Q(t_c) \succ 0$, then the adaptation law

$$\dot{\hat{\theta}} = \Gamma(C - Q\hat{\theta}), \quad \hat{\theta}(t_0) = \theta^0 \tag{9.6}$$

with $\Gamma = \Gamma^T \succ 0$ guarantees that $\|\tilde{\theta}\| = \|\theta - \hat{\theta}\|$ is non-increasing for $t_0 \le t \le t_c$ and converges to zero exponentially fast, starting from t_c. Moreover, the convergence rate is lower bounded by $\mathcal{E}(t) = \lambda_{\min}(\Gamma Q(t))$.

Proof: Consider a Lyapunov function

$$V_{\tilde{\theta}} = \frac{1}{2}\tilde{\theta}^T\tilde{\theta}, \tag{9.7}$$

it follows from (9.6) that

$$\dot{V}_{\tilde{\theta}}(t) = -\tilde{\theta}^T(t)\,\Gamma\Big(C(t) - Q(t)\hat{\theta}(t)\Big). \tag{9.8}$$

Since $w\theta = w\theta^0 + x - \hat{x} - \eta$ (from (9.2)), then

$$C(t) = \int_{t_0}^{t} \dot{C}(\tau)\,d\tau = \int_{t_0}^{t} w^T(\tau)w(\tau)\,d\tau\;\theta = Q(t)\,\theta \tag{9.9}$$

and (9.8) becomes

$$\dot{V}_{\tilde{\theta}}(t) = -\tilde{\theta}^T(t)\,\Gamma\, Q(t)\,\tilde{\theta}(t) \tag{9.10}$$

$$\le -\mathcal{E}(t)\, V_{\tilde{\theta}}(t) \tag{9.11}$$

This implies non-increase of $\|\tilde{\theta}\|$ for $t \ge t_0$ and the exponential claim follows from the fact that $\Gamma\, Q(t) = \Gamma \int_{t_0}^{t} w(\tau)^T w(\tau) d\tau$ is positive definite for all $t \ge t_c$. The convergence rate is shown by noting that

$$\dot{V}_{\tilde{\theta}}(t) = -\tilde{\theta}^T(t)\,\Gamma\left(Q(t_c) + \int_{t_c}^{t} w(\tau)^T\, w(\tau)\,d\tau\right)\tilde{\theta}(t), \quad \forall t \ge t_c \tag{9.12}$$

$$\le -\tilde{\theta}^T(t)\,\Gamma\, Q(t_c)\,\tilde{\theta}(t) \le -\mathcal{E}(t_c)\, V(t) \tag{9.13}$$

which implies

$$\|\tilde{\theta}(t)\| \leq \exp^{-\mathcal{E}(t_c)(t-t_0)} \|\tilde{\theta}(t_0)\|, \quad \forall t \geq t_c \tag{9.14}$$

■

Both the FT identification (8.9) and the adaptive compensator (9.6) use the static relationship developed between the unknown parameter θ and some measurable matrix signals C, that is, $Q\theta = C$. However, instead of computing the parameter values at a known FT by inverting matrix Q, the adaptive compensator is driven by the estimation error $C - Q\hat{\theta} = Q\tilde{\theta}$.

9.3 Incorporating adaptive compensator for performance improvement

It is assumed that the given control law u and stabilizing update law (herein denoted as $\dot{\hat{\theta}}^s$) result in closed-loop error system

$$\dot{Z} = AZ + \Phi^T \tilde{\theta}^s \tag{9.15a}$$

$$\dot{\hat{\theta}}^s = -\Gamma \Phi Z \tag{9.15b}$$

where the matrix A is such that $A + A^T < -2\,k_A I < 0$, Φ is a bounded matrix function of the regressor vectors, $\tilde{\theta}^s = \theta - \hat{\theta}^s$ and $Z = [z_1, z_2, \ldots, z_{n_x}]^T$ is a vector function of the tracking error with $z_1 = y - y^r$. This implies that the adaptive controller guarantees uniform boundedness of the estimation error $\tilde{\theta}^s$ and asymptotic convergence of the tracking error Z dynamics. Such adaptive controllers are very common in the literature. Examples include linearized control laws [117] and controllers designed via backstepping [98, 110].

Given the stabilizing adaptation law $\dot{\hat{\theta}}^s$, we propose the following update law which is a combination of the stabilizing update law (9.15b) and the adaptive compensator (9.6)

$$\dot{\hat{\theta}} = \Gamma(\Phi Z + C - Q\hat{\theta}). \tag{9.16}$$

Since $C(t) = Q(t)\theta$, the resulting error equations becomes

$$\begin{bmatrix} \dot{Z} \\ \dot{\tilde{\theta}} \end{bmatrix} = \begin{bmatrix} A & \Phi^T \\ -\Gamma\Phi & -\Gamma Q \end{bmatrix} \begin{bmatrix} Z \\ \tilde{\theta} \end{bmatrix}. \tag{9.17}$$

Considering the Lyapunov function $V = \frac{1}{2}(z^T z + \tilde{\theta}^T \Gamma^{-1} \tilde{\theta})$, and differentiating along (9.17) we have

$$\dot{V} = \frac{1}{2} z^T (A + A^T) z - \tilde{\theta}^T Q \tilde{\theta} \leq -k_A z^T z - \tilde{\theta}^T Q \tilde{\theta} \tag{9.18}$$

Hence $\tilde{\theta} \to 0$ exponentially for $t \geq t_c$ and the initial asymptotic convergence of Z is strengthened to exponential convergence.

For feedback linearizable systems

$$\begin{aligned} \dot{x}_i &= x_{i+1} \quad 1 \le i \le n-1 \\ \dot{x}_n &= f_1(x) + f_2(x)u + \theta^T g_n(x) \\ y &= x_1 \end{aligned}$$

the PE condition $Q(t_c) \succ 0$ translates to *a priori* verifiable sufficient condition on the reference setpoint. It requires the rows of the regressor vector $g_n(x)$ to be linearly independent along a desired trajectory $x^r(t)$ on any finite interval $t \in [t_1, t_2)$, $t_1 < t_2 < \infty$. This condition is less restrictive than the one given in Reference 79 for the same class of system. This is because the linear independence requirement herein is only required over a finite interval and it can be satisfied by a non-periodic reference trajectory while the asymptotic stability result in Reference 79 relies on a T-periodic reference setpoint. Moreover, exponential rather than asymptotic stability of the parametric equilibrium is achieved.

9.4 Dither signal update

Perturbation signal is usually added to the desired reference setpoint or trajectory to guarantee the convergence of system parameters to their true values. To reduce the variability of the closed-loop system, the added PE signal must be systematically removed in a way that sustains parameter convergence.

Suppose the dither signal $d(t)$ is selected as a linear combination of sinusoidal functions as detailed in Section 8.5. Let a^0 be the vector of the selected dither amplitude and let $T > 0$ be the first instant for which $d(T) = 0$, the amplitude of the excitation signal is updated as follows:

$$a(t) = \begin{cases} a^0, & t \in [0,\, T) \\ \exp^{-\gamma \bar{\mathcal{E}}\, T} a(j-1)T, & t \in [jT,\, (j+1)T), \qquad j \ge 1 \end{cases} \tag{9.19}$$

where the gain $\gamma > 0$ is a design parameter, $a(0) = a^0$ and

$$\begin{aligned} \mathcal{E}(0) &= 0, \qquad \mathcal{E}(\tau) = \lambda_{\min}(Q(\tau)) \\ \bar{\mathcal{E}} &= \max\{\mathcal{E}(jT),\ \mathcal{E}((j-1)T)\}. \end{aligned}$$

It follows from (9.19) that the reference setpoint will be subject to PE with constant amplitude a^0 if $t \in [0,\, T)$. After which the trajectory of $a(t)$ will be dictated by the filtered regressor matrix Q. The amplitude vector $a(t)$ will start to decay exponentially when $Q(t)$ becomes positive definite. Note that parameter convergence will be achieved regardless of the value of the gain γ selected as the only requirement for convergence is $Q(t) \succ 0$.

Remark 9.4.1. *The other major approach used in traditional adaptive control is parameter estimation-based design. A well-designed estimation-based adaptive control method achieves modularity of the controller – identifier pair. For nonlinear systems, the controller module must possess strong parametric robustness properties* *while the identifier module must guarantee certain boundedness properties independent of the control module. Assuming the existence of a bounded controller that is robust with respect to $(\tilde{\theta}, \dot{\hat{\theta}})$, the adaptive compensator (9.6) serves as a suitable identifier for modular adaptive control design.*

9.5 Simulation example

To demonstrate the effectiveness of the adaptive compensator, we consider the example in Section 8.6 for both the nominal system (8.36) and the system under additive disturbance (8.37). The simulation is performed for the same reference setpoint $y_r = 1$, disturbance vector $\vartheta = [0.1\sin(2\pi t/5), 0.2\cos(\pi t)]^T$, parameter values $\theta = [-1, -2, 1, 2, 3]$ and zero initial conditions.

The adaptive controller presented in Reference 110 is also used for the simulation. We modify the given stabilizing update law by adding the adaptive compensator (9.6) to it. The modification significantly improve upon the performance of the standard adaptation mechanism as shown in Figures 9.1 and 9.2. All the parameters converged

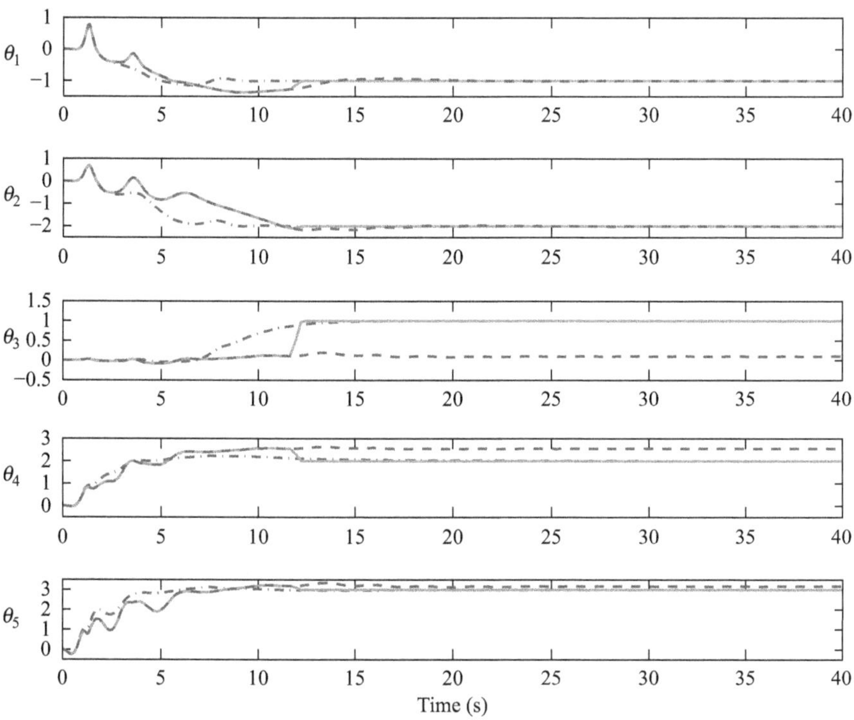

Figure 9.1 Trajectories of parameter estimates. Solid(−): compensated estimates; dashdot(− ·): FT estimates; dashed(− −): standard estimates [110]

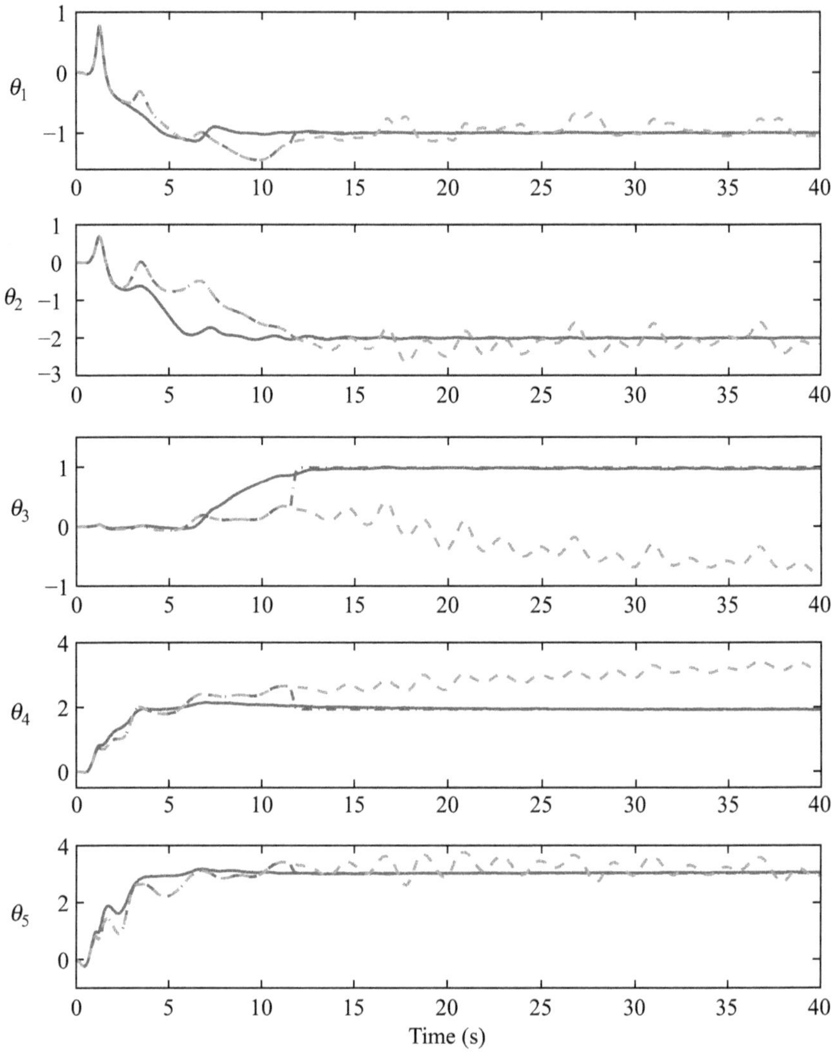

Figure 9.2 *Trajectories of parameter estimates under additive disturbances. Solid(–): compensated estimates; dashdot(– ·): FT estimates; dashed(– –): standard estimates* [110]

to their values and we recover the performance of the FTI (8.14). Figures 9.3 and 9.4 depict the performance of the output and the input trajectories. While the transient behavior of the output and input trajectories is slightly improved for the nominal adaptive system, a significant improvement is obtained for the system subject to additive disturbances.

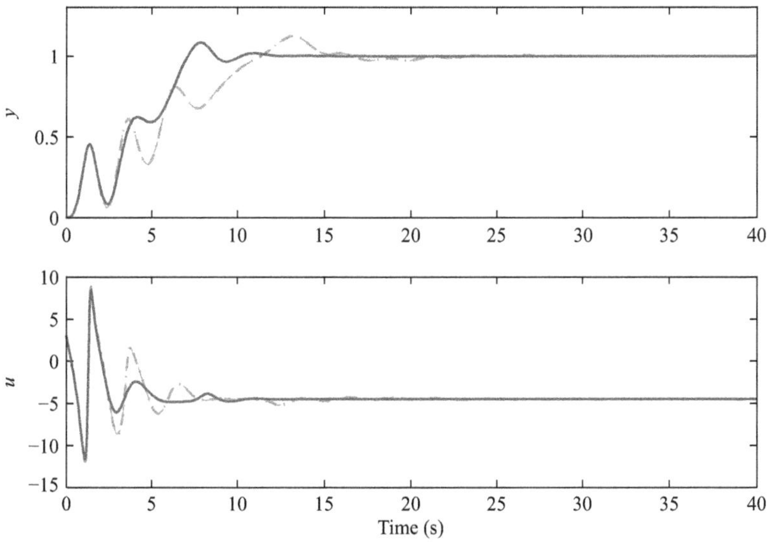

Figure 9.3 Trajectories of system's output and input for different adaptation laws. Solid(−): compensated estimates; dashdot(− ·): FT estimates; dashed(− −): standard estimates [110]

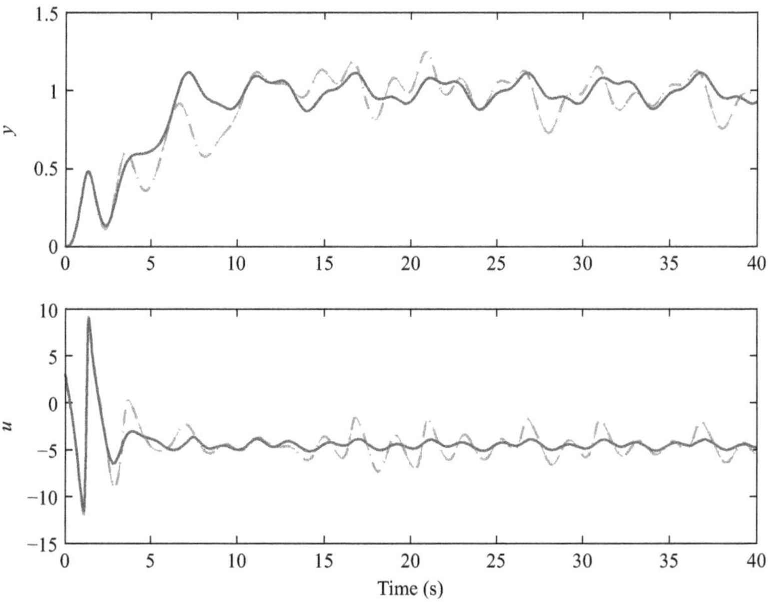

Figure 9.4 Trajectories of system's output and input under additive disturbances for different adaptation laws. Solid(−): compensated estimates; dashdot(−.): FT estimates; dashed(− −): standard estimates [110]

9.6 Summary

This chapter demonstrates how the FT identification procedure can be used to improve the overall performance (both transient and steady state) of adaptive control systems in a very appealing manner. First, we develop an adaptive compensator which guarantees exponential convergence of the estimation error provided the integral of a filtered regressor matrix is positive definite. The approach does not involve online checking of matrix invertibility and computation of matrix inverse or switching between parameter estimation methods. The convergence rate of the parameter estimator is directly proportional to the adaptation gain and a measure of the system's excitation. The adaptive compensator is then combined with existing adaptive controllers to guarantee exponential stability of the closed-loop system. The application reported in Section 9.3 is just an example, the adaptive compensator can easily be incorporated into other adaptive control algorithms.

Chapter 10

Adaptive MPC for constrained nonlinear systems

10.1 Introduction

This chapter is inspired by References 47 and 49. While the focus in References 47 and 49 is on the use of adaptation to reduce the conservatism of robust MPC controller, this study addresses the problem of adaptive MPC and incorporates robust features to guarantee closed-loop stability and constraint satisfaction. Simplicity is achieved here-in by generating a parameter estimator for the unknown parameter vector and parameterizing the control policy in terms of these estimates rather than adapting a parameter uncertainty set directly.

First, a min–max feedback nonlinear MPC scheme is combined with the adaptation mechanism developed in Chapter 8. The parameter estimation routine employed guarantees non-increase of the norm of the estimation error vector and provides exponential parameter convergence when an excitation condition is satisfied. The estimates are used to update the parameter uncertainty set, at every time step, in a manner that guarantees non-expansion of the set leading to a gradual reduction in the conservativeness or computational demands of the algorithms. The min–max formulation explicitly accounts for the effect of future parameter estimation and automatically injects some useful excitation into the closed-loop system to aid in parameter identification.

Second, the technique is extended to a less computationally demanding robust-MPC algorithm. The nominal model rather than the unknown bounded system state is controlled, subject to conditions that ensure that given constraints are satisfied for all possible uncertainties. State prediction error bound is determined based on assumed Lipschitz continuity of the model. Using a nominal model prediction, it is impossible to predict the actual future behavior of the parameter estimation error as was possible in the min–max framework. It is shown how the future model improvement over the prediction horizon can be considered by developing a worst-case upper bound on the future parameter estimation error. The conservativeness of the algorithm reduces as the error bound decreases monotonically over time.

Finally, it is shown how the FTI developed in Chapter 8 can be incorporated in the proposed adaptive MPC algorithms. The true value of the uncertain parameter vector is recovered in a known FT when an excitation condition is satisfied. Subsequently, the adaptive and robustness features of the MPC is eliminated and the complexity of the resultant controller reduces to that of nominal MPC.

10.2 Problem description

The system considered is the following nonlinear parameter affine system

$$\dot{x} = f(x,u) + g(x,u)\theta \triangleq \mathcal{F}(x,u,\theta) \tag{10.1}$$

$\theta \in \mathbb{R}^{n_\theta}$ is the unknown parameter vector whose entries may represent physically meaningful unknown model parameters or could be associated with any finite set of universal basis functions. It is assumed that θ is uniquely identifiable and lie within an initially known compact set $\Theta^0 \triangleq B(\theta^0, z_\theta^0)$, a ball described by an initial nominal estimate θ^0 and associated error bound $z_\theta^0 = \sup_{s\in\Theta^0} \|s - \theta^0\|$. The mapping $\mathcal{F}: \mathbb{R}^{n_x} \times \mathbb{R}^{n_u} \times \mathbb{R}^{n_\theta} \to \mathbb{R}^{n_x}$ is assumed to be locally Lipschitz with respect to its arguments. The state and the control input trajectories are assumed to be subject to pointwise constraints $x \in \mathcal{X} \in \mathbb{R}^{n_x}$ and $u \in \mathcal{U} \in \mathbb{R}^{n_u}$, respectively. The objective of the study is to (robustly) stabilize the plant by means of state feedback adaptive MPC. Optimality of the resulting trajectories are measured with respect to the accumulation of some stage cost $L(x,u) \geq 0$. The cost is assumed to be continuous, $L(0,0) = 0$, and $L(x,u) \geq \mu_L(\|x,u\|)$, where μ_L is a $\mathcal{K}_\infty$ function.[1]

10.3 Estimation of uncertainty

10.3.1 Parameter adaptation

Since parameter convergence is fundamental to the overall goal of this book, the estimation algorithm presented in Chapter 9 is used. The adaptive update law, driven by the parameter estimation error $\tilde{\theta} = \theta - \hat{\theta}$, remains active until parameter convergence is achieved and results in faster convergence than any traditional update laws that depends only on tracking or prediction error.

For ease of reference, the adaptive law is given by

$$\dot{\hat{\theta}} = \Gamma(C - Q\hat{\theta}), \quad \hat{\theta}(t_0) = \theta^0 \tag{10.2}$$

where the adaptive gain $\Gamma = \Gamma^T \succ 0$ and

$$\dot{Q} = w^T w, \qquad Q(t_0) = 0 \tag{10.3a}$$

$$\dot{C} = w^T(w\theta^0 + x - \hat{x} - \eta), \qquad C(t_0) = 0 \tag{10.3b}$$

$$\dot{\hat{x}} = f(x,u) + g(x,u)\theta^0 + k_w(x - \hat{x}), \quad \hat{x}(t_0) = x(t_0) \tag{10.3c}$$

$$\eta = x - \hat{x} - w(\theta - \theta^0) \tag{10.3d}$$

$$\dot{w} = g(x,u) - k_w w, \qquad w(t_0) = 0 \tag{10.3e}$$

$$\dot{\eta} = -k_w \eta, \qquad \eta(t_0) = e(t_0). \tag{10.3f}$$

[1] A continuous function $\mu : \mathbb{R}^+ \to \mathbb{R}^+$ is of class $\mathcal{K}_\infty$ if $\mu(0) = 0$, $\mu(\cdot)$ is strictly increasing on $\mathbb{R}^+$ and is radially unbounded.

The vector $\hat{x}$ is the adaptive predictor for (10.1), the constant $k_w > 0$ is the filter gain, θ^0 is the nominal initial estimate of θ, w is a first-order filter, η is an auxiliary variable defined to provide a direct relationship between the parameter estimation error $\tilde{\theta}$ and the prediction error $x - \hat{x}$. As shown in Section 9.2, the parameter estimation error $\|\tilde{\theta}\|$ is non-increasing for $t \leq t_c$ and converges to zero exponentially for $t \geq t_c$ where t_c is the time at which the matrix $Q(t_c) = \int_{t_0}^{t_c} w^T(\tau)w(\tau)\,d\tau > 0$. This was achieved by defining a Lyapunov function

$$V_{\tilde{\theta}} = \frac{1}{2}\tilde{\theta}^T\tilde{\theta} \tag{10.4}$$

and using the fact that $C(t) = Q(t)\,\theta$, to show

$$\dot{V}_{\tilde{\theta}}(t) = -\tilde{\theta}^T(t)\,\Gamma\,Q(t)\,\tilde{\theta}(t) \leq -\mathcal{E}(t)\,V_{\tilde{\theta}}(t) \tag{10.5}$$

where

$$\mathcal{E}(t) = \lambda_{\min}(\Gamma Q(t))\,.$$

10.3.2 Set adaptation

The uncertainty set $\Theta \triangleq B(\hat{\theta}, z_\theta)$ is updated online by updating the parameter estimate $\hat{\theta}$ and its associated error bound $z_\theta = \sup_{s\in\Theta} \|s - \hat{\theta}\|$. The vector $\hat{\theta}$ is updated via (10.2) while z_θ is updated based on the observed system's excitation contained in $\mathcal{E}(t)$ according to the following algorithm.

Algorithm 10.3.1. *Let $\mathcal{E}(\sigma) = \lambda_{\min}(\Gamma Q(\sigma))$, beginning from time $t_{i-1} = t_0$, the parameter and set adaptation is implemented iteratively as follows:*

1. ***Initialize*** $z_\theta(t_0) = z_\theta^0$, $\Theta(t_0) = B(\hat{\theta}(t_0), z_\theta(t_0))$, $\bar{\mathcal{E}} = \mathcal{E}(t_0) = 0$
2. ***Implement*** *the following adaptation law over the interval* $\tau \in [t_{i-1}, t_i)$

$$\dot{z}_\theta(\tau) = -\bar{\mathcal{E}} z_\theta(\tau) \tag{10.6}$$

3. *At time t_i, **perform** the updates*

$$\bar{\mathcal{E}} = \begin{cases} \mathcal{E}(t_i), & \text{if } \mathcal{E}(t_i) \geq \mathcal{E}(t_{i-1}) \\ \mathcal{E}(t_{i-1}), & \text{otherwise} \end{cases} \tag{10.7}$$

$$\left(\hat{\theta}, \Theta\right) = \begin{cases} \left(\hat{\theta}(t_i), \Theta(t_i)\right), & \text{if } z_\theta(t_i) \leq z_\theta(t_{i-1}) - \|\hat{\theta}(t_i) - \hat{\theta}(t_{i-1})\| \\ \left(\hat{\theta}(t_{i-1}), \Theta(t_{i-1})\right), & \text{otherwise} \end{cases} \tag{10.8}$$

4. ***Iterate*** *back to Step 2, **incrementing*** $i = i + 1$.

The advantage of updating z_θ according to (10.6) is that contraction of z_θ can be triggered even when the actual parameter estimation error is zero. The uncertainty set Θ when implemented according to Algorithm 10.3.1 contracts in a strictly nested fashion without excluding θ as shown in the following lemma.

Lemma 10.3.2. *The evolution of* $\Theta \triangleq B(\hat{\theta}, z_\theta)$ *under* (10.2), (10.6), *and Algorithm* 10.3.1 *is such that*

i. $\Theta(t_2) \subseteq \Theta(t_1), \quad t_0 \le t_1 \le t_2$
ii. $\theta \in \Theta(t_0) \Rightarrow \theta \in \Theta(t), \quad \forall t \ge t_0$

Proof:

i. If $\Theta(t_{i+1}) \nsubseteq \Theta(t_i)$, then

$$\sup_{s\in\Theta(t_{i+1})} \|s - \hat{\theta}(t_i)\| \ge z_\theta(t_i). \tag{10.9}$$

However, it follows from triangle inequality and Algorithm 10.3.1 that Θ, at update times, obeys

$$\begin{aligned}\sup_{s\in\Theta(t_{i+1})} \|s - \hat{\theta}(t_i)\| &\le \sup_{s\in\Theta(t_{i+1})} \|s - \hat{\theta}(t_{i+1})\| + \|\hat{\theta}(t_{i+1}) - \hat{\theta}(t_i)\| \\ &\le z_\theta(t_{i+1}) + \|\hat{\theta}(t_{i+1}) - \hat{\theta}(t_i)\| \le z_\theta(t_i),\end{aligned}$$

which contradicts (10.9). Hence, Θ update guarantees $\Theta(t_{i+1}) \subseteq \Theta(t_i)$ and the strict contraction claim follows from the fact that Θ is held constant over update intervals $\tau \in (t_i, t_{i+1})$.

ii. The θ inclusion claim is proven by showing that

$$\|\tilde{\theta}(t)\| \le z_\theta(t), \qquad \forall t \ge t_0 \tag{10.10}$$

which in turn establish that $\theta \in \Theta(t_0) \Rightarrow \theta \in B(\hat{\theta}(t), z_\theta(t)), \ \forall t \ge t_0$. We know that $\|\tilde{\theta}(t_0)\| \le z_\theta(t_0)$ (by definition). It follows from (10.5) that

$$\begin{aligned}\dot{V}_{\tilde{\theta}}(\tau) &= -\tilde{\theta}^T(\tau)\Gamma\left(Q(t_i) + \int_{t_i}^{\tau} w(\sigma)^T w(\sigma)d\sigma\right)\tilde{\theta}(\tau), \qquad \tau \in [t_i, \ t_{i+1}) \\ &\le -\tilde{\theta}^T(\tau)\,\Gamma\, Q(t_i)\,\tilde{\theta}(\tau) \le -\mathcal{E}(t_i)\, V_{\tilde{\theta}}(\tau).\end{aligned} \tag{10.11}$$

and using (10.4), (10.6), and (10.11) we have $\|\dot{\tilde{\theta}}(t)\| \le \dot{z}_\theta(t)$. Hence, by comparison lemma (89, Lemma 3.4) we have (10.10). ■

10.4 Robust adaptive MPC—a min–max approach

In this section, the concept of min–max robust MPC is employed to provide robustness for the MPC controller during the adaptation phase. The resulting optimization problem can either be solved in open-loop or closed-loop. In the approach proposed, we choose the least conservative option by performing optimization with respect to closed-loop strategies. As in typical feedback-MPC fashion, the controller chooses input u as a function of the current states. The formulation consists of maximizing a cost function with respect to θ and minimizing over feedback control policies κ.

The receding horizon control law is defined by

$$u = \kappa_{mpc}(x, \hat{\theta}, z_\theta) \triangleq \kappa^*(0, x, \hat{\theta}, z_\theta) \tag{10.12a}$$

$$\kappa^* \triangleq \arg \min_{\kappa(\tau, x^p, \hat{\theta}^p, z_\theta)} J(x, \hat{\theta}, z_\theta, \kappa) \tag{10.12b}$$

where $J(x, \hat{\theta}, z_\theta, \kappa)$ is the (worst-case) cost associated with the optimal control problem:

$$J(x, \hat{\theta}, z_\theta, \kappa) \triangleq \max_{\theta \in \Theta \triangleq B(\hat{\theta}, z_\theta)} \int_0^T L(x^p, u^p) d\tau + W(x^p(T), \tilde{\theta}^p(T)) \tag{10.13a}$$

$$\text{s.t. } \forall\ \tau \in [0, T]$$

$$\dot{x^p} = f(x^p, u^p) + g(x^p, u^p)\theta, \quad x^p(0) = x \tag{10.13b}$$

$$\dot{w^p} = g^T(x^p, u^p) - kw^p, \quad w^p(0) = w \tag{10.13c}$$

$$\dot{Q^p} = {w^p}^T w^p, \quad Q^p(0) = Q \tag{10.13d}$$

$$\dot{\hat{\theta}^p} = \Gamma Q^p \tilde{\theta}^p, \quad \tilde{\theta}^p = \theta - \hat{\theta}^p, \quad \hat{\theta}^p(0) = \hat{\theta} \tag{10.13e}$$

$$u^p(\tau) \triangleq \kappa(\tau, x^p(\tau), \quad \hat{\theta}^p(\tau)) \in \mathbb{U} \tag{10.13f}$$

$$x^p(\tau) \in \mathbb{X}, \quad x^p(T) \in \mathbb{X}_f(\tilde{\theta}^p(T)) \tag{10.13g}$$

In the formulation, the effect of future parameter adaptation is accounted for, which results in less conservative worst-case predictions. Also, the conservativeness of the terminal cost is reduced by parameterizing both W and $\mathbb{X}_f$ as functions of $\tilde{\theta}(T)$. Parameterizing the terminal penalty as a function of $\tilde{\theta}$ ensures that the algorithm will seek to reduce the parameter error in the process of optimizing the cost function J. This may require the algorithm to automatically inject some useful excitation into the closed-loop system.

10.4.1 Implementation algorithm

Algorithm 10.4.1. *The min–max MPC algorithm performs as follows: At sampling instant t_i*

1. ***Measure** the current state of the plant x and obtain the current values of matrices w and Q from* (10.3e) *and* (10.3a), *respectively.*

2. ***Update*** *the parameter estimates* $\hat{\theta}$ *and the uncertainty set* $\Theta(t) \triangleq B(\hat{\theta}(t), z_\theta(t))$ *using* (10.2) *and Algorithm 10.3.1.*
3. ***Solve*** *the optimization problem* (10.12, 10.13) *and apply the resulting feedback control law to the plant until the next sampling instant.*
4. ***Repeat*** *the procedure from Step* 1 *for the next sampling instant, incrementing* $i = i + 1$.

An alternative to the Θ update (10.8) is to define the uncertainty set

$$\Theta(t) \triangleq \bigcap_{\tau \in [t_0, t]} B(\hat{\theta}(\tau), z_\theta(\tau)), \tag{10.14}$$

and replace Step 2 of Algorithm 10.4.1 with the following alternative Step 2:

2. *Obtain the current value of parameter estimates* $\hat{\theta}$ *and uncertainty bound* z_θ *from* (10.2) *and* (10.6). *Then update the MPC quantities* Θ *and* $\hat{\theta}$ *as*

$$\Theta = \Theta(t_i) \triangleq \Theta(t_{i-1}) \bigcap B(\hat{\theta}(t_i), z_\theta(t_i)) \tag{10.15a}$$

$$\hat{\theta} = \bar{\theta} \tag{10.15b}$$

where $\bar{\theta}$ *is any point in* Θ.

We note that the evolution of Θ when updated according to (10.15) satisfies the main requirement for the MPC performance. The set contracts in a strictly nested fashion without excluding θ. The set contraction follows by definition:

$$\Theta(t_2) = \Theta(t_1) \bigcap B\big(\hat{\theta}(t_2),\ z_\theta(t_1)\big) \subseteq \Theta(t_1), \quad \forall t_2 \geq t_1. \tag{10.16}$$

Moreover, since z_θ is such that $\|\tilde{\theta}(t)\| \leq z_\theta(t)\ \forall t \geq t_0$, the θ inclusion claim follows by noting that $\theta \in \Theta(t_0) \Rightarrow \theta \in B(\hat{\theta}(t),\ z_\theta(t)),\ \forall t \geq t_0$. Hence, $\theta \in \Theta(t_j) := \bigcap_{i=0}^{j} B(\hat{\theta}(t_i), z_\theta(t_i))$. The benefit of using (10.15) is that the size of the uncertainty description Θ reduces faster over time but this is achieved at the expense of increased online computation due to the additional task of calculating the intersection of sets.

In the remainder of this section, we drop the explicit constraint (10.13g) by using the convention that if some of the constraints are not satisfied, then the value of J is $+\infty$, that is,

$$L(x,u) = \begin{cases} L(x,u) < \infty & \text{if } (x,u) \in \mathbb{X} \times \mathbb{U} \\ +\infty & \text{otherwise} \end{cases}$$

$$W(x,\tilde{\theta}) = \begin{cases} W(x,\tilde{\theta}) < \infty & \text{if } x \in \mathbb{X}_f(\tilde{\theta}) \\ +\infty & \text{otherwise} \end{cases}$$

10.4.2 Closed-loop robust stability

Robust stability is guaranteed if predicted state at terminal time belong to a robustly invariant set for all possible uncertainties. Let $\tilde{\Theta}^0 = \{\tilde{\theta} : \|\tilde{\theta}\| \leq z_\theta^0)\}$, a sufficient conditions for the robust MPC (10.12) to guarantee stabilization of the origin is outlined below.

Criterion 10.4.2. *The terminal penalty function* $W : \mathbb{X}_f \times \tilde{\Theta}^0 \to [0, +\infty]$ *and the terminal constraint function* $\mathbb{X}_f : \tilde{\Theta}^0 \to \mathbb{X}$ *are such that for each* $(\theta, \hat{\theta}, \tilde{\theta}) \in (\Theta^0 \times \Theta^0 \times \tilde{\Theta}^0)$, *there exists a feedback* $k_f(\cdot, \hat{\theta}) : \mathbb{X}_f \to \mathbb{U}$ *satisfying*

1. $0 \in \mathbb{X}_f(\tilde{\theta}) \subseteq \mathbb{X}$, $\mathbb{X}_f(\tilde{\theta})$ *closed*
2. $k_f(x, \hat{\theta}) \in \mathbb{U}$, $\forall x \in \mathbb{X}_f(\tilde{\theta})$
3. $W(x, \tilde{\theta})$ *is continuous with respect to* $x \in \mathbb{R}^{n_x}$
4. $\forall x \in \mathbb{X}_f(\tilde{\theta})$, $\mathbb{X}_f(\tilde{\theta})$ *is strongly positively invariant under* $k_f(x, \hat{\theta})$ *with respect to the differential inclusion* $\dot{x} \in f(x, k_f(x, \hat{\theta})) + g(x, k_f(x, \hat{\theta}))\Theta$
5. $W(x(t+\delta), \tilde{\theta}(t)) - W(x(t), \tilde{\theta}(t)) \leq -\int_t^{t+\delta} L(x, k_f(x, \hat{\theta}))d\tau$, $\forall x \in \mathbb{X}_f(\tilde{\theta})$.

In addition to Criterion (10.4.2), the $\tilde{\theta}$ dependence of W and $\mathbb{X}_f$ is required to satisfy the following.

Criterion 10.4.3. *For any* $\tilde{\theta}_1, \tilde{\theta}_2 \in \tilde{\Theta}^0$ *s.t.* $\|\tilde{\theta}_2\| \leq \|\tilde{\theta}_1\|$,

1. $W(x, \tilde{\theta}_2) \leq W(x, \tilde{\theta}_1)$, $\forall x \in \mathbb{X}_f(\tilde{\theta}_1)$
2. $\mathbb{X}_f(\tilde{\theta}_2) \supseteq \mathbb{X}_f(\tilde{\theta}_1)$

Note that Criterion (10.4.2) requires only the existence, not knowledge, of $k_f(x, \hat{\theta})$ and the stability condition requires the terminal penalty function $W(x, \tilde{\theta})$ to be a robust-CLF on the domain $\mathbb{X}_f(\tilde{\theta})$. Criterion (10.4.3) requires W to decrease and the domain $\mathbb{X}_f$ to enlarge with decreased parametric uncertainty as expected.

Theorem 10.4.4. *Let* $X_0 \triangleq X_0(\Theta^0) \subseteq \mathbb{X}$ *denote the set of initial states with uncertainty* Θ^0 *for which* (10.12) *has a solution. Assuming Criteria* 10.4.2 *and* 10.4.3 *are satisfied, then the closed-loop system state x, given by* (10.1, 10.2, 10.6, 10.12), *originating from any* $x_0 \in X_0$ *feasibly approaches the origin as* $t \to +\infty$. ■

The proof of the theorem is given in Section 10.9.

10.5 Robust adaptive MPC—a Lipschitz-based approach

Due to the computational complexity associated with (feedback) min–max optimization problem for nonlinear systems, it is (sometimes) more practical to use more conservative but computationally efficient methods. Examples of such approaches include Lipschitz-based methods [120, 128] and those based on the concept of reachable sets [108].

In this section, we present a Lipschitz-based method whereby the nominal model rather than the unknown bounded system state is controlled, subject to conditions that ensure that given constraints are satisfied for all possible uncertainties. State prediction error bound is determined based on the Lipschitz continuity of the model. A knowledge of appropriate Lipschitz bounds for the x-dependence of the dynamics $f(x,u)$ and $g(x,u)$ are assumed as follows.

Assumption 10.5.1. *A set of functions* $\mathcal{L}_j : \mathbb{X} \times \mathbb{U} \to \mathbb{R}^+$, $j \in \{f, g\}$ *are known which satisfy*

$$\mathcal{L}_j(\mathbb{X}, u) \geq \min\Big\{\mathcal{L}_j \;\Big|\; \sup_{x_1,x_2 \in \mathbb{X}} \big(\|j(x_1,u) - j(x_2,u)\| - \mathcal{L}_j\|x_1 - x_2\|\big) \leq 0\Big\},$$

where for $j \equiv g$ *is interpreted as an induced norm since* $g(x,u)$ *is a matrix.* ■

10.5.1 Prediction of state error bound

In order to consider the effect of the uncertainty $\tilde{\theta}$ in the controller synthesis, we have to compute a bound on the difference between the nominal state trajectory and the solution of the actual system. To this end, consider the actual system

$$\dot{x} = f(x,u) + g(x,u)\theta, \tag{10.18}$$

and the nominal model controlled by the same input u

$$\dot{x}^p = f(x^p,u) + g(x^p,u)\hat{\theta}, \tag{10.19}$$

it follows that

$$\begin{aligned} \|\dot{x} - \dot{x}^p\| &\leq \|f(x,u) - f(x^p,u)\| + \|g(x,u)\theta - g(x^p,u)\theta\| \\ &\quad + \|g(x^p,u)\theta - g(x^p,u)\theta^p\| \\ &\leq \mathcal{L}_f\|x - x^p\| + \mathcal{L}_g\|\theta\|\,\|x - x^p\| + \|g(x^p,u)\|\,\|\theta - \hat{\theta}\|. \end{aligned}$$

Therefore, a worst-case deviation $z_x^p \geq \max_{\theta\in\Theta}\|x - x^p\|$ can be generated from

$$\dot{z}_x^p = (\mathcal{L}_f + \mathcal{L}_g\Pi)z_x^p + \|g(x^p,u)\|z_\theta, \quad z_x^p(t_0) = 0 \tag{10.20}$$

where $\Pi = z_\theta + \|\hat{\theta}\|$.

10.5.2 Lipschitz-based finite horizon optimal control problem

The model predictive feedback is defined as

$$u = \kappa_{mpc}(x,\hat{\theta},z_\theta) = u^*(0) \tag{10.21a}$$

$$u^*(\cdot) \triangleq \arg\min_{u^p_{[0,T]}} J(x,\hat{\theta},z_\theta,u^p) \tag{10.21b}$$

where $J(x, \hat{\theta}, z_\theta, u^p)$ is given by the optimal control problem:

$$J(x, \hat{\theta}, z_\theta, u^p) = \int_0^T L(x^p, u^p)d\tau + W(x^p(T), z_\theta^p(T)) \tag{10.22a}$$

$$\text{s.t. } \forall\tau \in [0,\ T]$$

$$\dot{x^p} = f(x^p, u^p) + g(x^p, u^p)\hat{\theta}, \quad x^p(0) = x \tag{10.22b}$$

$$\dot{z}_x^p = (\mathcal{L}_f + \mathcal{L}_g\Pi)z_x^p + \|g(x^p, u^p)\|z_\theta, \quad z_x^p(0) = 0 \tag{10.22c}$$

$$X^p(\tau) \triangleq B(x^p(\tau), z_x^p(\tau)) \subseteq \mathbb{X}, \quad u^p(\tau) \in \mathbb{U} \tag{10.22d}$$

$$X^p(T) \subseteq \mathbb{X}_f(z_\theta^p(T)) \tag{10.22e}$$

In the proposed formulation, the parameter estimate $\hat{\theta}$ and the uncertainty radius z_θ which appears in (10.22b) and (10.22c) are updated at every sampling instant and held constant over the prediction horizon. However, the effect of the future model improvement along the prediction horizon is incorporated in the formulation by parameterizing the terminal expressions in (10.22a) and (10.22e) as a function of $z_\theta(T)$. This enlarges the terminal domain and hence reduces the conservatism of the robust MPC. Using a nominal model prediction, it is impossible to predict the actual future behavior of the parameter estimation error as was possible in the min–max framework. However, based upon the excitation of the real system at sampling instants t_i, one can generate an upper bound on the future parameter estimation error according to Algorithm 10.3.1, (10.6), that is,

$$z_\theta^p(\tau) = \exp^{-\bar{\mathcal{E}}(\tau - t_i)} z_\theta(t_i) \quad \tau \in [t_i,\ t_i + T) \tag{10.23}$$

where

$$\bar{\mathcal{E}} \geq \mathcal{E}(t_i) = \lambda_{\min}(\Gamma Q(t_i))$$

10.5.3 Implementation algorithm

Algorithm 10.5.2. *The Lipschitz-based MPC algorithm is implemented as follows: At sampling instant t_i*

1. ***Measure*** *the current state of the plant* $x = x(t_i)$.
2. ***Update*** *the parameter estimates* $\hat{\theta} = \hat{\theta}(t_i)$ *and uncertainty bounds* $z_\theta = z_\theta(t_i)$ *and* $z_\theta^p(T) = z_\theta^p(t_i + T)$ *via* (10.2), (10.6), *and* (10.23), *respectively.*
3. ***Solve*** *the optimization problem* (10.21, 10.22) *and apply the resulting feedback control law to the plant until the next sampling instant.*
4. ***Repeat*** *the procedure from Step* 1 *for the next sampling instant, incrementing* $i = i + 1$.

The conservatism of the Lipschitz-based approach is mainly due to the computation of the uncertainty cone $B(x^p, z_x^p)$ around the nominal trajectory. The rate at which the cone expands over the prediction horizon reduces at each sampling instant as z_θ reduces. When z_θ is zero, the effect of parameter uncertainty on the state prediction can be totally eliminated from the adaptive framework by replacing the error dynamic (10.22c) with $\dot{z}_x^p = 0$ when $z_\theta \approx 0$.

Theorem 10.5.3. *Let* $X_0' \triangleq X_0'(\Theta^0) \subseteq \mathbb{X}$ *denote the set of initial states for which* (10.21) *has a solution. Assuming Assumption* 10.5.1 *and Criteria* 10.4.2 *and* 10.4.3 *are satisfied, then the origin of the closed-loop system given by* (10.1, 10.2, 10.6, 10.21) *is feasibly asymptotically stabilized from any* $x_0 \in X_0'$. ■

The proof can be found in Section 10.9.

10.6 Incorporating FTI

The performance and computational demand of the adaptive MPC schemes developed depend on the performance of the parameter and set adaptation mechanism employed. An identification mechanism that provides faster convergence of $\tilde{\theta}$ to zero (in a known time) is beneficial. In this section, we employ the FTI, presented in Chapter 8, in developing an adaptive predictive control structure that reduces to a nominal MPC problem when exact parameter estimates are obtained.

Let the parameter estimate $\hat{\theta}$, matrices Q and C be generated from (10.2), (10.3a), and (10.3b), respectively. Also, let t_c be a time such that $Q(t_c)$ is invertible, the FTI is given by

$$\hat{\theta}^c(t) = \begin{cases} \hat{\theta}(t), & \text{if } t < t_c \\ Q(t_c)^{-1}\, C(t_c), & \text{if } t \geq t_c. \end{cases} \tag{10.24}$$

The revised algorithm based on the FTI is given in the following.

10.6.1 FTI-based min–max approach

Let the filter (10.13c) and excitation dynamics (10.13d) be replaced by

$$\dot{w}^p = \beta\big(g^T(x^p, u^p) - kw^p\big), \quad w^p(0) = w \tag{10.25}$$

$$\dot{Q}^p = \beta(w^{p^T} w^p), \quad Q^p(0) = Q \tag{10.26}$$

with $\beta \in \{0, 1\}$ a design parameter. The proposed FTI-based algorithm is as follows:

Algorithm 10.6.1. *FT min–max MPC algorithm: At sampling instant* t_i

1. *Measure the current state of the plant* x.
2. ***Obtain** the current value of matrices* Q *and* C *from* (10.3a) *and* (10.3b) *respectively.*
3. ***If*** $det(Q) = 0$ *or* $cond(Q)$ *is not satisfactory update the parameter estimates* $\hat{\theta}$ *and the uncertainty set* $\Theta(t) \triangleq B(\hat{\theta}(t), z_\theta(t))$ *according to Algorithm* 10.3.1.
   Else if $det(Q) > 0$ *and* $cond(Q)$ *is satisfactory, set* $\beta = 0$ *and update*

   $$\hat{\theta} = Q^{-1}(t_i)C(t_i), \quad z_\theta = 0$$

   End

4. *Solve the optimization problem* (10.12, 10.13) *and apply the resulting feedback control law to the plant until the next sampling instant.*
5. ***Increment*** $i = i + 1$. ***If*** $z_\theta > 0$, *repeat the procedure from Step* 1 *for the next sampling instant.* ***Otherwise***, *repeat only Steps* 1 *and* 3 *for the next sampling instant.*

Implementing the adaptive MPC controller according to Algorithm 10.6.1 guarantees that the uncertainty ball $\Theta \triangleq B(\hat{\theta}, z_\theta)$ is contained in the previous one, that is, $\Theta(t_i) \subseteq \Theta(t_{i-1})$. Hence, a successive reduction in the computational requirement of (10.12) is ensured. Moreover, when the parameter estimate θ^c becomes available, the uncertainty set Θ reduces to a single point with $\tilde{\theta} = 0$ and the predictive robust control structure becomes that of a nominal MPC:

$$u = \kappa_{mpc}(x) \triangleq \kappa^*(0, x) \tag{10.27a}$$

$$\kappa^* \triangleq \arg\min_{\kappa(\cdot,\cdot)} J(x, \kappa) \triangleq \int_0^T L(x^p, u^p) d\tau + W(x^p(T)) \tag{10.27b}$$

$$\text{s.t.} \quad \forall \tau \in [0, T]$$

$$\dot{x}^p = f(x^p, u^p) + g(x^p, u^p)\theta^c, \quad x^p(0) = x \tag{10.27c}$$

$$u^p(\tau) \triangleq \kappa(\tau, x^p(\tau)) \in \mathbb{U}, \quad x^p(\tau) \in \mathbb{X}, \quad x^p(T) \in \mathbb{X}_f \tag{10.27d}$$

10.6.2 FTI-based Lipshitz-bound approach

For the Lipshitz-based approach, the error bound dynamic (10.22c) is replaced by

$$\dot{z}_x^p = \beta(\mathcal{L}_f + \mathcal{L}_g \Pi) z_x^p + \|g^p\| z_\theta, \quad z_x^p(0) = 0, \tag{10.28}$$

with $\beta \in \{0, 1\}$ and the controller is implemented according to the following algorithm.

Algorithm 10.6.2. *FT Lipschitz-based MPC algorithm: At sampling instant* t_i

1. ***Measure*** *the current state of the plant* x.
2. ***Obtain*** *the current value of matrices* Q *and* C *from* (10.3a) *and* (10.3b) *respectively.*
3. ***If*** $\det(Q) = 0$ *or* $\mathrm{cond}(Q)$ *is not satisfactory, set* $\beta = 1$ *and update the parameter estimates* $\hat{\theta} = \hat{\theta}(t_i)$ *and uncertainty bounds* $z_\theta = z_\theta(t_i)$ *and* $z_\theta^p(T) = z_\theta^p(t_i + T)$ *via* (10.2), (10.6), *and* (10.23), *respectively.*
   Else if $\det(Q) > 0$ *and* $\mathrm{cond}(Q)$ *is satisfactory, set* $\beta = 0$ *and update*

   $$\hat{\theta} = Q^{-1}(t_i)C(t_i), \quad z_\theta = 0$$

   End
4. ***Solve*** *the optimization problem* (10.21, 10.22) *and apply the resulting feedback control law to the plant until the next sampling instant.*
5. ***Increment*** $i = i + 1$. ***If*** $z_\theta > 0$, *repeat the procedure from Step* 1 *for the next sampling instant.* ***Otherwise***, *repeat only Steps* 1 *and* 3 *for the next sampling instant.*

Implementing Algorithm 10.6.2 ensures that the size of the uncertainty cone around the nominal state trajectory reduces as z_θ shrinks and when exact parameter estimate vector θ^c is obtained, $z_x^p = 0$, which implies that the problem becomes that of a nominal MPC (10.27).

10.7 Simulation example

Consider the regulation of a continuous stirred tank reactor where a first order, irreversible exothermic reaction $\mathbf{A} \to \mathbf{B}$ is carried out. Assuming constant liquid level, the reaction is described by the following dynamic model [112]:

$$\begin{aligned}\dot{C}_A &= \frac{q}{V}(C_{Ain} - C_A) - k_0 \exp\left(\frac{-E}{RT_r}\right)C_A \\ \dot{T}_r &= \frac{q}{V}(T_{in} - T_r) - \frac{\Delta H}{\rho c_p} k_0 \exp\left(\frac{-E}{RT_r}\right)C_A + \frac{UA}{\rho c_p V}(T_c - T_r)\end{aligned}$$

The states C_A and T_r are the concentrations of components A and the reactor temperature, respectively. The manipulated variable T_c is temperature of the coolant stream.

It is assumed that reaction kinetic constant k_0 and heat of reaction ΔH are only nominally known and parameterized as $k_0 = \theta_1 \times 10^{10}$ min^{-1} and $\Delta H\ k_0 = -\theta_2 \times 10^{15}$ J/mol min with the parameters satisfying $0.1 \leq \theta_1 \leq 10$ and $0.1 \leq \theta_2 \leq 10$. The objective is to adaptively regulate the unstable equilibrium $C_A^{eq} = 0.5$ mol/l, $T_r^{eq} = 350$ K, $T_c^{eq} = 300$ K while satisfying the constraints $0 \leq C_A \leq 1$, $280 \leq T_r \leq 370$, and $280 \leq T_c \leq 370$. The nominal operating conditions, which corresponds to the given unstable equilibrium are taken from Reference 112: $q = 100$ l/min, $V = 100$ l, $\rho = 1000$ g/l, $c_p = 0.239$ J/g K, E/R = 8750 K, $UA = 5 \times 10^4$ J/min K, $C_{Ain} = 1$ mol/l, and $T_{in} = 350$ K.

Defining $x = \left[\frac{C_A - C_A^{eq}}{0.5}, \frac{T_r - T_r^{eq}}{20}\right]'$, $u = \frac{T_c - T_c^{eq}}{20}$, the stage cost $L(x, u)$ was selected as a quadratic function of its arguments:

$$L(x, u) = x^T Q_x x + u^T R_u u \tag{10.29a}$$

$$Q_x = \begin{bmatrix} 0.5 & 0 \\ 0 & 1.1429 \end{bmatrix} \quad R_u = 1.333. \tag{10.29b}$$

The terminal penalty function used is a quadratic parameter-dependent Lyapunov function $W(x, \theta) = x^T P(\theta) x$ for the linearized system. Denoting the closed-loop system under a local robust stabilizing controller $u = k_f(\theta)x$ as $\dot{x} = A_{cl}(\theta)x$. The matrix $P(\theta) := P_0 + \theta_1 P_1 + \theta_2 P_2 + \cdots \theta_{n\theta} P_{n\theta}$ was selected to satisfy the Lyapunov system of linear matrix inequalities (LMIs)

$$\begin{aligned}&P(\theta) > 0 \\ &A_{cl}(\theta)^T P(\theta) + P(\theta) A_{cl}(\theta) < 0\end{aligned}$$

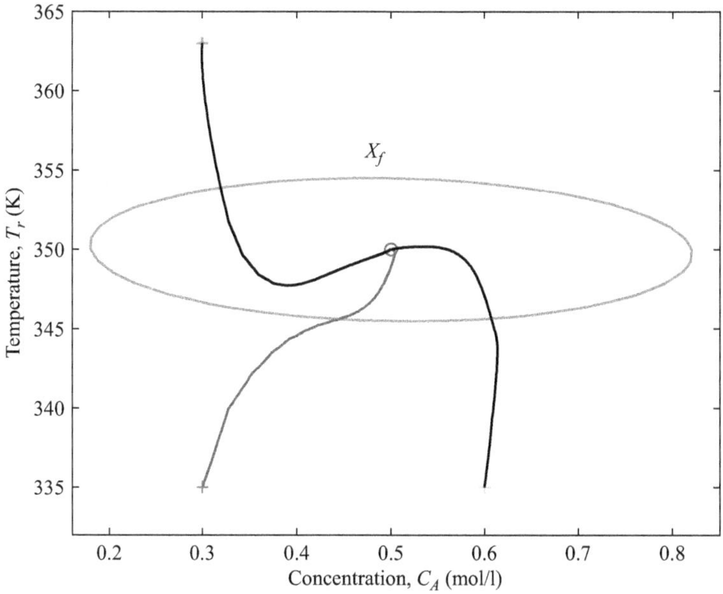

Figure 10.1 Closed-loop reactor state trajectories

for all admissible values of θ. Since θ lie between known extrema values, the task of finding $P(\theta)$ reduces to solving a finite set of LMIs by introducing additional constraints [66]. For the initial nominal estimate $\theta^0 = 5.05$ and $z_\theta^0 = 4.95$, the matrix $P(\theta^0)$ obtained is

$$P(\theta^0) = \begin{bmatrix} 0.6089 & 0.1134 \\ 0.1134 & 4.9122 \end{bmatrix} \tag{10.30}$$

and the corresponding terminal region is

$$\mathbb{X}_f = \{x : x^T P(\theta^0) x \leq 0.25\}. \tag{10.31}$$

For simulation purposes, the true values of the unknown parameters were chosen as $k_0 = 7.2 \times 10^{10}\,\text{min}^{-1}$ and $\Delta H = -5.0 \times 10^4$ J/mol which implies $\theta_1 = 7.2$ and $\theta_2 = 3.6$. The Lipschitz-based approach was used for the controller calculations and the result was implemented according to Algorithm 10.5.2. Since the regressor matrix for this reactor model is diagonal, we define uncertainty bound z_θ for each parameter estimate and adapt the pairs $(\hat{\theta}_1, z_{\theta_1})$ and $(\hat{\theta}_2, z_{\theta_2})$ separately.

The system was simulated from three different initial states $(C_A(0),\ T_r(0)) = (0.3,\ 335)$, $(0.6,\ 335)$ and $(0.3,\ 363)$. The closed-loop trajectories are reported in Figures 10.1–10.4. The results demonstrate that the adaptive MPC regulates the system states to the open loop unstable equilibrium values and satisfies the imposed state and input constraints. The parameter estimates converge to the true values and the uncertainty bound z_θ reduces over time.

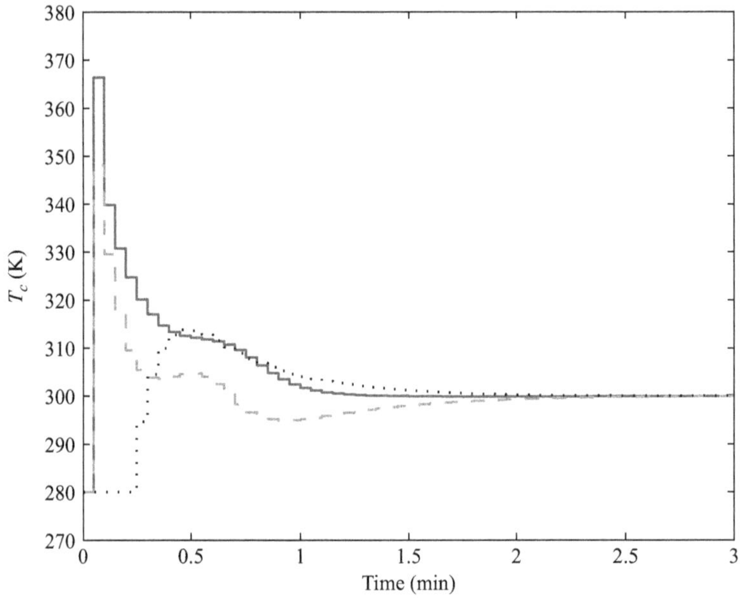

Figure 10.2 Closed-loop reactor input profiles for states starting at different initial conditions ($C_A(0)$, $T_r(0)$): (0.3, 335) is the solid line, (0.6, 335) is the dashed line, and (0.3, 363) is the dotted line

10.8 Conclusions

In this chapter, we presented an adaptive MPC design technique for constrained nonlinear systems with parametric uncertainties. The system's performance is improved over time as the adaptive control updates the model online. The controller parameters is updated only when an improved parameter estimate is obtained. Robustly stabilizing MPC schemes are incorporated to ensure robustness of the algorithm to parameter estimation error during the adaptation phase. The two robust approaches, min–max and Lipschitz-based method, presented provides a tradeoff between computational complexity and conservatism of the solutions. In both cases, the controller is designed in such a way that the computational requirement/conservativeness of the robust adaptive MPC reduces with reduction in parameter uncertainty. Moreover, the complexity of the resultant controller reduces to that of nominal MPC when a FTI is employed and an excitation condition is satisfied.

10.9 Proofs of main results

10.9.1 Proof of Theorem 10.4.4

Feasibility: The closed-loop stability is based upon the feasibility of the control action at each sample time. Assuming, at time t, that an optimal solution $u^p_{[0,T]}$ to

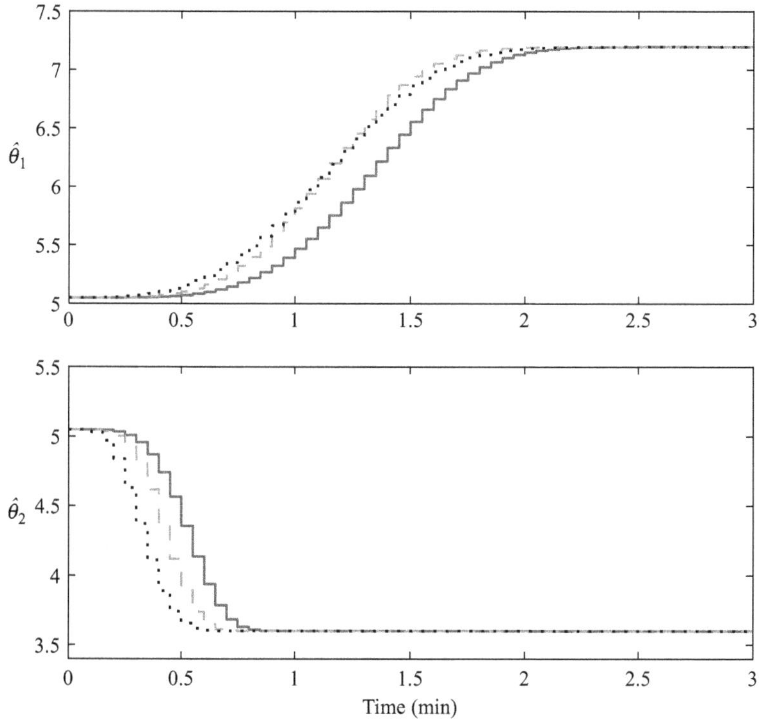

Figure 10.3 Closed-loop parameter estimates profile for states starting at different initial conditions ($C_A(0)$, $T_r(0)$): (0.3, 335) is the solid line, (0.6, 335) is the dashed line, and (0.3, 363) is the dotted line

the optimization problem (10.12) exist and is found. Let Θ^p denote the estimated uncertainty set at time t and Θ^v denote the set at time $t+\delta$ that would result with the feedback implementation of $u_{[t,t+\delta]} = u^p_{[0,\delta]}$. Also, let x^p represent the worst-case state trajectory originating from $x^p(0) = x(t)$ and x^v represents the trajectory originating from $x^v(0) = x + \delta v$ under the same feasible control input $u^v_{[\delta,T]} = u^p_{[\delta,T]}$. Moreover, let $X^a_{\Theta^b} \triangleq \{x^a | \dot{x}^a \in \mathcal{F}(x^a, u^p, \Theta^b) \triangleq f(x^a, u^p) + g(x^a, u^p)\Theta^b\}$.

Since the $u^p_{[0,T]}$ is optimal with respect to the worst-case uncertainty scenario, it is suffice to say that $u^p_{[0,T]}$ drives any trajectory $x^p \in X^p_{\Theta^p}$ into the terminal region $\mathbb{X}^p_f$. Since Θ is non-expanding over time, we have $\Theta^v \subseteq \Theta^p$ implying $x^v \in X^p_{\Theta^v} \subseteq X^p_{\Theta^p}$. The terminal region $\mathbb{X}^p_f$ is strongly positively invariant for the nonlinear system (10.1) under the feedback $k_f(\cdots)$, the input constraint is satisfied in $\mathbb{X}^p_f$ and $\mathbb{X}^v_f \supseteq \mathbb{X}^p_f$ by Criteria 2.2, 2.4, and 3.2, respectively. Hence, the input $u = [u^p_{[\delta,T]}, k_{f[T,T+\delta]}]$ is a feasible solution of (10.12) at time $t+\delta$ and by induction, the optimization problem is feasible for all $t \geq 0$.

Stability: The stability of the closed-loop system is established by proving strict decrease of the optimal cost $J^*(x, \hat{\theta}, z_\theta) \triangleq J(x, \hat{\theta}, z_\theta, \kappa^*)$. Let the trajectories

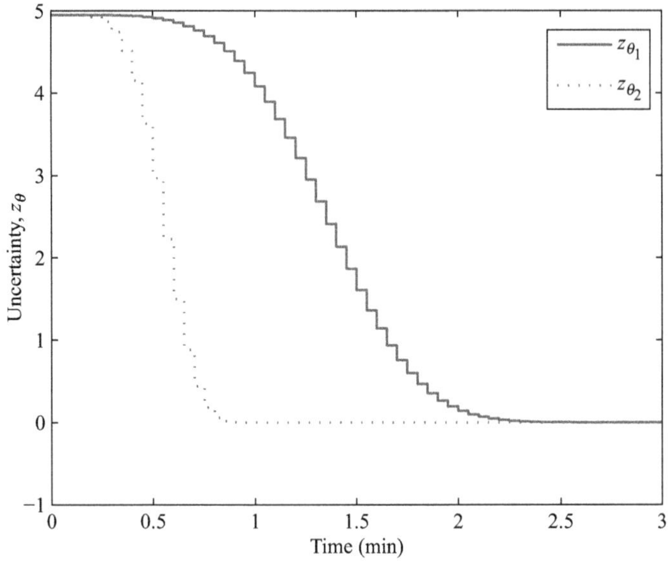

Figure 10.4 Closed-loop uncertainty bound trajectories for initial condition $(C_A,\ T_r) = (0.3,\ 335)$

$(x^p,\ \hat{\theta}^p,\ \tilde{\theta}^p, z_\theta^p)$ and control u^p correspond to any worst-case minimizing solution of $J^*(x, \hat{\theta}, z_\theta)$. If $x^p_{[0,T]}$ were extended to $\tau \in [0,\ T+\delta]$ by implementing the feedback $u(\tau) = k_f(x^p(\tau), \hat{\theta}^p(\tau))$ on $\tau \in [T, T+\delta]$, then criterion 10.4.2(5) guarantees the inequality

$$\int_T^{T+\delta} L(x^p, k_f(x^p, \hat{\theta}^p))d\tau + W(x^p_{T+\delta}, \tilde{\theta}^p_T) - W(x^p_T, \tilde{\theta}^p_T) \le 0 \tag{10.32}$$

where in (10.32) and in the remainder of the proof, $x^p_\sigma \triangleq x^p(\sigma)$, $\tilde{\theta}^p_\sigma \triangleq \tilde{\theta}^p(\sigma)$, for $\sigma = T,\ T+\delta$.

The optimal cost $J^*(x, \hat{\theta}, z_\theta)$

$$= \int_0^T L(x^p, u^p)d\tau + W(x^p_T, \tilde{\theta}^p_T)$$

$$\ge \int_0^T L(x^p, u^p)d\tau + W(x^p_T, \tilde{\theta}^p_T) + \int_T^{T+\delta} L(x^p, k_f(x^p, \hat{\theta}^p))d\tau + W(x^p_{T+\delta}, \tilde{\theta}^p_T) - W(x^p_T, \tilde{\theta}^p_T) \tag{10.33}$$

$$\ge \int_0^\delta L(x^p, u^p)d\tau + \int_\delta^T L(x^p, u^p)d\tau + \int_T^{T+\delta} L(x^p, k_f(x^p, \hat{\theta}^p))d\tau + W(x^p_{T+\delta}, \tilde{\theta}^p_{T+\delta}) \tag{10.34}$$

$$\ge \int_0^\delta L(x^p, u^p)d\tau + J^*(x(\delta), \hat{\theta}(\delta), z_\theta(\delta)) \tag{10.35}$$

Then, it follows from (10.35) that

$$J^*(x(\delta),\hat{\theta}(\delta),z_\theta(\delta)) - J^*(x,\hat{\theta},z_\theta) \le -\int_0^\delta L(x^p,u^p)d\tau$$
$$\le -\int_0^\delta \mu_L(\|x,u\|)d\tau. \qquad (10.36)$$

where μ_L is a class $\mathcal{K}_\infty$ function. Hence $x(t) \to 0$ asymptotically.

Remark 10.9.1. *In the above proof,*

- (10.33) *is obtained using inequality* (10.32)
- (10.34) *follows from Criterion* 10.4.3.1 *and the fact that* $\|\tilde{\theta}\|$ *is non-increasing*
- (10.35) *follows by noting that the last* 3 *terms in* (10.32) *is a (potentially) suboptimal cost on the interval* $[\delta, T+\delta]$ *starting from the point* $(x^p(\delta),\hat{\theta}^p(\delta))$ *with associated uncertainty set* $B(\hat{\theta}^p(\delta), z_\theta^p(\delta))$.

10.9.2 Proof of Theorem 10.5.3

Feasibility: Let $u^p_{[0,T]}$ denotes the initial optimal or feasible solution of (10.21) and let $X^p \triangleq B(x^p, z_x^p)$ denotes the corresponding predicted ball of possible trajectories that starts at $(x^p, z_x^p, z_\theta^p)|_{\tau=0} = (x(t), 0, z_\theta(t))$. Similarly, let $X^v \triangleq B(x^v, z_x^v)$ denote the resulting cone originating from $(x^v, z_x^v, z_\theta^v)|_{\tau=\delta} = (x+\delta v, 0, z_\theta(t+\delta))$, under the same feasible control input $u^v_{[\delta,T]} = u^p_{[\delta,T]}$.

At time $\tau = 0$, it follows from (10.22b) and (10.22c) that $\dot{z}_x^p$ provides an upper bound on $\|\dot{x}^p - \dot{x}\|$, for $\dot{x} \in \mathcal{F}(x, u^p, \Theta)$. Thus the continuity of the trajectories $x^p(\tau)$ and $z_x^p(\tau)$ ensures that for a small enough $\delta > 0$, $x + \delta v \in X^p(\delta)$. Therefore, we assume, without loss of generality, that $x^v(\delta) \in X^p(\delta)$. Since $z_x^v(\delta) = 0$, it follows that $X^v(\delta) \subseteq X^p(\delta)$ and $\|x^p(\delta) - x^v(\delta)\| \le z_x^p(\delta) - z_x^v(\delta)$.

Next, we establish that the inclusion $X^v(\tau) \subseteq X^p(\tau)$ holds for all $\tau \in [\delta, T]$ by showing that $\|x^p(\tau) - x^v(\tau)\| \le z_x^p(\tau) - z_x^v(\tau)$, $\forall \tau \in [\delta, T]$. Defining the variable $e_z \triangleq z_x^p - z_x^v - \|x^p - x^v\|$, the dynamics of e_z satisfy

$$\begin{aligned}
\dot{e}_z &= \dot{z}_x^p - \dot{z}_x^v - \|\dot{x}^p - \dot{x}^v\| \\
&= (\mathcal{L}_f + \mathcal{L}_g\Pi)z_x^p - (\mathcal{L}_f + \mathcal{L}_g\Pi)z_x^v - \|f(x^p,u) - f(x^v,u) \\
&\quad + (g(x^p,u) - g(x^v,u))\hat{\theta}\| \\
&\ge (\mathcal{L}_f + \mathcal{L}_g\Pi)(z_x^p - z_x^v) - \left(\mathcal{L}_f + \mathcal{L}_g\|\hat{\theta}\|\right)\|x^p - x^v\| \\
&\ge (\mathcal{L}_f + \mathcal{L}_g\Pi)e_z,
\end{aligned}$$

from which the initial condition $e_z(\delta) \ge 0$ guarantees that $e_z(\tau) \ge 0$, $\forall \tau \in [\delta, T]$. This implies that $X^v(\tau) \subseteq X^p(\tau) \subseteq \mathbb{X}$ and $X^v(T) \subseteq X^p(T) \subseteq \mathbb{X}_f^p$. Moreover, from the non-increase of the uncertainty bound z_θ, we have $z_\theta^v \leq z_\theta^p$ which implies that $\mathbb{X}_f^p \subseteq \mathbb{X}_f^v$. Therefore, the input $u = [u_{[\delta,T]}^p, k_{f[T,T+\delta]}]$ serves as a feasible solution for (10.21) at time $t + \delta$ and the feasibility result can be achieved by induction.

Stability: closed-loop stability is established by showing that the optimal value function is non-increasing. The proof is similar to that of Theorem 10.4.4.

Chapter 11
Adaptive MPC with disturbance attenuation

11.1 Introduction

In general, modeling error consists of parametric and non-parametric uncertainties and the system dynamics can be influenced by exogenous disturbances as well. In this chapter, we extend the adaptive MPC framework presented in Chapter 10 to nonlinear systems with both constant parametric uncertainty and additive exogenous disturbances.

Intuitively, an adaptive controller should lead to controller with better robustness properties than their non-adaptive counterpart since they use more information on the systems uncertainties. However, this is not generally the case. Under external disturbance input, adaptive controllers can lead to inferior transient behavior, infinite parameter drift, and burstiness in the closed-loop system. To address these problems, parameter projection [98] is used to ensure the estimate remains in a convex set and the parameter estimates are updated only when an improved estimate is obtained. The formulation provides robustness to parameter estimation error and bounded disturbances $\vartheta \in \mathcal{D}$. While the disturbance set $\mathcal{D}$ remains unchanged over time, the parametric uncertainty set Θ is adapted in such a way that guarantees its contraction.

11.2 Revised problem set-up

Consider the uncertain nonlinear system

$$\dot{x} = f(x,u) + g(x,u)\theta + \vartheta \triangleq \mathcal{F}(x,u,\theta,\vartheta) \tag{11.1}$$

where the disturbance $\vartheta \in \mathcal{D} \subset \mathbb{R}^{n_d}$ is assumed to satisfy a known upper bound $\|\vartheta(t)\| \leq M_\vartheta < \infty$. The objective of the study is to (robustly) stabilize the plant to some target set $\Xi \subset \mathbb{R}^{n_x}$ while satisfying the pointwise constraints $x \in \mathcal{X} \in \mathbb{R}^{n_x}$ and $u \in \mathcal{U} \in \mathbb{R}^{n_u}$. The target set is a compact set, contains the origin, and is robustly invariant under no control.

11.3 Parameter and uncertainty set estimation

11.3.1 Preamble

Consider the dynamical system (14.1) and assume we use the same adaptive compensator (10.2) and (10.3). Since ϑ is not known, the true η dynamic is

$$\dot{\eta}_\vartheta = -k_w \eta_\vartheta + \vartheta, \quad \eta_\vartheta(t_0) = e(t_0) \tag{11.2}$$

which results in the estimation error $\tilde{\eta} = \eta_\vartheta - \eta$ and dynamic

$$\dot{\tilde{\eta}} = -k_w \tilde{\eta} + \vartheta, \quad \tilde{\eta}(t_0) = 0. \tag{11.3}$$

Considering the Lyapunov function

$$V_{\tilde{\theta}} = \frac{1}{2}\tilde{\theta}^T \tilde{\theta} \tag{11.4}$$

Since $w\theta = w\theta^0 + x - \hat{x} - \eta + \tilde{\eta}$ in this case, we have

$$C(t) = Q(t)\theta + \int_{t_0}^{t} w^T(\sigma)\tilde{\eta}(\sigma)\, d\sigma \tag{11.5}$$

hence, it follows that

$$\dot{V}_{\tilde{\theta}}(t) = -\tilde{\theta}^T(t)\,\Gamma\, Q(t)\tilde{\theta}(t) - \tilde{\theta}^T(t)\,\Gamma \int_{t_0}^{t} w^T(\sigma)\tilde{\eta}(\sigma)\, d\sigma, \tag{11.6}$$

which guarantees boundedness of the parameter estimation error. To compute z_θ, the upper bound on the estimation error that must depend on measurable signal, we replace (11.6) with

$$\dot{V}_{\tilde{\theta}}(t) \leq -\mathcal{E}(t)V_{\tilde{\theta}}(t) + k_d\sqrt{V_{\tilde{\theta}}(t)} \int_{t_0}^{t} \|w^T(\sigma)\| d\sigma \tag{11.7}$$

where

$$\mathcal{E}(t) = \lambda_{\min}(\Gamma Q(t)) \quad \text{and} \quad k_d = \lambda_{\max}(\Gamma)\frac{M_\vartheta}{k_w}.$$

Though the adaptive compensator gives a stronger convergence result for systems subject to uncertainties, its usefulness in developing robust adaptive MPC for systems subject to disturbances is limited. Updating the uncertainty bound z_θ based on (11.7) would result in a very conservative design, mainly because of the integral in the positive term. To obtain a tighter parameter estimation error bound, we present an alternative update law that is based on the closed-loop system states and M_ϑ.

11.3.2 Parameter adaptation

Let the estimator model for 14.1 be selected as

$$\dot{\hat{x}} = f(x,u) + g(x,u)\hat{\theta} + k_w e + w\dot{\hat{\theta}}, \quad k_w > 0 \tag{11.8}$$

$$\dot{w} = g(x,u) - k_w w, \quad w(t_0) = 0. \tag{11.9}$$

resulting in state prediction error $e = x - \hat{x}$ and auxiliary variable $\eta = e - w\tilde{\theta}$ dynamics:

$$\dot{e} = g(x,u)\tilde{\theta} - k_w e - w\dot{\hat{\theta}} + \vartheta \quad e(t_0) = x(t_0) - \hat{x}(t_0) \tag{11.10}$$

$$\dot{\eta} = -k_w \eta + \vartheta, \quad \eta(t_0) = e(t_0). \tag{11.11}$$

Since ϑ is not known, an estimate of η is generated from

$$\dot{\hat{\eta}} = -k_w \hat{\eta}, \quad \hat{\eta}(t_0) = e(t_0). \tag{11.12}$$

with resulting estimation error $\tilde{\eta} = \eta - \hat{\eta}$ dynamics

$$\dot{\tilde{\eta}} = -k_w \tilde{\eta} + \vartheta, \quad \tilde{\eta}(t_0) = 0. \tag{11.13}$$

Let $\Sigma \in \mathbb{R}^{n_\theta \times n_\theta}$ be generated from

$$\dot{\Sigma} = w^T w, \quad \Sigma(t_0) = \alpha I \succ 0, \tag{11.14}$$

based on (12.3), (12.4), and (11.12), our preferred parameter update law is given by

$$\dot{\Sigma}^{-1} = -\Sigma^{-1} w^T w \Sigma^{-1}, \quad \Sigma^{-1}(t_0) = \frac{1}{\alpha} I \tag{11.15a}$$

$$\dot{\hat{\theta}} = \text{Proj}\left\{\gamma\, \Sigma^{-1} w^T (e - \hat{\eta}),\ \hat{\theta}\right\}, \quad \hat{\theta}(t_0) = \theta^0 \in \Theta^0 \tag{11.15b}$$

where $\gamma = \gamma^T > 0$ and $\text{Proj}\{\phi,\ \hat{\theta}\}$ denotes a Lipschitz projection operator such that

$$-\text{Proj}\{\phi,\ \hat{\theta}\}^T \tilde{\theta} \leq -\phi^T \tilde{\theta}, \tag{11.16}$$

$$\hat{\theta}(t_0) \in \Theta^0 \Rightarrow \hat{\theta}(t) \in \Theta^0_\epsilon, \quad \forall t \geq t_0. \tag{11.17}$$

where $\Theta^0_\epsilon \triangleq B(\theta^0,\ z^0_\theta + \epsilon), \epsilon > 0$. More details on parameter projection can be found in Reference 98.

Lemma 11.3.1. *The identifier* (11.15a) *is such that the estimation error* $\tilde{\theta} = \theta - \hat{\theta}$ *is bounded. Moreover, if*

$$\vartheta \in \mathcal{L}_2 \quad or \quad \int_{t_0}^{\infty} \left[\|\tilde{\eta}\|^2 - \underline{\gamma}\, \|e - \hat{\eta}\|^2\right] d\tau < +\infty \tag{11.18}$$

with $\underline{\gamma} = \lambda_{\min}(\gamma)$ *and the strong condition*

$$\lim_{t \to \infty} \lambda_{\min}(\Sigma) = \infty \tag{11.19}$$

is satisfied, then $\tilde{\theta}$ *converges to zero asymptotically.*

Proof: Let $V_{\tilde{\theta}} = \tilde{\theta}^T \Sigma\, \tilde{\theta}$, it follows from (11.15a) and the relationship $w\tilde{\theta} = e - \hat{\eta} - \tilde{\eta}$ that

$$\begin{aligned} \dot{V}_{\tilde{\theta}} &\leq -2\underline{\gamma}\, \tilde{\theta}^T w^T (e - \hat{\eta}) + \tilde{\theta}^T w^T w \tilde{\theta} \\ &= -\underline{\gamma}\, (e - \hat{\eta})^T (e - \hat{\eta}) + \|\tilde{\eta}\|^2, \end{aligned} \tag{11.20}$$

implying that $\tilde{\theta}$ is bounded. Moreover, it follows from (11.20) that

$$V_{\tilde{\theta}}(t) = V_{\tilde{\theta}}(t_0) + \int_{t_0}^{t} \dot{V}_{\tilde{\theta}}(\tau)d\tau \tag{11.21}$$

$$\leq V_{\tilde{\theta}}(t_0) - \underline{\gamma} \int_{t_0}^{t} \|e - \hat{\eta}\|^2 \, d\tau + \int_{t_0}^{t} \|\tilde{\eta}\|^2 \, d\tau \tag{11.22}$$

Considering the dynamics of (11.13), if $\vartheta \in \mathcal{L}_2$, then $\tilde{\eta} \in \mathcal{L}_2$ (Lemma 8.4.2). Hence, the right-hand side of (11.22) is finite in view of (11.18), and by (11.19) we have $\lim_{t\to\infty} \tilde{\theta}(t) = 0$. ∎

11.3.3 Set adaptation

An update law that measures the worst-case progress of the parameter identifier in the presence of disturbance is given by:

$$z_\theta = \sqrt{\frac{V_{z\theta}}{\lambda_{\min}(\Sigma)}} \tag{11.23a}$$

$$V_{z\theta}(t_0) = \lambda_{\max}\big(\Sigma(t_0)\big)\,(z_\theta^0)^2 \tag{11.23b}$$

$$\dot{V}_{z\theta} = -\underline{\gamma}\,(e - \hat{\eta})^T(e - \hat{\eta}) + \left(\frac{M_\vartheta}{k_w}\right)^2. \tag{11.23c}$$

Using the parameter estimator (11.15) and its error bound z_θ (11.16), the uncertain ball $\Theta \triangleq B(\hat{\theta}, z_\theta)$ is adapted online according to the following algorithm.

Algorithm 11.3.2. *Beginning from time $t_{i-1} = t_0$, the parameter and set adaptation is implemented iteratively as follows:*

1. ***Initialize*** $z_\theta(t_{i-1}) = z_\theta^0$, $\hat{\theta}(t_{i-1}) = \hat{\theta}^0$ *and* $\Theta(t_{i-1}) = B(\hat{\theta}(t_{i-1}), z_\theta(t_{i-1}))$.
2. *At time t_i, using* (11.15) *and* (12.16) ***perform*** *the update*

$$\left(\hat{\theta},\, \Theta\right) = \begin{cases} (\hat{\theta}(t_i),\ \Theta(t_i)), & \text{if } z_\theta(t_i) \leq z_\theta(t_{i-1}) - \|\hat{\theta}(t_i) - \hat{\theta}(t_{i-1})\| \\ (\hat{\theta}(t_{i-1}),\ \Theta(t_{i-1})), & \text{otherwise} \end{cases} \tag{11.24}$$

3. ***Iterate back*** *to Step* 2, ***incrementing*** $i = i + 1$.

The algorithm ensures that Θ is only updated when z_θ value has decreased by an amount which guarantees a contraction of the set. Moreover z_θ evolution as given in (12.16) ensures non-exclusion of θ as shown below.

Lemma 11.3.3. *The evolution of $\Theta = B(\hat{\theta}, z_\theta)$ under (11.15), (11.16), and Algorithm* 12.3.2 *is such that*

i. $\Theta(t_2) \subseteq \Theta(t_1),\ t_0 \leq t_1 \leq t_2$
ii. $\theta \in \Theta(t_0) \Rightarrow \theta \in \Theta(t),\ \forall t \geq t_0$

Proof:

i. The proof of the first claim is the same as that of Lemma 10.3.2i
ii. We know that $V_{\tilde{\theta}}(t_0) \leq V_{z\theta}(t_0)$ (by definition) and it follows from (11.20) and (12.16c) that $\dot{V}_{\tilde{\theta}}(t) \leq \dot{V}_{z_\theta}(t)$. Hence, by the comparison lemma, we have

$$V_{\tilde{\theta}}(t) \leq V_{z\theta}(t), \quad \forall t \geq t_0. \tag{11.25}$$

and since $V_{\tilde{\theta}} = \tilde{\theta}^T \Sigma \tilde{\theta}$, it follows that

$$\|\tilde{\theta}(t)\|^2 \leq \frac{V_{z\theta}(t)}{\lambda_{\min}(\Sigma(t))} = z_\theta^2(t), \quad \forall t \geq t_0. \tag{11.26}$$

Hence, if $\theta \in \Theta(t_0)$, then $\theta \in B(\hat{\theta}(t), z_\theta(t)), \ \forall t \geq t_0$.

■

11.4 Robust adaptive MPC

11.4.1 Min–max approach

The formulation of the min–max MPC consists of maximizing a cost function with respect to $\theta \in \Theta$, $\vartheta \in \mathcal{D}$ and minimizing over feedback control policies κ. The robust receding horizon control law is

$$u = \kappa_{mpc}(x, \hat{\theta}, z_\theta) \triangleq \kappa^*(0, x, \hat{\theta}, z_\theta) \tag{11.27a}$$

$$\kappa^* \triangleq \arg \min_{\kappa(\cdot,\cdot,\cdot,\cdot)} J(x, \hat{\theta}, z_\theta, \kappa) \tag{11.27b}$$

where

$$J(x, \hat{\theta}, z_\theta, \kappa) \triangleq \max_{\theta\in\Theta,\ \vartheta\in\mathcal{D}} \int_0^T L(x^p, u^p)d\tau + W(x^p(T), \tilde{\theta}^p(T)) \tag{11.28a}$$

$$\text{s.t. } \forall \ \tau \in [0, T]$$

$$\dot{x}^p = f(x^p, u^p) + g(x^p, u^p)\theta + \vartheta, \quad x^p(0) = x \tag{11.28b}$$

$$\dot{w}^p = g^T(x^p, u^p) - k_w w^p, \quad w^p(0) = w \tag{11.28c}$$

$$(\dot{\Sigma}^{-1})^p = -(\Sigma^{-1})^p w^T w(\Sigma^{-1})^p, \quad (\Sigma^{-1})^p(0) = \Sigma^{-1} \tag{11.28d}$$

$$\dot{\hat{\theta}}^p = \text{Proj}\left\{\gamma\,(\Sigma^{-1})^p w^T(e - \hat{\eta}),\ \hat{\theta}\right\} \quad \tilde{\theta}^p = \theta - \hat{\theta}^p, \quad \hat{\theta}^p(0) = \hat{\theta} \tag{11.28e}$$

$$u^p(\tau) \triangleq \kappa(\tau, x^p(\tau), \hat{\theta}^p(\tau)) \in \mathbb{U} \tag{11.28f}$$

$$x^p(\tau) \in \mathbb{X}, \quad x^p(T) \in \mathbb{X}_f(\tilde{\theta}^p(T)) \tag{11.28g}$$

The effect of future parameter adaptation is also accounted for in this formulation. The conservativeness of the algorithm is reduced by parameterizing both W and $\mathbb{X}_f$ as functions of $\tilde{\theta}(T)$. While it is possible for the set Θ to contract upon θ over time, the robustness feature due to $\vartheta \in \mathcal{D}$ will still remain.

Algorithm 11.4.1. *The MPC algorithm performs as follows: At sampling instant t_i*

1. ***Measure*** *the current state of the plant $x(t)$ and obtain the current value of matrices w and Σ^{-1} from* (11.9) *and* (11.15a), *respectively.*

2. ***Obtain*** *the current value of parameter estimates* $\hat{\theta}$ *and uncertainty bound* z_θ *from* (11.15a) *and* (11.23)*, respectively*

   If $z_\theta(t_i) \leq z_\theta(t_{i-1}) - \|\hat{\theta}(t_i) - \hat{\theta}(t_{i-1})\|$

   $$\hat{\theta} = \hat{\theta}(t_i), \quad z_\theta = z_\theta(t_i)$$

   Else

   $$\hat{\theta} = \hat{\theta}(t_{i-1}), \quad z_\theta = z_\theta(t_{i-1})$$

   End

3. ***Solve*** *the optimization problem* (11.28) *and apply the resulting feedback control law to the plant until the next sampling instant.*
4. ***Increment*** $i = i + 1$. *Repeat the procedure from Step* 11.4.3 *for the next sampling instant.*

11.4.2 Lipschitz-based approach

Assuming a knowledge of the Lipschitz bounds for the x-dependence of the dynamics $f(x, u)$ and $g(x, u)$ as given as follows:

Assumption 11.4.2. *A set of functions* $\mathcal{L}_j : \mathbb{X} \times \mathbb{U} \rightarrow \mathbb{R}^+, j \in \{f, g\}$ *are known which satisfy*

$$\mathcal{L}_j(\mathbb{X}, u) \geq \min\Big\{\mathcal{L}_j \;\Big|\; \sup_{x_1, x_2 \in \mathbb{X}} \big(\|j(x_1, u) - j(x_2, u)\| - \mathcal{L}_j\|x_1 - x_2\|\big) \leq 0\Big\},$$

where for $j \equiv g$ *is interpreted as an induced norm since* $g(x, u)$ *is a matrix.* ■

Let $\Pi = z_\theta + \|\hat{\theta}\|$, a worst-case deviation $z_x^p \geq \max_{\theta \in \Theta} \|x - x^p\|$ can be generated from

$$\dot{z}_x^p = (\mathcal{L}_f + \mathcal{L}_g\Pi)z_x^p + \|g(x^p, u)\|z_\theta + M_\vartheta, \quad z_x^p(t_0) = 0. \tag{11.30}$$

Using this error bound, the robust Lipschitz-based MPC is given by

$$u = \kappa_{mpc}(x, \hat{\theta}, z_\theta) = u^*(0) \tag{11.31a}$$

$$u^*(\cdot) \triangleq \arg\min_{u^p_{[0,T]}} J(x, \hat{\theta}, z_\theta, u^p) \tag{11.31b}$$

where

$$J(x, \hat{\theta}, z_\theta, u^p) = \int_0^T L(x^p, u^p)d\tau + W(x^p(T), z_\theta^p) \tag{11.32a}$$

$$\text{s.t. } \forall \; \tau \in [0, \; T]$$

$$\dot{x}^p = f(x^p, u^p) + g(x^p, u^p)\hat{\theta}, \quad x^p(0) = x \tag{11.32b}$$

$$\dot{z}_x^p = (\mathcal{L}_f + \mathcal{L}_g\Pi)z_x^p + \|g^p\|z_\theta + M_\vartheta, \quad z_x^p(0) = 0 \tag{11.32c}$$

$$X^p(\tau) \triangleq B(x^p(\tau), z_x^p(\tau)) \subseteq \mathbb{X}, \quad u^p(\tau) \in \mathbb{U} \tag{11.32d}$$

$$X^p(T) \subseteq \mathbb{X}_f(z_\theta^p) \tag{11.32e}$$

The effect of the disturbance is built into the uncertainty cone $B(x^p(\tau), z_x^p(\tau))$ via (11.31c). Since the uncertainty bound is no more monotonically decreasing in this case, the uncertainty radius z_θ which appears in (11.31c) and in the terminal expressions of (11.31a) and (11.31e) are held constant over the prediction horizon. However, the fact that they are updated at sampling instants when z_θ shrinks reduces the conservatism of the robust MPC and enlarges the terminal domain that would otherwise have been designed based on a large initial uncertainty $z_\theta(t_0)$.

Algorithm 11.4.3. *The Lipschitz-based MPC algorithm performs as follows: At sampling instant t_i*

1. ***Measure*** *the current state of the plant $x = x(t_i)$*
2. ***Obtain*** *the current value of the parameter estimates $\hat{\theta}$ and uncertainty bound z_θ from* (11.15a) *and* (11.23a), *respectively,*

   If $z_\theta(t_i) \leq z_\theta(t_{i-1})$

   $$\hat{\theta} = \hat{\theta}(t_i), \quad z_\theta = z_\theta(t_i)$$

   Else

   $$\hat{\theta} = \hat{\theta}(t_{i-1}), \quad z_\theta = z_\theta(t_{i-1})$$

   End

3. ***Solve*** *the optimization problem* (11.31a) *and apply the resulting feedback control law to the plant until the next sampling instant*
4. ***Increment*** *i:=i+1; repeat the procedure from Step* 1 *for the next sampling instant.*

11.5 Closed-loop robust stability

Robust stabilization to the target set Ξ is guaranteed by appropriate selection of the design parameters W and X_f. The robust stability conditions require the satisfaction of Criteria 10.4.2 and 10.4.3, with Criteria 10.4.2.4 strengthened to account for the effect of the disturbance $\vartheta \in \mathcal{D}$. The criteria are given below for ease of reference.

Criterion 11.5.1. *The terminal penalty function $W : \mathbb{X}_f \times \tilde{\Theta}^0 \to [0, +\infty]$ and the terminal constraint function $\mathbb{X}_f : \tilde{\Theta}^0 \to \mathbb{X}$ are such that for each $(\theta, \hat{\theta}, \tilde{\theta}) \in (\Theta^0 \times \Theta^0 \times \tilde{\Theta}^0_\epsilon)$, there exists a feedback $k_f(., \hat{\theta}) : \mathbb{X}_f \to \mathbb{U}$ satisfying*

1. $0 \in \mathbb{X}_f(\tilde{\theta}) \subseteq \mathbb{X}$, $\mathbb{X}_f(\tilde{\theta})$ *closed*
2. $k_f(x, \hat{\theta}) \in \mathbb{U}$, $\forall x \in \mathbb{X}_f(\tilde{\theta})$
3. $W(x, \tilde{\theta})$ *is continuous with respect to* $x \in \mathbb{R}^{n_x}$
4. $\forall x \in \mathbb{X}_f(\tilde{\theta})\backslash\Xi$, $\mathbb{X}_f(\tilde{\theta})$ *is strongly positively invariant under $k_f(x, \hat{\theta})$ with respect to $\dot{x} \in f(x, k_f(x, \hat{\theta})) + g(x, k_f(x, \hat{\theta}))\Theta + \mathcal{D}$*
5. $L(x, k_f(x, \hat{\theta})) + \frac{\partial W}{\partial x}\mathcal{F}(x, k_f(x, \hat{\theta}), \theta, \vartheta) \leq 0$, $\forall x \in \mathbb{X}_f(\tilde{\theta})\backslash\Xi$.

Criterion 11.5.2. *For any* $\tilde{\theta}_1,\ \tilde{\theta}_2 \in \tilde{\Theta}^0$ *s.t.* $\|\tilde{\theta}_2\| \leq \|\tilde{\theta}_1\|$,

1. $W(x,\tilde{\theta}_2) \leq W(x,\tilde{\theta}_1), \quad \forall x \in \mathbb{X}_f(\tilde{\theta}_1)$
2. $\mathbb{X}_f(\tilde{\theta}_2) \supseteq \mathbb{X}_f(\tilde{\theta}_1)$

The revised condition 11.5.1.5 requires W to be a local robust CLF for the uncertain system 11.1 with respect to $\theta \in \Theta$ and $\vartheta \in \mathcal{D}$.

11.5.1 Main results

Theorem 11.5.3. *Let* $X_{d0} \triangleq X_{d0}(\Theta^0) \subseteq \mathbb{X}$ *denote the set of initial states with uncertainty* Θ^0 *for which* (11.27) *has a solution. Assuming Criteria* (11.5.1) *and* (11.5.2) *are satisfied, then the closed-loop system state x, given by* (11.1,11.15,11.23,14.14), *originating from any* $x_0 \in X_{d0}$ *feasibly approaches the target set* Ξ *as* $t \to +\infty$.

Proof: The closed-loop stability is established by the feasibility of the control action at each sample time and the strict decrease of the optimal cost J^*. The proof follows from that of Theorem 10.4.4 since the control law is optimal with respect to the worst case uncertainty $(\theta,\ \vartheta) \in (\Theta,\ \mathcal{D})$ scenario and the terminal region $\mathbb{X}_f^p$ is strongly positively invariant for 14.1 under the (local) feedback $k_f(\cdot,\cdot)$. ■

Theorem 11.5.4. *Let* $X'_{d0} \triangleq X'_{d0}(\Theta^0) \subseteq \mathbb{X}$ *denote the set of initial states for which* (14.18) *has a solution. Assuming Assumption* (14.4.2) *and Criteria* (11.5.1) *and* (11.5.2) *are satisfied, then the origin of the closed-loop system given by* (14.1,11.15,11.23,11.30) *is feasibly asymptotically stabilized from any* $x_0 \in X'_{d0}$ *to the target set* Ξ. ■

The proof of the Lipschitz-based control law follows from that of Theorem 10.5.3.

11.6 Simulation example

To illustrate the effectiveness of the proposed design, we consider the regulation of the CSTR in Example 10.7, subject to an additional disturbance on the temperature dynamic:

$$\dot{C}_A = \frac{q}{V}(C_{Ain} - C_A) - k_0 \exp\left(\frac{-E}{R\,T_r}\right) C_A$$

$$\dot{T}_r = \frac{q}{V}(T_{in} - T_r) - \frac{\Delta H}{\rho\, c_p} k_0 \exp\left(\frac{-E}{R\,T_r}\right) C_A + \frac{UA}{\rho\, c_p\, V}(T_c - T_r) + \vartheta$$

where $\vartheta(t)$ is an unknown function of time. We also assume that the reaction kinetic constant k_0 and ΔH are only nominally known. The operating conditions and system constraints are as detailed in Section 10.7. The control objective is to robustly regulate the reactor temperature and concentration to the (open loop) unstable equilibrium $C_A^{eq} = 0.5\,\mathrm{mol}/l$, $T_r^{eq} = 350\,\mathrm{K}$, $T_c^{eq} = 300\,\mathrm{K}$ by manipulating the temperature of the coolant stream T_c.

For simulation purposes, the disturbance is selected as a fluctuation of the inlet temperature $\vartheta(t) = 0.01\, T_{in}\sin(3t)$ and the true values of the unknown parameters

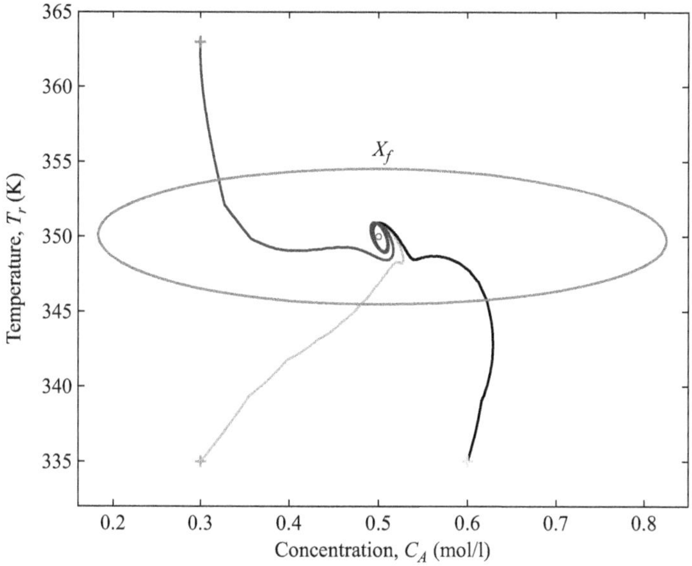

Figure 11.1 Closed-loop reactor trajectories under additive disturbance $\vartheta(t)$

were also chosen as $k_0 = 7.2 \times 10^{10}\,\text{min}^{-1}$ and $\Delta H = -5.0 \times 10^4$ J/mol. The stage cost (10.29), terminal penalty (10.30), and terminal region (10.31) were used. The Lipschitz-based approach was used for the controller calculations and the result was implemented according to Algorithm 11.4.3. As depicted in Figures 11.1–11.3, the robust adaptive MPC drives the system to a neighborhood of the equilibrium while satisfying the imposed constraints and achieves parameter convergence. Figure 11.4 shows that the uncertainty bound z_θ also reduces over time, although at much more conservative rate compared to Figure 10.4 obtained for systems with no disturbances.

11.7 Conclusions

The adaptive MPC design technique is extended to constrained nonlinear systems with both parametric and time-varying disturbances. The proposed robust controller updates the plant model online when model improvement is guaranteed. The embedded adaptation mechanism enables us to construct less conservative terminal design parameters based upon subsets of the original parametric uncertainty. While the introduced conservatism/computation complexity due to the parametric uncertainty reduces over time, the portion due to the disturbance $\vartheta \in \mathcal{D}$ remains active for all time.

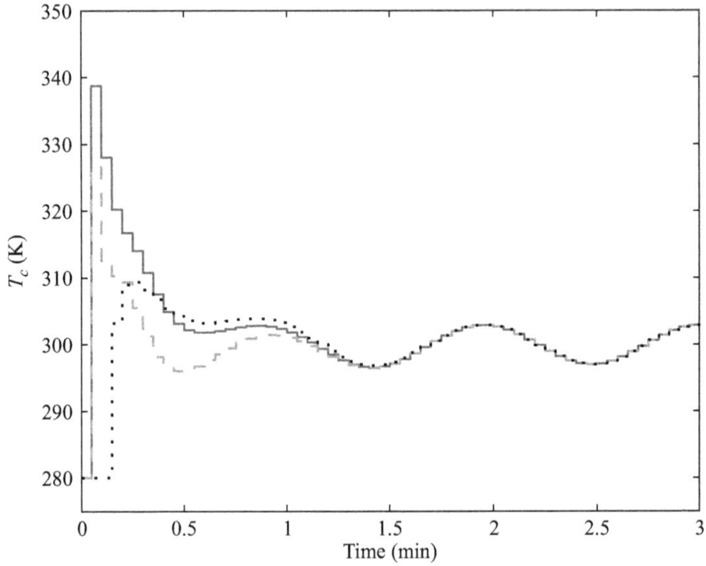

Figure 11.2 Closed-loop input profiles for states starting at different initial conditions ($C_A(0)$, $T_r(0)$): (0.3, 335) is solid line, (0.6, 335) is dashed line, and (0.3, 363) is the dotted line

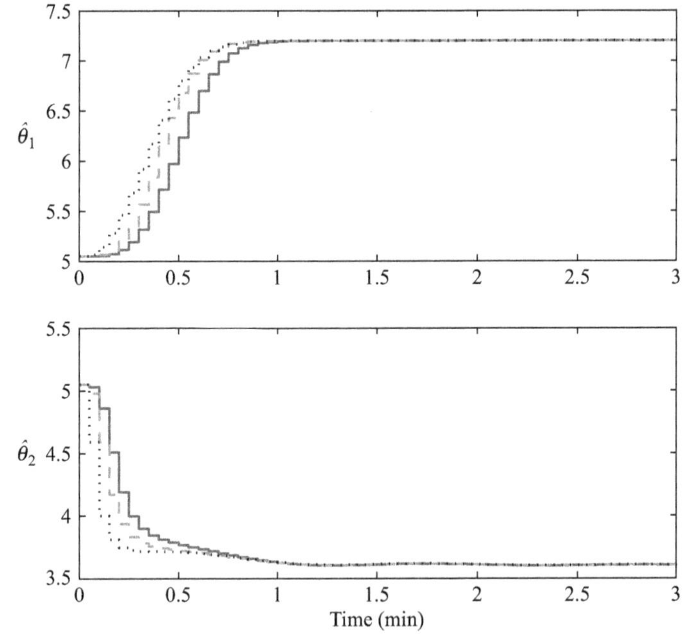

Figure 11.3 Closed-loop parameter estimates profile for states starting at different initial conditions ($C_A(0)$, $T_r(0)$): (0.3, 335) is solid line, (0.6, 335) is dashed line, and (0.3, 363) is the dotted line

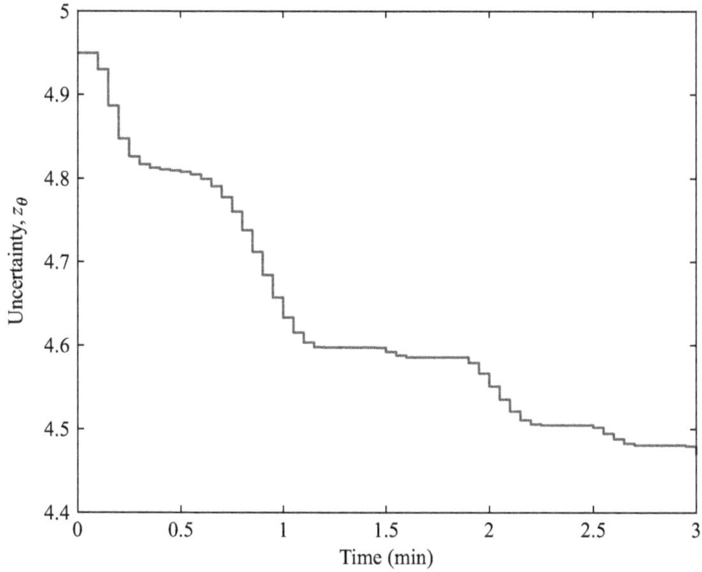

Figure 11.4 *Closed-loop uncertainty bound trajectories for initial condition $(C_A,\ T_r) = (0.3,\ 335)$*

Chapter 12

Robust adaptive economic MPC

In this chapter, we propose the design of economic MPC systems based on a single-step approach of the adaptive MPC technique proposed for a class of uncertain nonlinear systems subject to parametric uncertainties and exogenous variables. The framework considered assumes that the economic function is a known function of constrained system's states, parameterized by unknown parameters. The objective and constraint functions may explicitly depend on time, which means that our proposed method is applicable to both dynamic and steady-state economic optimization. A simulation example is used to demonstrate the effectiveness of the design technique.

12.1 Introduction

One of the key challenges in the process industry is how to best operate plants in light of varying processing and economic conditions. If one focusses on economic considerations, it may be extremely difficult to know, *a priori*, what the optimal operating conditions may be. Several technologies have been developed to address this problem. The leading and most popular approach remains RTO [18, 102], which refers to the online economic optimization of a process plant. RTO attempts to optimize process performance (usually measured in terms of profit or operating cost) thereby enabling companies to push the profitability of their processes to their true potential as operating conditions change. RTO systems are usually designed to solve steady-state optimization problems. It is therefore generally assumed that the process dynamics can be neglected if the optimization execution time interval is long enough to allow the process to reach and maintain steady state.

The integration of RTO and control for an optimal plant operation is an active area of research. RTO is in the family of adaptive extremum-seeking control techniques. Many techniques have been developed in the literature to address the regulation of processes to optimal (unknown) setpoints. The main challenge has been to address the combined task of steady-state optimization and transient performance. One approach proposed in Reference 75 is to use the profit (or cost) function to construct a Lyapunov function. Since the profit function is not generally measured and may depend on unknown model parameters, an adaptive control approach is usually required to ensure that the control system can reach the true unknown optimum without any bias associated with model and parametric uncertainties. In the context of MPC, the problem has been treated in Reference 8 where the integration of RTO and MPC is considered. The technique in Reference 8 is particularly well suited as it provides a robust adaptive integrated control approach that can effectively deal with uncertainties and economic objectives. It is based on the robust adaptive MPC techniques proposed in Reference 9 and provides robustness to parametric uncertainties. The approach of Reference 8 proposes integration of a nonlinear adaptive model predictive controller and an adaptive RTO approach. The adaptive real-dynamic optimization routine provides the best reference trajectory, given estimates of the parameters, that meets process constraints and asymptotically converges to the unknown optimal setpoint. Under the assumption that there exists an input trajectory that can steer the system to this optimal steady-state point, an adaptive MPC design technique is proposed that can robustly stabilize the nonlinear system about the unknown setpoint.

Another approach that has been proposed in the literature is a framework for integration of RTO and MPC called economic MPC. The leading approach proposed in [13]. In contrast to the two step approach proposed in [8], the economic MPC approach proposes to use the economic objective cost in the stage cost of the MPC. Assuming that the cost function is exactly known, the resulting MPC can be shown to asymptotically convergence to a neighborhood of the best feasible operating conditions for the control system. Questions associated with transient performance and stability of the control system remain open. Stability results are limited to linear systems subject to convex cost and convex constraints. Recent studies [14, 53] provide a summary of results for economic MPC for nonlinear systems.

In this chapter, we propose the design of economic MPC systems based on a single-step approach of the adaptive MPC technique proposed for a class of uncertain nonlinear systems subject to parametric uncertainties and exogenous variables. The framework considered assumes that the economic function is a known function of constrained system's states, parameterized by unknown parameters. The objective and constraint functions may explicitly depend on time, which means that our proposed method is applicable to both dynamic and steady-state economic optimization. The control objective is to simultaneously identify and regulate the system to the operating point that optimizes the economic function. The control input and state trajectories of the closed-loop system may also be required to satisfy some constraints. The approach proposed in this paper generalizes the results in Reference 8 by incorporating the cost information directly in the MPC stage cost. The cost information is integrated in a very specific way that allows one to address stability and robustness issues for the class of economic MPC control systems. The approach is based on a set-based robust adaptive estimation method that provides estimates of the unknown parameters along with an uncertain region guaranteed to contain the true value. The main result states that the economic MPC technique robustly stabilizes a neighborhood of the minimizer of the cost function.

12.2 Problem description

Consider a constrained optimization problem of the form

$$\min_{x\in\mathbb{R}^n} p(x,\theta) \tag{12.1a}$$

$$\text{s.t. } c_j(x) \leq 0 \quad j = 1\ldots n_c \tag{12.1b}$$

with θ representing unknown parameters, assumed to be uniquely identifiable and lie within an initially known convex set $\Theta^0 \triangleq B(\theta^0, z_\theta^0)$ where $B(\theta^0, z_\theta^0)$ denotes the ball of radius z_0^θ centered at θ^0. The functions p and c_j are assumed to be C^2 in all of their arguments (with locally Lipschitz second derivatives). The constraint $c_j \leq 0$ must be satisfied along the system's state trajectory $x(t)$.

In contrast to existing economic MPC techniques, it is assumed that the optimum solution depends on unknown parameters, θ. As a result, the cost function $p(x,\theta)$ is not known exactly and its value must be inferred from plant measurements. This is in line with standard RTO where plant data must be used to update unknown optimal operating conditions.

Assumption 12.2.1. *The following assumptions are made about* (12.1).

1. *There exists* $\varepsilon_0 > 0$ *such that* $\frac{\partial^2 p}{\partial x \partial x^T} \geq \varepsilon_0 I$ *and* $\frac{\partial^2 c}{\partial x \partial x^T} \geq 0$ *for all* $(x,\theta) \in (\mathbb{R}^n \times \Theta^\epsilon)$, *where* Θ^ϵ *is an* ϵ *neighborhood of* Θ.
2. *The feasible set*

$$\mathbb{X} = \{x \in \mathbb{R}^{n_x} \mid \max_j c_j(x) \leq 0\},$$

has a nonempty interior. ■

Assumption 12.2.1 states that the cost surface is strictly convex in x and $\mathbb{X}$ is a non-empty convex set. Standard nonlinear optimization results guarantee the existence of a unique minimizer $x^*(x,\theta) \in \mathbb{X}$ to problem 12.1. In the case of non-convex cost surface, only local attraction to an extremum can be guaranteed.

Consider the uncertain nonlinear system

$$\dot{x} = f(x) + \sum_{i=1}^{m} g_i(x)u_i + \sum_{j=1}^{p} q_j(x)\theta_i + \vartheta \triangleq \mathcal{F}(x,u,\theta,\vartheta) \tag{12.2}$$

where $x \in \mathbb{R}^n$ are the state variables, $u \in \mathbb{R}^m$ are the input variables, $\theta \in \mathbb{R}^p$ are unknown system parameters, and $\vartheta \in \mathbb{R}^n$ is a vector of unstructured exogenous disturbances. It is assumed that all vector fields $f(x)$, $g_i(x)$ $(i = 1,\ldots,m)$ and $q_j(x)$ $(j = 1,\ldots,p)$ are smooth vector valued functions. It is assumed that θ is uniquely identifiable and lie within an initially known compact set $\Theta^0 = B(\theta_0, z_\theta)$ where θ_0 is a nominal parameter value, z_θ is the radius of the parameter uncertainty set. The disturbance $\vartheta \in \mathcal{D} \subset \mathbb{R}^{n_d}$ is assumed to satisfy a known upper bound $\|\vartheta(t)\| \leq M_\vartheta < \infty$.

Remark 12.2.2. *In this study, the exogenous variable ϑ represents an unstructured bounded time-varying uncertainty.*

The control objective is to stabilize the nonlinear system (12.2) to the optimum operating point or trajectory given by the solution of (12.1) while obeying the input constraint $u \in \mathcal{U} \in \mathbb{R}^m$ in addition to the state constraint $x \in \mathcal{X} \in \mathbb{R}^n$.

In this next section, we present the first element of the proposed design method. It consists of a set-based adaptive parameter estimation technique that can be used to estimate the unknown parameters while simultaneously monitoring the size of the parameter uncertainty set.

12.3 Set-based parameter estimation routine

The set-based parameter estimation technique proposed requires two adaptive estimation mechanisms to estimate the uncertainty set $B(\theta_0, z_0^\theta)$. A suitable adaptive parameter estimation technique is first proposed to estimate the center of the parameter uncertainty set, θ_0. A update algorithm is then proposed to estimate the radius of the uncertainty ball, z_0^θ.

12.3.1 Adaptive parameter estimation

Using a standard estimation technique, we first consider the state prediction:

$$\dot{\hat{x}} = f(x) + g(x)u + q(x)\hat{\theta} + k_w(x - \hat{x}) + w\dot{\hat{\theta}}, \tag{12.3}$$

where x and u are the state and input variables from the system (12.2), $\hat{\theta}$ are the parameter estimates, $g(x) = [g_1(x), \ldots, g_m(x)]$ and $q(x) = [q_1(x), \ldots, q_p(x)]$, $k_w > 0$ are a strictly positive constant gain to be assigned, and w is a n by p matrix solution of the matrix differential equation,

$$\dot{w} = q(x) - k_w w \tag{12.4}$$

with initial condition $w(0) = 0_{n\times p}$.

Let us define the state prediction error $e = x - \hat{x}$ and the parameter estimation error $\dot{\hat{\theta}} = \theta - \tilde{\theta}$. For the specific choice of the state prediction dynamics (12.3), the error dynamics are given by:

$$\dot{e} = q(x)\tilde{\theta} - k_w e - w\dot{\hat{\theta}} + \vartheta. \tag{12.5}$$

We then define the vector of auxiliary variable $\eta = e - w\tilde{\theta}$. The η dynamics are as follows:

$$\dot{\eta} = -k_w\eta + \vartheta. \tag{12.6}$$

By definition, the initial conditions of η are $\eta(0) = e(0)$.

In the proposed estimation technique, an estimate of η, $\hat{\eta}$, is required to filter the impact of the uncertainties ϑ on η. The $\hat{\eta}$ dynamics are chosen as follows:

$$\dot{\hat{\eta}} = -k_w\hat{\eta} \tag{12.7}$$

with initial condition $\hat{\eta}(0) = e(0)$. As a result, the η estimation error $\tilde{\eta} = \eta - \hat{\eta}$ has dynamics of the form:

$$\dot{\tilde{\eta}} = -k_w \tilde{\eta} + \vartheta \tag{12.8}$$

with $\tilde{\eta}(0) = 0$.

Next, we present the proposed parameter estimation update. One key element to monitor the parametric uncertainty is the definition of the matrix $\Sigma \in \mathbb{R}^{p \times p}$. The matrix Σ is defined as the solution of the matrix differential equation:

$$\dot{\Sigma} = w^T w, \qquad \Sigma(0) = \alpha I \succ 0, \tag{12.9}$$

where $\alpha > 0$ is a positive constant to be assigned. It follows that the inverse of Σ, Σ^{-1} is defined as the solution of the matrix differential equation:

$$\dot{\Sigma}^{-1} = -\Sigma^{-1} w^T w \Sigma^{-1}, \quad \Sigma^{-1}(0) = \frac{1}{\alpha} I. \tag{12.10}$$

Finally, the preferred parameter update law is defined as:

$$\dot{\hat{\theta}} = \text{Proj}\left\{\Sigma^{-1} w^T (e - \hat{\eta}), \hat{\theta}\right\}, \quad \hat{\theta}(0) = \theta^0. \tag{12.11}$$

where $\theta^0 \in \Theta^0$ is the center of the initial parameter uncertainty set. The notation $\text{Proj}\{\phi, \ \hat{\theta}\}$ denotes a Lipschitz projection operator. This operator is defined such that

$$-\text{Proj}\{\phi, \ \hat{\theta}\}^T \tilde{\theta} \le -\phi^T \tilde{\theta}, \tag{12.12}$$

$$\hat{\theta}(t_0) \in \Theta^0 \Rightarrow \hat{\theta}(t) \in \Theta^0_\epsilon, \quad \forall t \ge t_0. \tag{12.13}$$

where $\Theta^0_\epsilon \triangleq B(\theta^0, z^0_\theta + \epsilon), \epsilon > 0$. More details on the choice of parameter projection operators and their properties can be found in Reference 98. The main property of the proposed estimation technique can be summarized in the following lemma due to Reference 99.

Lemma 12.3.1. *The identifier* (12.10), (12.11) *is such that the estimation error* $\tilde{\theta} = \theta - \hat{\theta}$ *is bounded. Moreover, if*

$$\vartheta \in \mathcal{L}_2 \qquad or \qquad \int_{t_0}^{\infty} \left[\|\tilde{\eta}\|^2 - \|e - \hat{\eta}\|^2\right] d\tau < +\infty \tag{12.14}$$

with the strong condition

$$\lim_{t \to \infty} \lambda_{\min}(\Sigma) = \infty \tag{12.15}$$

is satisfied, then $\tilde{\theta}$ *converges to zero asymptotically.*

12.3.2 Set adaptation

An update law that measures the worst-case progress of the parameter identifier in the presence of disturbance is given as follows. We let z_θ represent the current estimate of the radius of the uncertainty. The update law is given by:

$$z_\theta = \sqrt{\frac{V_{z\theta}(t)}{\lambda_{\min}(\Sigma)}} \tag{12.16a}$$

$$V_{z\theta}(t_0) = \lambda_{\max}\left(\Sigma(t_0)\right)(z_\theta^0)^2 \tag{12.16b}$$

$$\dot{V}_{z\theta} = -(e-\hat{\eta})^T(e-\hat{\eta}) + \left(\frac{M_\vartheta}{k_w}\right)^2 \tag{12.16c}$$

where $V_{z\theta}(t)$ represents the solution of the ordinary differential equation (12.16c) with initial condition (12.16b). The function V_{z_θ} is used as an upper bound of the rate of change of the function $V_{\tilde{\theta}} = \frac{1}{2}\tilde{\theta}^T\Sigma\tilde{\theta}$. The purpose of the update law (12.16a) is to estimate the radius of the uncertainty. The next step yields to provide a mechanism to update uncertainty set with desirable invariance properties.

Using the parameter estimator (12.11) and its error bound z_θ (12.16a), the uncertain ball $\Theta \triangleq B(\hat{\theta}, z_\theta)$ is adapted online according to the following algorithm.

Algorithm 12.3.2. *The parameter and set adaptation is implemented iteratively as follows:*

1. ***Initialize*** $z_\theta(0) = z_\theta^0$, $\hat{\theta}(0) = \hat{\theta}^0$ *and* $\Theta(0) = B(\hat{\theta}(0), z_\theta(0))$.
2. *At time* $t_{i-1} \le t \le t_i$, *using* (12.11) *and* (12.16a) ***perform*** *the update*

$$\left(t, \hat{\theta}(t), \Theta(t)\right) = \begin{cases} \left(t_i, \hat{\theta}(t_i), \Theta(t_i)\right), & \text{if } z_\theta(t_i) \le z_\theta(t_{i-1}) - \|\hat{\theta}(t_i) - \hat{\theta}(t_{i-1})\| \\ \left(t_{i-1}, \hat{\theta}(t_{i-1}), \Theta(t_{i-1})\right), & \text{otherwise} \end{cases} \tag{12.17}$$

3. ***Iterate back*** *to Step* 2, ***incrementing***, $i = i+1$, *if* $t = t_i$.

The algorithm ensure, that Θ is only updated when z_θ value has decreased by an amount which guarantees a contraction of the set. Moreover z_θ evolution as given in (12.16a) ensures non-exclusion of θ as shown below.

Lemma 12.3.3. [10] *The evolution of* $\Theta = B(\hat{\theta}, z_\theta)$ *under* (12.11), (12.16a) *and Algorithm* 12.3.2 *is such that*

i) $\Theta(t_2) \subseteq \Theta(t_1), \quad 0 \le t_1 \le t_2$
ii) $\theta \in \Theta(0) \Rightarrow \theta \in \Theta(t), \quad \forall t \ge 0$

Lemma 12.3.3 establishes the key properties of the update algorithm. The algorithm guarantees that the uncertainty set estimate always contains the true value of the parameters. It also ensures that each set $\Theta(t_i)$ is always contained in the previous uncertainty set $\Theta(t_{i-1})$. It provides an effective mechanism for the uncertainty in the parameter estimate that reflects the information content of the dynamical system trajectories. The next section proposes an adaptive MPC algorithm that utilizes the set-based estimation technique to achieve the extremum-seeking objective.

12.4 Robust adaptive economic MPC implementation

In this section, we propose a design technique to achieve the integrated RTO/MPC task using a one-step approach. As reported in Reference 13, RTO objectives can be integrated by incorporating the cost function directly in the stage cost for the MPC. The main disadvantage of this technique is that the problem of RTO is transformed artificially to a dynamic optimization problem.

12.4.1 Alternative stage cost in economic MPC

One alternative to economic MPC is to consider stage cost that is associated with the best possible transient performance achievable for a gradient system. Consider the cost to be minimized $y = p(x)$ and assume that the closed-loop system is such that:

$$\dot{x} = -\frac{\partial p}{\partial x}$$

Then the rate of change of the cost is given by:

$$\dot{y} = -\left\| \frac{\partial p}{\partial x} \right\|^2.$$

Note that if the hessian of $p(x)$ is such that

$$\frac{\partial^2 p(x)}{\partial x \partial x^T} \geq \alpha I_n, \quad \forall x \in \mathbb{X}$$

then the closed-loop system would converge exponentially to the local minimum of $p(x)$. This simple observation would suggest that a suitable stage cost to the combined problem would be:

$$L(x, u) = \left\| \frac{\partial p}{\partial x} \right\|^2$$

The ultimate objective in the integration of RTO and MPC is to achieve a closed-loop system that behaves like a gradient system with respect to the cost function $p(x)$.

In practice, one must also contend with the presence of constraints that must be enforced by the control system. For the solution of the constrained problem (12.1), we propose the application of an interior point method. To achieve this, we consider the modified cost:

$$p_m(x, \mu) = p(x) - \mu \sum_{i=1}^{n_c} \lambda_i \psi(-\mu^{-1} c_i(x) + 1) \tag{12.18}$$

where $\mu > 0$ is a positive constant and λ_i acts as Lagrange multipliers. The function $\psi(\cdot)$ is a barrier function. Standard barrier function candidates include logarithmic barrier function, $\psi(c) = \ln(c)$, or the inverse function $\psi(c) = \frac{1}{c}$. In this study, we primarily focus on logarithm barrier functions.

If $p(x)$ and all constraints $c_i(x)$ are convex, it follows by standard arguments that the modified cost is also convex. Let $x^*(\mu)$ denote unique point such that:

$$\nabla_x p_m(x^*(\mu),\mu) = \nabla_x p(x^*(\mu)) + \sum_{i=1}^{n} \frac{1}{-\mu^{-1}c_i(x^*(\mu)) + 1}\nabla_x c_i(x^*(\mu)) = 0$$

The unconstrained minimization of (12.18) provides an $\mathcal{O}(\mu)$ approximation of the optimum x^*. Thus, as μ decreases, the approximate solution $x^*(\mu)$ of problem (12.18) approaches the optimal solution x^* of (12.1).

In order to avoid numerical problems associated with barrier functions, a standard approach is to approximate the barrier by a quadratic function close to the boundary of the feasible region. Thus for each constraint $c_i(x)$ such that $c_i(x) \geq \epsilon$, we use the quadratic approximation:

$$\psi(c_i) = a_1 + b_1(c_i(x) - \epsilon) + \frac{1}{2}q_1(c_i(x) - \epsilon)^2$$

where $a_1 = \psi(\epsilon)$, $b_1 = \psi'(\epsilon)$, and $q_1 = \psi''(\epsilon)$, ψ' is first derivative of ψ with respect to its argument, $c_i(x)$.

As a simple demonstration, we consider the following simple example.

Consider the quadratic optimization problem:

$$\min_x [-2, 0]x + \frac{1}{2}x^T \begin{bmatrix} 1 & 0 \\ 0 & \frac{1}{10} \end{bmatrix} x$$

subj. to.

$$\begin{aligned} &x_1 \geq 0, \quad x_2 \geq 0 \\ &1 - x_1 - x_2 \geq 0 \end{aligned}$$

The optimal solution occurs at $x^* = [1,\ 0]^T$. We fix $\mu = 0.01$, $\epsilon = 10^{-4}$ and consider the modified cost in the gradient descent formula:

$$\dot{x} = -k\nabla_x p_m(x,\mu). \tag{12.19}$$

where $k = 10$.

Figure 12.1 shows the corresponding trajectory system (12.19) for the choice of tuning parameters, with initial conditions $x(0) = [0.5, 0.5]^T$. Figure 12.2 shows the performance of the same update formula starting from the infeasible initial condition $x(0) = [2, -2]$, the use of the quadratic approximation described above allows one to enter the feasible region in order to eventually converge to the correct constrained optimum. The barrier function approach provides a very effective mechanism to encode constraints in RTO approaches. The design of barrier functions in the formulation of model predictive controllers has been investigated in References 50 and 161. It has also been considered in the solution of extremum-seeking control problems [48]. Within the context of the current study, the barrier function approach is used to address the design of integrated RTO/MPC systems. The integration is handled by posing the MPC using the stage cost

$$L(x,u) = \left\| \frac{\partial p_m(x,\mu)}{\partial x} \right\|^2. \tag{12.20}$$

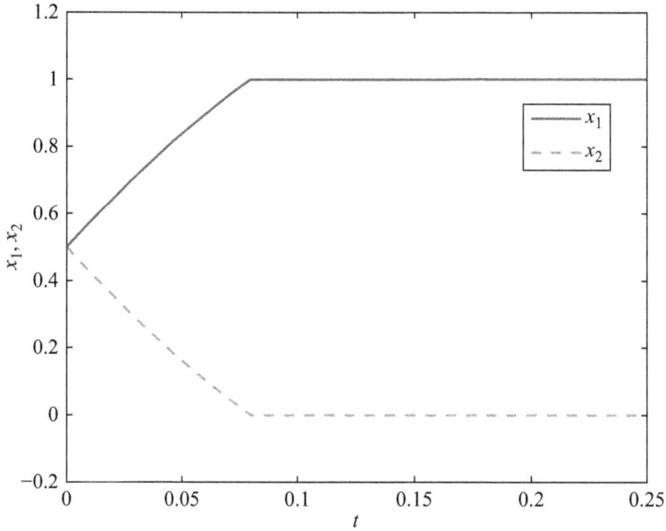

Figure 12.1 Trajectories of the gradient descent for the barrier function method (12.19) with initial condition x(0) = [0.5, −0.5]

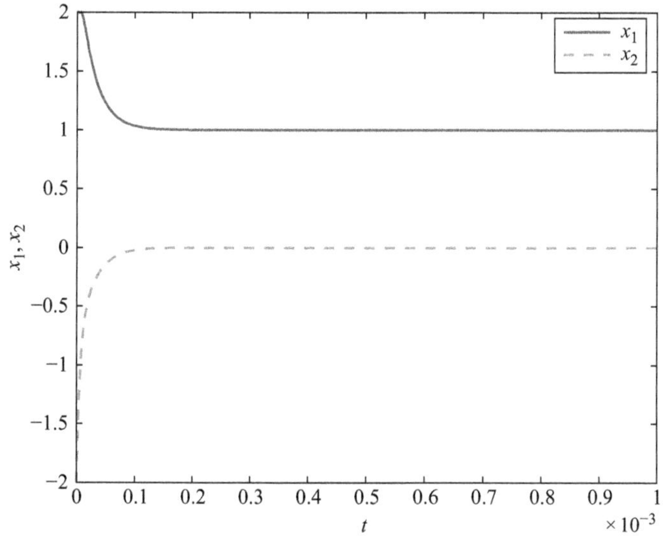

Figure 12.2 Trajectories of the gradient descent for the barrier function method 12.19 with initial condition x(0) = [2, −2]

Throughout this chapter, it is understood that the MPC system using the modified stage cost (12.20) is operated for a constant value of the parameter μ leading to an $\mathcal{O}(\mu)$ approximation of the optimum x^*. In what follows, we propose two formulations of the RTO/MPC system that can handle systems subject to parametric uncertainties and disturbances.

12.4.2 A min–max approach

Most robust-MPC techniques consider some form of min–max approach to handle the effect of uncertainties. When one considers adaptive MPC, one must also consider the effect of parametric uncertainties in addition to the impact of bounded model uncertainties such as exogenous disturbances. In this section, we propose a minmax MPC approach that implements a one-step RTO procedure. The approach guarantees robust performance of the closed-loop to a neighborhood of the optimum x^* of the optimization problem (12.1).

The formulation of the min–max MPC consists of maximizing a cost function with respect to $\theta \in \Theta$, $\vartheta \in \mathcal{D}$ and minimizing over feedback control policies κ. The feedback control policy is assumed to be a time-varying function of the state variables, and the parameter estimates can be viewed as a robust adaptive feedback. To account for the uncertainty in the parameter estimates, the robust-MPC formulation takes into account the predicted change in the parameter estimates subject to the predicted state trajectories. This allows one to incorporate the impact of the state trajectories on the uncertainty reduction in the parameter estimates. One monitors the impact of the parametric uncertainties by using a form of adaptive CLF as a terminal cost. Doing so, one is able to assign a cost to the parametric uncertainty that can be adjusted by the MPC without the need for external perturbations or dither signals. Thus, we treat both parametric uncertainties and disturbances but only the impact of parametric uncertainties can be accounted for in the formulation since no disturbance model is assumed.

The robust receding horizon control law is given as follows:

$$u = \kappa_{mpc}(x, \hat{\theta}, z_\theta) \triangleq \kappa^*(0, x, \hat{\theta}, z_\theta) \tag{12.21a}$$

$$\kappa^* \triangleq \arg \min_{\kappa(\cdot,\cdot,\cdot)} J(x, \hat{\theta}, z_\theta, \kappa) \tag{12.21b}$$

where

$$J(x, \hat{\theta}, z_\theta, \kappa) \triangleq \max_{\theta\in\Theta, \vartheta\in\mathcal{D}} \int_0^T L(\|\gamma^p\|, u^p)\, d\tau + W(x^p(T), \tilde{\theta}^p(T)) \tag{12.22a}$$

$$\text{s.t. } \forall\ \tau \in [0,\ T]$$

$$\dot{x}^p = f(x^p) + g(x^p)u^p + q(x^p)\theta + \vartheta, \quad x^p(0) = x \tag{12.22b}$$

$$\gamma^p = \frac{\partial p(x^p, \hat{\theta}^p)}{\partial x} \tag{12.22c}$$

$$\dot{\hat{x}}^p = f(x^p) + g(x^p)u^p + q(x^p)\hat{\theta}^p + k_w(x^p - \hat{x}^p) + w\dot{\hat{\theta}}, \quad \hat{x}^p(0) = \hat{x}$$

$$\dot{\hat{\eta}}^p = -k_w \hat{\eta}^p, \quad \hat{\eta}^p(0) = \hat{\eta} \tag{12.22d}$$

$$\dot{w}^p = g^T(x^p, u^p) - k_w w^p, \quad w^p(0) = w \tag{12.22e}$$

$$(\dot{\Sigma}^{-1})^p = -(\Sigma^{-1})^p (w^p)^T (w^p)(\Sigma^{-1})^p, \quad (\Sigma^{-1})^p(0) = \Sigma^{-1} \tag{12.22f}$$

$$\dot{\hat{\theta}}^p = \text{Proj}\Big\{(\Sigma^{-1})^p w^T (x^p - \hat{x}^p - \hat{\eta}^p)\hat{\theta}\Big\},$$

$$\tilde{\theta}^p = \theta - \hat{\theta}^p, \quad \hat{\theta}^p(0) = \hat{\theta} \tag{12.22g}$$

$$u^p(\tau) \triangleq \kappa(\tau, x^p(\tau), \hat{\theta}^p(\tau)) \in \mathbb{U} \tag{12.22h}$$

$$x^p(\tau) \in \mathbb{X}, \quad x^p(T) \in \mathbb{X}_f(\tilde{\theta}^p(T)) \tag{12.22i}$$

The effect of future parameter adaptation is also accounted for in this formulation. Constraints (12.22d)–(12.22h) are used to model the effect of the parameter estimation routine. The conservativeness of the algorithm is reduced by parameterizing the terminal cost W and the terminal set $\mathbb{X}_f$ as functions of the predicted parameter estimation error $\tilde{\theta}(T)$. The maximization step utilizes the current value of the parametric uncertainty set $\Theta(t)$. As a result, the proposed MPC will eventually reduce to a robust MPC as the set Θ will contract around the true unknown parameter vector θ over time. The robustness feature due to $\vartheta \in \mathcal{D}$ is retained. The stage cost $L(\|\gamma^p\|, u_p)$ is assumed to be a function of the predicted gradient γ^p, given by (12.22d), of the modified RTO cost as presented in the previous section. It is also a function of the input energy, if required. The design criterion for this MPC formulation is given in the following. The state constraints $\mathcal{X}$ encodes the constraints of the system. If the set $\mathcal{X}$ is identified with the feasible set of the problem (12.1) then one can rely on the barrier function entering the gradient γ^p in the stage cost to ensure feasiblity of $\mathcal{X}$.

We first consider the implementation algorithm. A standard receding horizon implementation is assumed in which only the current value of the control is fed back to the process.

Algorithm 12.4.1. *The MPC algorithm performs as follows: At sampling instant* t_i

1. ***Measure*** *the current state of the plant* $x(t)$ *and obtain the current value of matrices* w *and* Σ^{-1} *from* (12.4) *and* (12.10), *respectively.*
2. ***Obtain*** *the current value of parameter estimates* $\hat{\theta}$ *and uncertainty bound* z_θ *from* (12.11) *and* (12.16a), *respectively. Update the uncertainty sets following* (12.17).
3. ***Solve*** *the optimization problem* (12.21) *and apply the resulting feedback control law to the plant until the next sampling instant.*
4. ***Increment*** $i = i + 1$. *Repeat the procedure from Step* 1 *for the next sampling instant.*

The following criteria are required to guarantee robust stability of the unknown approximate optimal equilibrium $x^*(\mu)$ of the closed-loop MPC system. We let $\tilde{\Theta}^0_\epsilon$ denote an ϵ inner approximation of the set of parameter estimation error $\tilde{\theta}(= \theta - \hat{\theta})$. This set can be approximated by a ball of radius $2z^\theta$ centered at the origin. The set Ξ represents the target set which is a forward invariant set containing the unknown optimum x^*. The unknown optimum is written as $x^*(\theta)$ to indicate its dependence on the unknown parameters.

Criterion 12.4.2. *The terminal penalty function* $W : \mathbb{X}_f \times \tilde{\Theta}^0 \rightarrow [0, +\infty]$ *and the terminal constraint function* $\mathbb{X}_f : \tilde{\Theta}^0 \rightarrow \mathbb{X}$ *are such that for each* $(\theta, \hat{\theta}, \tilde{\theta}) \in (\Theta^0 \times \Theta^0 \times \tilde{\Theta}^0_\epsilon)$, *there exists a feedback* $k_f(\cdot, \hat{\theta}) : \mathbb{X}_f \rightarrow \mathbb{U}$ *satisfying*

1. $x^*(\theta) \in \mathbb{X}_f(\tilde{\theta}) \subseteq \mathbb{X}$, $\mathbb{X}_f(\tilde{\theta})$ *closed*
2. $k_f(x, \hat{\theta}) \in \mathbb{U}$, $\forall x \in \mathbb{X}_f(\tilde{\theta})$

3. $W(x, \tilde{\theta})$ *is continuous with respect to* $x \in \mathbb{R}^{n_x}$
4. $\forall x \in \mathbb{X}_f(\tilde{\theta})\backslash\Xi$, $\mathbb{X}_f(\tilde{\theta})$ *is strongly positively invariant under* $k_f(x, \hat{\theta})$ *with respect to* $\dot{x} \in f(x) + g(x)k_f(x, \hat{\theta})) + q(x)\Theta + \mathcal{D}$
5. $L\left(\gamma, k_f(x, \hat{\theta})\right) + \frac{\partial W}{\partial x}\mathcal{F}(x, k_f(x, \hat{\theta}), \theta, \vartheta) \leq 0,$
   $\forall x \in \mathbb{X}_f(\tilde{\theta})\backslash\Xi$, $\gamma = \left\| \frac{\partial p(x,\hat{\theta})}{\partial x} \right\|.$

We also require the following assumption concerning the impact of parametric uncertainty on the terminal cost and the terminal set.

Criterion 12.4.3. *For any* $\tilde{\theta}_1,\ \tilde{\theta}_2 \in \tilde{\Theta}^0$ *s.t.* $\|\tilde{\theta}_2\| \leq \|\tilde{\theta}_1\|$,

1. $W(x, \tilde{\theta}_2) \leq W(x, \tilde{\theta}_1), \quad \forall x \in \mathbb{X}_f(\tilde{\theta}_1)$
2. $\mathbb{X}_f(\tilde{\theta}_2) \supseteq \mathbb{X}_f(\tilde{\theta}_1)$

The target set Ξ has the same significance as in other standard robust-MPC approaches. In this case, $\Xi \in \mathbb{X}$ is a set containing the unknown optimal setpoint x^*. The use of the gradient of $p(x)$ in the stage cost allows one to force convergence of the closed-loop system in a neighborhood of an estimated optimal setpoint $x^*(\hat{\theta})$. Convergence to the set Ξ comes as a result of the set-based parameter identification routine proposed which guarantees convergence of the parameter estimates $\hat{\theta}$ to a neighborhood of the true value, θ. The choice of stage cost proposed attempts to combine the goals of MPC and RTO. By minimizing a measure of the estimated gradient of the objective function, one can guarantee that the MPC controller achieves the RTO while ensuring some degree of transient performance. The robust stabilization of the target set required a revised condition C5 stating that the function W is a local robust CLF for the uncertain system (12.2) with respect to $\theta \in \Theta$ and $\vartheta \in \mathcal{D}$.

The main challenge with this formulation is that the resulting optimal equilibrium to be stabilized cannot be known in advance. This particular property of the proposed MPC formulation requires some care in the definition of a suitable terminal cost and terminal set.

We state the robust stability of the closed-loop MPC system to the target set Ξ, that is, a set containing the unknown optimal setpoint.

12.4.3 Main result

We now state the main result of the paper.

Theorem 12.4.4. *Let* $X_{d0} \triangleq X_{d0}(\Theta^0) \subseteq \mathbb{X}$ *denote the set of initial states with uncertainty* Θ^0 *for which* (12.21) *has a solution. Assuming Criteria* 12.4.2 *and* 12.4.3 *are satisfied, then the closed-loop system state* x*, given by* (12.2, 12.11, 12.16a, 12.21)*, originating from any* $x_0 \in X_{d0}$*, feasibly approaches the target set* Ξ *as* $t \to +\infty$.

Proof:
Feasibility: The closed-loop stability is based upon the feasibility of the control action at each sample time. Assuming, at time t, that an optimal solution $u^p_{[0,T]}$ to the optimization problem (12.21) exists and is found (where $u^p_{[0,T]}$ denotes the feedback policy defined in (12.22i)). Let Θ^p denote the estimated uncertainty set at time t and Θ^v denote the set at time $t+\delta$ that would result with the feedback implementation of $u_{[t,t+\delta]} = u^p_{[0,\delta]}$. Also, let x^p represents the worst-case state trajectory originating from $x^p(0) = x(t)$ and x^v represents the trajectory originating from $x^v(0) = x + \delta v$ under the same feasible control input $u^v_{[\delta,T]} = u^p_{[\delta,T]}$. Moreover, let $X^a_{\Theta^b} \triangleq \{x^a \,|\, \dot{x}^a \in \mathcal{F}(x^a, u^p, \Theta^b, \mathcal{D}) \triangleq f(x^a) + g(x^a)u^p + q(x^a)\Theta^b + \mathcal{D}\}$.

Since the $u^p_{[0,T]}$ is optimal with respect to the worst-case uncertainty scenario, it follows hat $u^p_{[0,T]}$ steers any trajectory $x^p \in X^p_{\Theta^p}$ to the terminal region $\mathbb{X}^p_f$. Since Θ is guaranteed not to increase in size over time, it follows that $\Theta^v \subseteq \Theta^p$. This, in turn, implies that $x^v \in X^p_{\Theta^v} \subseteq X^p_{\Theta^p}$. Since the terminal region $\mathbb{X}^p_f$ is strongly positively invariant for the nonlinear system (12.2) under the feedback $k_f(\cdot,\cdot)$ and since the input constraints are satisfied in $\mathbb{X}^p_f$ and $\mathbb{X}^v_f \supseteq \mathbb{X}^p_f$ by Criteria 2.2, 2.4, and 3.2 respectively, one can conclude that the input $u = [u^p_{[\delta,T]}, k_{f[T,T+\delta]}]$ is a feasible solution of (12.21) at time $t+\delta$. By induction, it follows that the dynamic optimization problem is feasible for all $t \geq 0$.

Stability: The stability of the closed-loop system is established by proving strict decrease of the optimal cost $J^*(x, \hat{\theta}, z_\theta) \triangleq J(x, \hat{\theta}, z_\theta, \kappa^*)$. Let the trajectories $(x^p,\ \hat{\theta}^p,\ \tilde{\theta}^p, z^p_\theta)$ and control u^p correspond to any worst-case minimizing solution of $J^*(x, \hat{\theta}, z_\theta)$. Let

$$\gamma^p = \left\| \frac{\partial p(x^p, \hat{\theta}^p)}{\partial x} \right\|.$$

If $x^p_{[0,T]}$ were extended to $\tau \in [0, T+\delta]$ by implementing the feedback $u(\tau) = k_f(x^p(\tau), \hat{\theta}^p(\tau))$ on $\tau \in [T, T+\delta]$, then Criterion 12.4.2(5) guarantees the inequality

$$\int_T^{T+\delta} L(\gamma^p, k_f(x^p, \hat{\theta}^p))d\tau + W(x^p_{T+\delta}, \tilde{\theta}^p_T) - W(x^p_T, \tilde{\theta}^p_T) \leq 0 \tag{12.23}$$

where in (12.23) and in the remainder of the proof, $x^p_\sigma \triangleq x^p(\sigma)$, $\tilde{\theta}^p_\sigma \triangleq \tilde{\theta}^p(\sigma)$, for $\sigma = T,\ T+\delta$.

The optimal cost

$$\begin{aligned} J^*(x, \hat{\theta}, z_\theta) &= \int_0^T L(\gamma^p, u^p)d\tau + W(x^p_T, \tilde{\theta}^p_T) \\ &\geq \int_0^T L(\gamma^p, u^p)d\tau + W(x^p_T, \tilde{\theta}^p_T) + \int_T^{T+\delta} L(\gamma^p, k_f(x^p, \hat{\theta}^p))d\tau \\ &\quad + W(x^p_{T+\delta}, \tilde{\theta}^p_T) - W(x^p_T, \tilde{\theta}^p_T) \end{aligned} \tag{12.24}$$

$$\geq \int_0^\delta L(\gamma^p, u^p)d\tau + \int_\delta^T L(x^p, u^p)d\tau + \int_T^{T+\delta} L(\gamma^p, k_f(x^p, \hat{\theta}^p))d\tau$$

$$+ W(x^p_{T+\delta}, \tilde{\theta}^p_{T+\delta}) \tag{12.25}$$

$$\geq \int_0^\delta L(\gamma^p, u^p)d\tau + J^*(x(\delta), \hat{\theta}(\delta), z_\theta(\delta)) \tag{12.26}$$

Then, it follows from (12.25) that

$$J^*(x(\delta), \hat{\theta}(\delta), z_\theta(\delta)) - J^*(x, \hat{\theta}, z_\theta) \leq - \int_0^\delta L(\gamma^p, u^p)\, d\tau \tag{12.27}$$

Since the stage cost is assumed to be such that $L(0, u^p) = 0$, and locally convex with respect to the gradient of $p(x^p, \hat{\theta}^p)$, it follows that $x(t)$ converges to a neighborhood of $x^*(\hat{\theta})$ asymptotically where $x^*(\hat{\theta})$ is the critical value of $p(x, \hat{\theta})$.

The closed-loop stability is established by the feasibility of the control action at each sample time and the strict decrease of the optimal cost J^*. The proof follows from the fact that the control law is optimal with respect to the worst-case uncertainty $(\theta,\ \vartheta) \in (\Theta,\ \mathcal{D})$ scenario and the terminal region $\mathbb{X}^p_f$ is strongly positively invariant for (12.2) under the (local) feedback $k_f(\cdot, \cdot)$.

If the conditions of Lemma 12.3.1 are met, it follows that $z_\theta \to 0$ and therefore the closed-loop system reaches a neighborhood of the true unknown setpoint $x^\star$ subject to the worst-case disturbance $\vartheta \in \mathcal{D}$. ■

12.4.4 Lipschitz-based approach

The min–max approach presented above can constitute an insurmountable computational problem. As shown in Reference 9, it is possible to substitute the min–max approach with a simplified, but more conservative, approach termed the Lipschitz-based method. In this approach, the nominal model rather than the unknown bounded system state is controlled, subject to conditions that ensure robust feasibility of given constraints. To this end, the uncertain state prediction error bound is approximated using Lipschitz bounds.

It is assumed that appropriate Lipschitz bounds for the x-dependence of the dynamics $f(x)$, $g(x)$ and $q(x)$ are known. They are given as follows:

Assumption 12.4.5. *A set of functions $\mathcal{L}_j : \mathbb{X} \to \mathbb{R}^+, j \in \{f, g, q\}$ are known which satisfy*

$$\mathcal{L}_j(\mathbb{X}) \geq \min\left\{\mathcal{L}_j \;\middle|\; \sup_{x_1, x_2 \in \mathbb{X}} \left(\|j(x_1) - j(x_2)\| - \mathcal{L}_j\|x_1 - x_2\|\right) \leq 0\right\}, \tag{12.28}$$

where for $j \equiv g$ and $j \equiv q$ is interpreted as an induced norm since $g(x)$ and $q(x)$ are matrices. ■

Assuming a knowledge of the Lipschitz bounds for the x-dependence of the dynamics $f(x)$, $g(x)$, and $q(x)$ as given in Assumption 12.4.5 and let $\Pi = z_\theta + \|\hat{\theta}\|$, a worst-case deviation $z_x^p \geq \max_{\theta \in \Theta} \|x - x^p\|$ can be generated from

$$\dot{z}_x^p = (\mathcal{L}_f + \mathcal{L}_g\|u\| + \mathcal{L}_q\Pi)z_x^p + \|q(x^p)\|z_\theta + M_\vartheta, \quad z_x^p(t_0) = 0. \tag{12.29}$$

Using this error bound, the robust Lipschitz-based MPC is given by

$$u = \kappa_{mpc}(x, \hat{\theta}, z_\theta) = u^*(0) \tag{12.30a}$$

$$u^*(\cdot) \triangleq \arg\min_{u^p_{[0,T]}} J(x, \hat{\theta}, z_\theta, u^p) \tag{12.30b}$$

where

$$J(x, \hat{\theta}, z_\theta, u^p) = \int_0^T L(\gamma^p, u^p)d\tau + W(x^p(T), z_\theta) \tag{12.31a}$$

$$\text{s.t. } \forall\ \tau \in [0,\ T]$$

$$\dot{x}^p = f(x^p, u^p) + g(x^p, u^p)\hat{\theta}, \quad x^p(0) = x \tag{12.31b}$$

$$\dot{z}_x^p = (\mathcal{L}_f + \mathcal{L}_g\Pi)z_x^p + \|g^p\|z_\theta + M_\vartheta, \quad z_x^p(0) = 0 \tag{12.31c}$$

$$X^p(\tau) \triangleq B(x^p(\tau), z_x^p(\tau)) \subseteq \mathbb{X}, \quad u^p(\tau) \in \mathbb{U} \tag{12.31d}$$

$$X^p(T) \subseteq \mathbb{X}_f(z_\theta^p) \tag{12.31e}$$

The effect of the additive disturbances on the predicted trajectories takes the form of the uncertainty cone $B(x^p(\tau), z_x^p(\tau))$ computed using the bound (12.31c). The uncertainty radius z_θ which appears in (12.31c) and in the terminal expressions of (12.31a) and (12.31e) are held throughout the prediction horizon. However, the update algorithm for the bounds z_θ provides a mechanism to reduce the conservatism of the robust MPC. The shrinking of this uncertainty enlarges the terminal region.

Algorithm 12.4.6. *The Lipschitz-based MPC algorithm performs as follows: At sampling instant* t_i

1. ***Measure** the current state of the plant* $x = x(t_i)$
2. ***Obtain** the current value of the parameter estimates* $\hat{\theta}$ *and uncertainty bound* z_θ *from* (12.11) *and* (12.16a), *respectively,*

If $z_\theta(t_i) \leq z_\theta(t_{i-1})$

$$\hat{\theta} = \hat{\theta}(t_i), \quad z_\theta = z_\theta(t_i)$$

Else

$$\hat{\theta} = \hat{\theta}(t_{i-1}), \quad z_\theta = z_\theta(t_{i-1})$$

End

3. ***Solve*** *the optimization problem* (12.30) *and apply the resulting feedback control law to the plant until the next sampling instant*
4. ***Increment*** *i*: $= i + 1$; *repeat the procedure from Step* 1 *for the next sampling instant.*

Theorem 12.4.7. *Let* $X'_{d0} \triangleq X'_{d0}(\Theta^0) \subseteq \mathbb{X}$ *denote the set of initial states for which* (12.30) *has a solution. Assuming Assumption* 12.4.5 *and Criteria* 12.4.2 *and* 12.4.3 *are satisfied, then the local minimizer of the cost function* $y = p(x, \theta)$ *is feasibly asymptotically stabilized from any* $x_0 \in X'_{d0}$ *to the target set* Ξ *for the closed-loop system given by* (12.2, 12.11, 12.16a, 12.30). ∎

The proof of the Lipschitz-based control law follows from that of Theorem 12.4.4.

12.5 Simulation example

Consider the parallel isothermal stirred-tank reactor in which reagent A forms product B and waste-product C (as presented in Reference 50). Let $x = [A_1,\ A_2]^T$, $\theta = [k_{11}, k_{12}, k_{21}, k_{22}]^T$ and $u = [F_1^{in},\ F_2^{in}]^T$ where A_i denote the concentration of chemical A in reactor i, k_{ij} are the reaction kinetic constants, which are only nominally known. The inlet flows F_i^{in} are the control inputs. The dynamics of the system can be expressed in the form:

$$\dot{x} = -\underbrace{\begin{bmatrix} \dfrac{x_1 k_{V1}(\xi_1 - V_1^0 + \xi_3)}{\xi_1} \\ \dfrac{x_2 k_{V2}(\xi_2 - V_2^0 + \xi_4)}{\xi_2} \end{bmatrix}}_{f} + \underbrace{\begin{bmatrix} \dfrac{A_{in}}{\xi_1} & 0 \\ 0 & \dfrac{A_{in}}{\xi_2} \end{bmatrix}}_{g} u - \underbrace{\begin{bmatrix} x_1 & 2x_1^2 & 0 & 0 \\ 0 & 0 & x_2 & 2x_2^2 \end{bmatrix}}_{q} \theta + \vartheta,$$

where ξ_1, ξ_2 are the two tank volumes and ξ_3, ξ_4 are the PI integrators. The system parameters are $V_1^0 = 0.9$, $V_2^0 = 1.5$, $k_{v_1} = k_{v_2} = 1$, $P_A = 5$, $P_B = 26$, $p_{11} = p_{21} = 3$ and $p_{12} = p_{22} = 1$. The disturbances are $\vartheta = [0.001\sin(0.1t),\ 0.001\sin(0.1t)]$.

The economic cost function is the net expense of operating the process at steady state.

$$p(A_i,\ s,\ \theta) = \sum_{i=1}^{2} [(p_{i1}s_i + P_A - P_B)k_{i1}A_iV_i^0 + (p_{i2}s_i + 2P_A)k_{i2}A_i^2V_i^0] \quad (12.32)$$

where P_A, P_B denote component prices, p_{ij} is the net operating cost of reaction j in reactor i. Disturbances s_1, s_2 reflect changes in the operating cost (utilities, etc.) of each reactor. The control objective is to robustly regulate the process to the optimal operating point that optimizes the economic cost (12.32) while satisfying the following state constraints $0 \leq A_i \leq 3$, $c_v = A_1^2V_1^0 + A_2^2V_2^0 - 15 \leq 0$ and input constraint $0.01 \leq F_i^{in} \leq 0.2$. The reaction kinetics are assumed to satisfy $0.01 \leq k_i \leq 0.2$. A logarithmic barrier function is first with parameters $\mu = 0.00001$ and $\epsilon = 10^{-7}$.

The sampling time is take to be 0.1 s. The robustness of the adaptive controller is guaranteed via the Lipschitz bound method. The stage cost is selected as a quadratic cost $L(\gamma, u) = \frac{1}{2}\sum_{i=1}^{2}\left(\frac{\partial p(A_i\, s, \hat{\theta})}{\partial A_i}\right)^2$.

12.5.1 Terminal penalty and terminal set design

A Lyapunov function for the terminal penalty is defined as the input to state stabilizing CLF (iss-clf):

$$W(x) = \frac{1}{2}\left\| \frac{\partial p(x, s, \hat{\theta})}{\partial x} \right\|^2 = \frac{1}{2}\|\gamma\|^2 \tag{12.33}$$

Let $\Gamma = \frac{\partial^2 p(x,s,\hat{\theta})}{\partial x \partial x^T}$ and $\Upsilon = \frac{\partial^2 p(x,s,\hat{\theta})}{\partial x \partial \hat{\theta}^T}$.

Choosing a terminal controller

$$u = k_f(x) = -g^{-1}\big(-f + q(x)\hat{\theta} + k_1\Gamma^{-1}\gamma + k_2 qq^T\,\Gamma\gamma + k_3\Gamma\gamma + \Upsilon\dot{\hat{\theta}}\big), \tag{12.34}$$

with design constants $k_1,\ k_2 > 0$, the time derivative of (12.33) becomes

$$\dot{W}(x) = -k_1\,\gamma^T\gamma - \gamma^T\Gamma g\tilde{\theta} - k_2\gamma^T\Gamma gg^T\Gamma\gamma - k_3\gamma^T\Gamma\Gamma\gamma \tag{12.35}$$

$$\leq -k_1\|\gamma\|^2 + \frac{1}{4k_2}\|\tilde{\theta}\|^2 + \frac{1}{4k_3}\|\vartheta\|^2 \tag{12.36}$$

Since the stability condition requires $\dot{W}(x(T)) + L(T) \leq 0$, we choose the weighting matrices of L as $Q = 0.5I$ and $R = 0$. The terminal state region is selected as

$$\mathbb{X}_f = \{x\ :\ W(x) \leq \alpha\} \tag{12.37}$$

such that

$$k_f(x) \in \mathbb{U}, \quad \dot{W}(T) + L(T) \leq 0, \quad \forall(\theta, x) \in (\Theta, \mathbb{X}_e) \tag{12.38}$$

Since $\tilde{\theta}$ and ϑ have known upper bounds, it follows that there exists k_1, k_2, and k_3 such that

$$\dot{W} + L = -(k_1 - 0.5)\|\gamma\|^2 - \gamma^T\Gamma\, g\,\tilde{\theta} - \gamma^T\Gamma\vartheta - k_2\gamma^T\Gamma\, g\, g^T\,\Gamma\gamma - k_3\gamma^T\Gamma\Gamma\gamma \leq 0$$

$\forall\ \theta \in \Theta$ and $\forall\ \vartheta \in \mathcal{D}$, outside a small neighborhood of x^*. The task of computing the terminal set is then reduced to computing the largest possible α such that for $k_f(\cdot) \in \mathbb{U}$ for all $x \in \mathbb{X}_f$ yields the terminal set. The terminal cost (12.33) is used for this simulation and the terminal set is re-computed at every sampling instant using the current setpoint value.

The system is first simulated subject to a ramping measured economic disturbance in s_2 from $t = 6$ to 10. The simulation results are presented in Figures 12.3–12.7. The phase trajectories displayed in Figure 12.3 show that the reactor states x_1 and x_2 obey the imposed constraints. The concentration of A in reactor x_1 is shown to approach

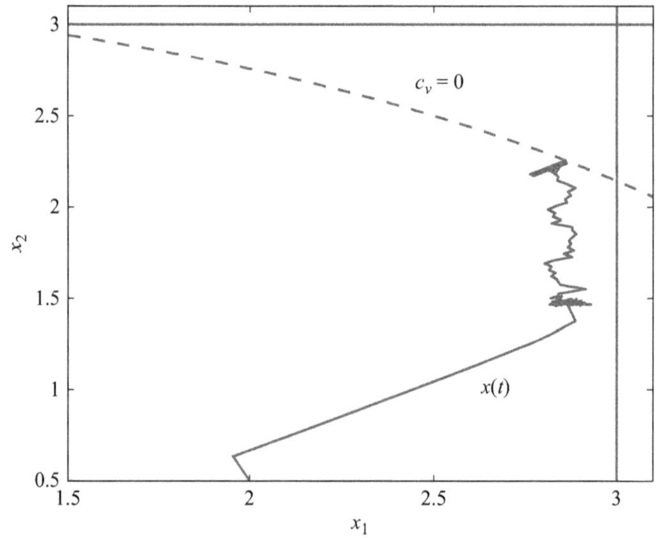

Figure 12.3 Phase diagram and feasible state region

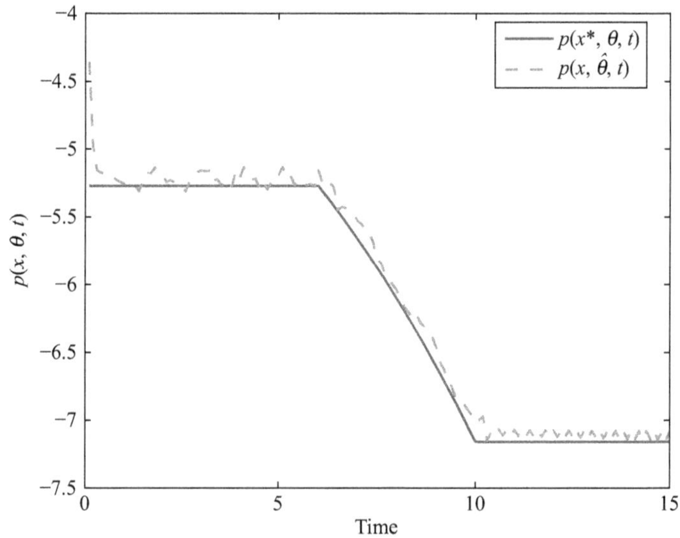

Figure 12.4 Optimal and actual profit functions

its upper bound. The system trajectories are also shown to approach the constraints toward the end of the simulation.

Figure 12.4 shows that the cost $p(t, x, \theta)$ converges to the optimal, unknown $p^*(t, x^*, \theta)$. The initial effect of the parameter estimation is observed initially but vanishes quickly. As soon as the parameter estimates reach their unknown true values,

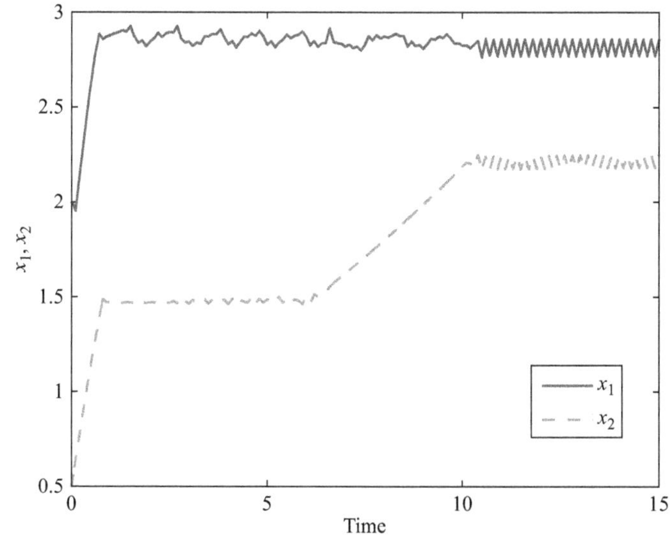

Figure 12.5 Closed-loop states trajectories

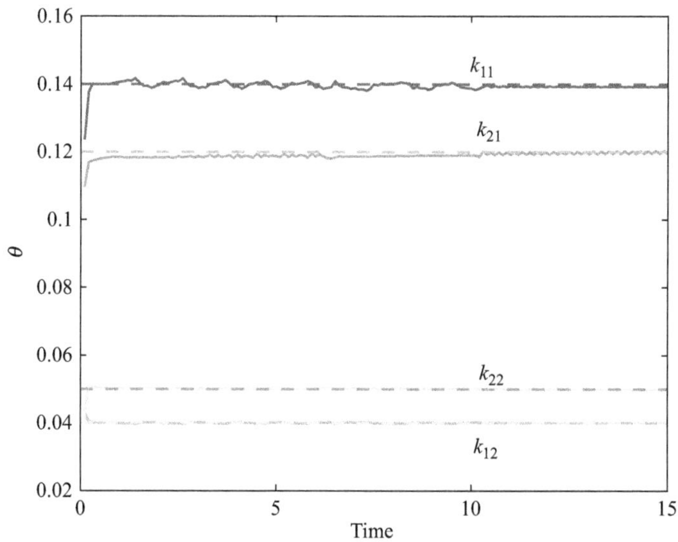

Figure 12.6 Unknown parameters and estimates

the economic MPC approach converges quickly to the unknown optimum, as expected. Figure 12.5 confirms the effectiveness of the adaptive MPC in tracking the desired setpoint. Figure 12.6 demonstrates that the convergence of the parameter estimates to their true values. Note that the adaptive MPC has a self-exciting feature that penalizes large estimation errors. The simulation results show that this approach is extremely

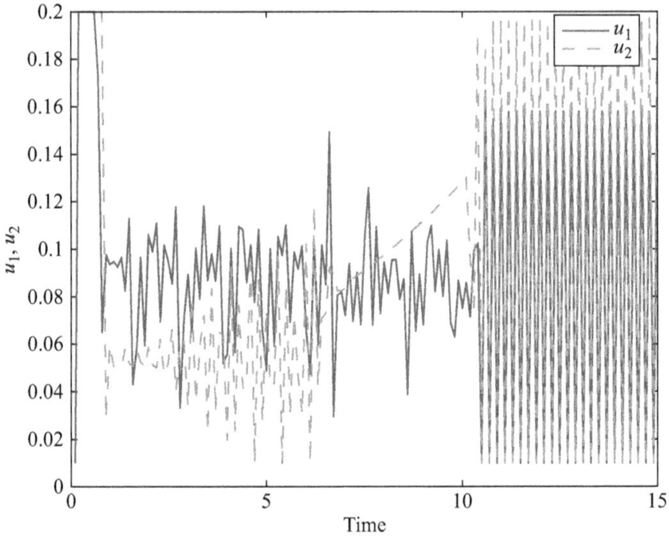

Figure 12.7 Closed-loop system's inputs

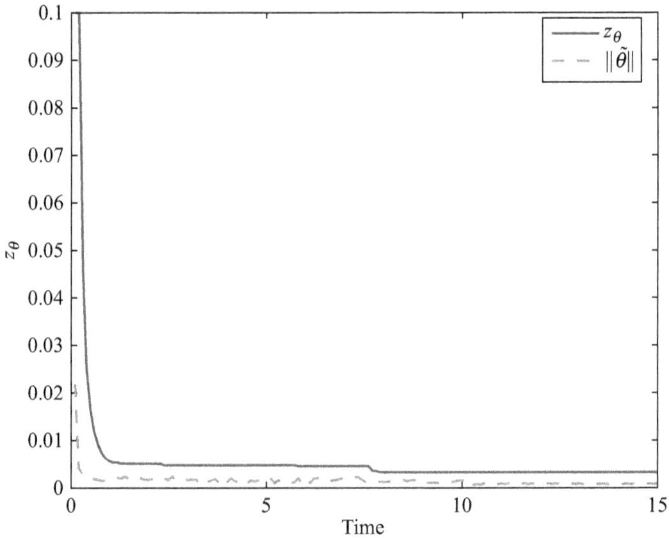

Figure 12.8 Uncertainty set radius and the true parameter estimate error norm

successful for the design of the adaptive control system. The control variables are shown in Figure 12.7. The required control action is implementable and satisfies the given constraints. Figure 12.8 compares the estimated uncertainty radius z_θ and the actual parameter estimation error $\|\tilde{\theta}\|$. As predicted, the estimated bound provides an accurate upper bound on the parameter estimation error. It also shows that the

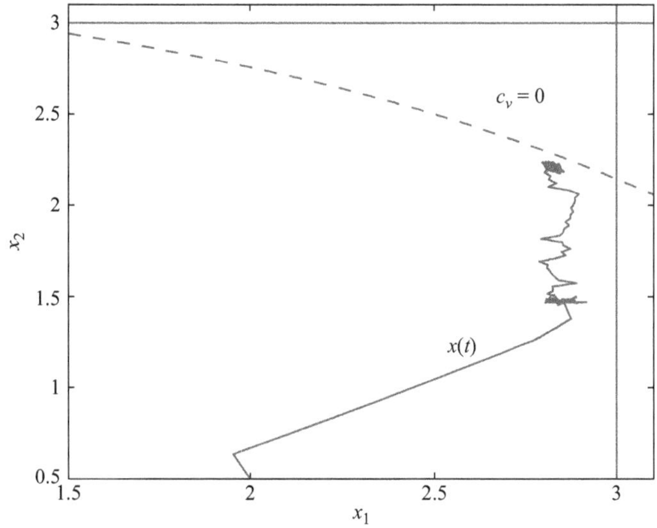

Figure 12.9 Phase diagram and feasible state region

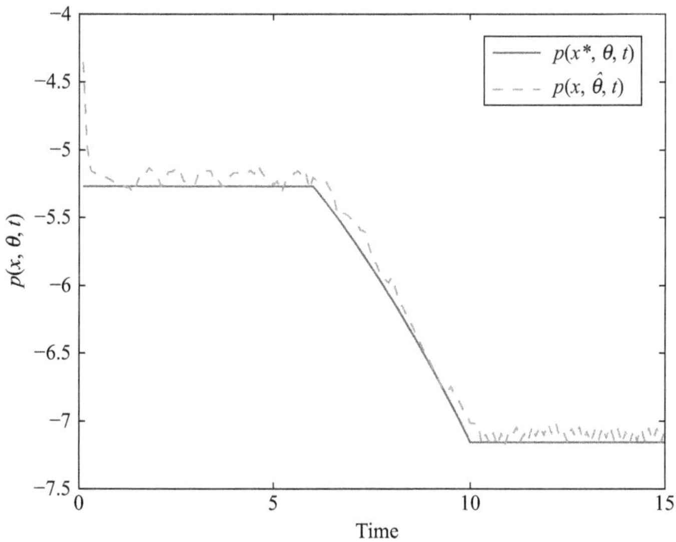

Figure 12.10 Optimal and actual profit functions

uncertainty set always contains the true value of the parameters. The proposed robust adaptive extremum-seeking model predictive controller performs extremely well for this specific problem, despite the absence of any external perturbation (apart from the contribution of ϑ) or dither signal.

In this specific case, the input trajectories demonstrate a considerable amount of chatter as the system approaches the constraints. One can reduce the sensitivity by increasing the value of μ. Figures 12.9–12.11 show the simulation results for

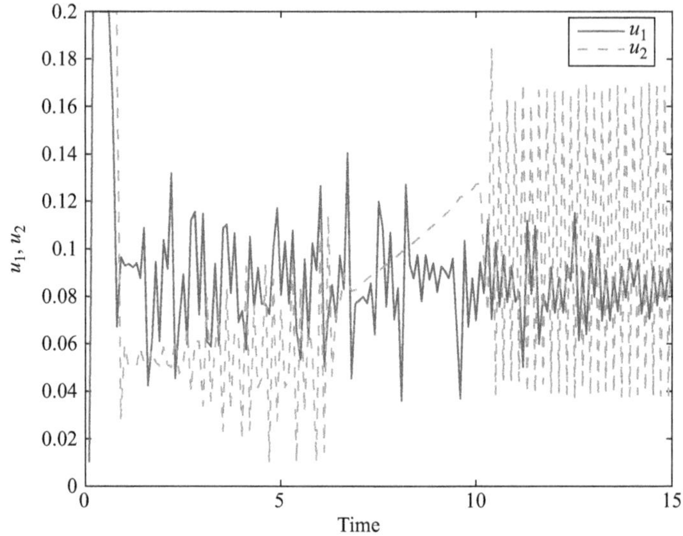

Figure 12.11 Closed-loop system's inputs

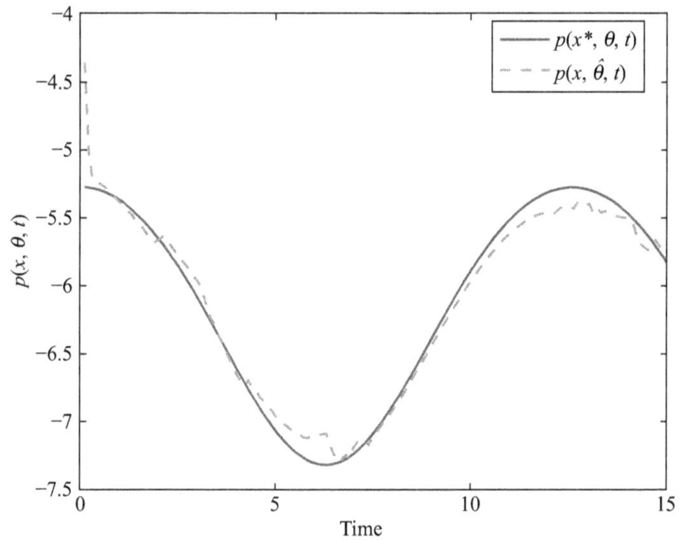

Figure 12.12 Optimal and actual profit functions for disturbance $s_2(t) = \frac{6}{5} + \frac{4}{5}\cos\left(\frac{1}{2}t\right)$

$\mu = 0.001$. The state trajectories are shown along with the constraints are shown in Figure 12.9. Figure 12.10 compares the true and estimated optimal profit trajectories. The input trajectories are shown in Figure 12.11. As expected, the input is shown to be less sensitive close to the constraints. However, the estimated profit function is shown to deviate from the the true profit. Nevertheless, the system is shown to perform adequately and the system converges effectively to the true unknown optimum.

The choice of s_2 considered in this simulation study is arbitrary. In order to demonstrate that this specific choice is not unique we consider a simulation where the value s_2 is chosen as

$$s_2(t) = \frac{6}{5} + \frac{4}{5}\cos\left(\frac{1}{2}t\right).$$

Figure 12.12 shows the resulting optimal profit function and the estimated profit. The MPC controller effectively steers the system to the optimum of the profit function. As in the previous case, all constraints are satisfied and the parameter estimates converge to their true values.

12.6 Conclusions

This chapter provides a formal design technique for solving economic optimization problems for a class of constrained nonlinear uncertain systems subject to parametric and disturbance uncertainties. The approach proposed implements a new direct adaptive extremum-seeking MPC approach for a class of uncertain nonlinear systems. The main advantage of the direct approach is that it naturally leads to a robust adaptive economic MPC approach that guarantees robust stabilization of the unknown optimal operating conditions. The technique is also readily implemented for discrete-time nonlinear systems.

Chapter 13
Set-based estimation in discrete-time systems

In this chapter, we present new techniques for parameter identification for nonlinear dynamical discrete-time systems. The methods presented are intended to improve the performance of adaptive control systems such as RTO schemes and adaptive extremum-seeking systems. Using recent results on FT adaptive control, we develop alternative techniques that can be used to guarantee the convergence of parameter estimates to their true values in the presence of model-mismatch and exogenous variables. Three methods are presented. The first two methods rely on system excitation and a regressor matrix, in either case, the true parameters are identified when the regressor matrix is of full rank and can be inverted. The third method is based on a novel set-based adaptive estimation method proposed in Chapter 10 to simultaneously estimate the parameters and the uncertainty associated with the true value. The uncertainty set is updated periodically when sufficient information has been obtained to shrink the uncertainty set around the true parameters. Each method guarantees convergence of the parameter estimation error, provided an appropriate PE condition is met. The effectiveness of each method is demonstrated using a simulation example, displaying convergence of the parameter error estimation error.

13.1 Introduction

Parameter identification is an important feature in many control situations. In many adaptive control algorithms, the reference trajectory is unknown and dependent on system dynamics, which can rely on a set of unknown system parameters. For example, in adaptive extremum-seeking control, the system is optimized using a cost function that may rely on unknown parameters [73, 160]. The performance of the system is dependent on the performance of the parameter identification method. It has been shown that efficient parameter convergence increases the robustness properties of closed-loop adaptive systems [109].

Few studies are available regarding parameter identification for discrete-time systems in the context of adaptive control. The problem of output feedback systems and strict feedback systems [165, 166] have been proposed where the parameter identification algorithm is separated from some control task. This algorithm uses a two-phase approach to identify parameters then apply some appropriate control to achieve a desired objective. This approach is limited by the assumption that the system is free from noise, though robustness is shown for a small additive random noise. Several recent studies solve similar identification problems with a variety of methods. The application of a neural network identification approach is shown in Reference 164. Results from this study demonstrate convergence of the internal state of the neural network to the true system state. A similar problem with the addition of time-varying parameters is solved in Reference 61.

This chapter adapts parameter estimation methods developed in Chapters 8, 9 and 10. The first method, the FT identification method, allows the direct and exact recovery of parameters immediately once a PE condition is met. This method requires the online inversion and computation of rank of a regressor matrix. Since the parameters are identified in FT, it is possible to remove the excitation signal, at the moment the parameters are recovered. Several results on intelligent excitation signals, where the magnitude of the excitation signal is adjusted as needed, have become available [2, 31]. Traditionally, the excitation signal is removed when the parameters are assumed to have converged. The method presented in this chapter allows the excitation signal to be removed once the parameters are guaranteed to have converged. The second method is a refinement of the first: it uses an adaptive compensator to eliminate the need for online inversion and rank computation of a matrix. The parameter estimation error can be shown to be non-increasing once a PE condition is met. The third method defines a parameter uncertainty set that evolves based on a worst-case estimate. Further, the parameter estimates are not allowed to fall outside the uncertainty set. This method ensures convergence of the parameter uncertainty set to the true parameters provided the true parameters fall within the initial uncertainty set, as the update algorithm ensures non-exclusion of the true parameter estimate.

13.2 Problem description

The parameter identification methods discussed in this paper are:

$$x_{k+1} = x_k + F(x_k, u_k) + G(x_k, u_k)\theta \tag{13.1}$$

where x_k is a state at some time step k, u_k is a control input at some time step k, and θ is a column vector of system parameters.

Assumption 13.2.1. *The state of the system x_k is available for measurement at any time step k.* ■

Assumption 13.2.2. *There is some known, bounded control law, u_k that achieves some control objective.* ■

Given the bounded control law, the objective of the three methods presented below is to determine the true values of the plant parameters.

13.3 FT parameter identification

Consider the following state predictor

$$\begin{aligned}\hat{x}_{k+1} &= \hat{x}_k + F(x_k, u_k) + G(x_k, u_k)\hat{\theta}_{k+1} \\ &\quad + K_k e_k - \omega_k(\hat{\theta}_k - \hat{\theta}_{k+1}) \\ &\quad + K_k \omega_k(\hat{\theta}_k - \hat{\theta}_{k+1}) \end{aligned} \tag{13.2}$$

where $\hat{\theta}_k$ is the vector of parameter estimates at time step k given by any update law, K_k is a correction factor at time step k, and $e_k = x_k - \hat{x}_k$ is the state estimation error at time step k. The variable ω_k is the following output filter at time step k

$$\omega_{k+1} = \omega_k + G(x_k, u_k) - K_k \omega_k, \qquad \omega_0 = 0 \tag{13.3}$$

Let the parameter estimation error at some time step k be $\tilde{\theta}_k = \theta - \hat{\theta}_k$. Now from (14.2) and (13.1) the state estimation error at time step $k+1$ is given by

$$\begin{aligned} e_{k+1} &= e_k + G(x_k, u_k)\tilde{\theta}_{k+1} - K_k e_k \\ &\quad + \omega_k(\hat{\theta}_k - \hat{\theta}_{k+1}) - K_k \omega_k(\hat{\theta}_k - \hat{\theta}_{k+1}). \end{aligned} \tag{13.4}$$

Define the auxiliary variable

$$\eta_k = e_k - \omega_k \tilde{\theta}_k. \tag{13.5}$$

From (14.3), (13.4), and (13.5) it follows that

$$\eta_{k+1} = \eta_k - K_k \eta_k, \qquad \eta_0 = e_0. \tag{13.6}$$

Let $Q \in \mathbb{R}^{p \times p}$ and $C \in \mathbb{R}^p$ be defined as

$$\begin{aligned} Q_{k+1} &= Q_k + \omega_k^T \omega_k \\ Q_0 &= 0 \end{aligned} \tag{13.7}$$

$$\begin{aligned} C_{k+1} &= C_k + \omega_k^T(\omega_k \hat{\theta}_k + e_k - \eta_k) \\ C_0 &= 0. \end{aligned} \tag{13.8}$$

Lemma 13.3.1. *If there exists some time step k_c such that Q_{k_c} is invertible, that is,*

$$Q_N = \sum_{i=0}^{N} \omega_i^T \omega_i \succ 0 \tag{13.9}$$

then the parameters are given by $\theta = Q_k^{-1} C_k$, $\forall k \geq k_c$.

Proof: This results can be shown from

$$Q_N \theta = \sum_{i=0}^{N} \omega_i^T \omega_i [\hat{\theta}_i + \tilde{\theta}_i]. \tag{13.10}$$

Upon substitution of (13.5), it follows that

$$\begin{aligned}\theta &= Q_k^{-1}\sum_{i=0}^{k}\omega_i^T(\omega_i\hat{\theta}_i - +e_i - \eta_i)\\ &= Q_k^{-1}C_k \quad \forall k \geq k_c, \end{aligned} \tag{13.11}$$

which proves the result. ■

13.4 Adaptive compensation design

Application of the FTI is problematic since it requires that one checks the nonsingularity of Q_k at all step k. In this section, an adaptive compensation design is proposed that recovers exponential stability of the parameter estimation error in FT without the need to test the matrix Q_k.

Consider the state predictor for system (13.1)

$$\begin{aligned}\hat{x}_{k+1} &= \hat{x}_k + F(x_k, u_k) + G(x_k, u_k)\theta^o\\ &\quad + K_k(x_k - \hat{x}_k)\end{aligned} \tag{13.12}$$

where $K_k > 0$ and θ^o is the vector of initial parameter estimates.

As above, define the auxiliary variable and the filter dynamic as:

$$\eta_k = x_k - \hat{x}_k - \omega_k(\theta - \theta^o) \tag{13.13}$$

$$\omega_{k+1} = \omega_k + G(x_k, u_k) - K_k\omega_k \quad \omega_0 = 0 \tag{13.14}$$

The auxiliary variable η_k can be generated from:

$$\eta_{k+1} = \eta_k - K_k\eta_k, \quad \eta_0 = e_0 \tag{13.15}$$

Let Q and C be generated by

$$Q_{k+1} = Q_k + \omega_k^T\omega_k, \quad Q_0 = 0, \tag{13.16}$$

$$C_{k+1} = C_k + \omega_k^T(\omega_k\theta^o + e_k - \eta_k), \quad C_0 = 0 \tag{13.17}$$

and let k_c be a time step at which $Q_{k_c} \succ 0$.

The parameter update law proposed in Reference 6 is given by:

$$\hat{\theta}_{k+1} = \hat{\theta}_k + \Gamma_k(C_k - Q_k\hat{\theta}_k). \tag{13.18}$$

It follows from (13.18) that the dynamics of the parameter estimation error are

$$\tilde{\theta}_{k+1} = \tilde{\theta}_k - \Gamma_k(C_k - Q_k\hat{\theta}_k). \tag{13.19}$$

For all time steps $k \geq k_c$, $Q_k\theta = C_k$, it follows that $\forall k \geq k_c$

$$\tilde{\theta}_{k+1} = \tilde{\theta}_k - \Gamma_k(Q_k\theta - Q_k\hat{\theta}_k). \tag{13.20}$$

Define the variable $\Gamma_k = \frac{1}{\|Q_k\|+\epsilon}$, where ϵ is some small positive number

$$\tilde{\theta}_{k+1} = \left(I - \frac{Q_k}{\|Q_k\| + \epsilon}\right)\tilde{\theta}_k \tag{13.21}$$

It follows from (13.21) that for all time steps $k \geq k_c$ that $\tilde{\theta}$ is non-increasing, and

$$\lim_{k\to\infty} \tilde{\theta} = 0. \tag{13.22}$$

13.5 Parameter uncertainty set estimation

The adaptive compensator design provides an effective mechanism to recover the FTI performance. However, the properties of this design can be lost in the presence of exogenous disturbance variables and model mismatch. In this section, a parameter estimation technique is proposed to handle nonlinear systems subject to exogenous variables. The technique relies on an uncertainty set update formulation that provides robust performance.

13.5.1 Parameter update

Consider the uncertain nonlinear system

$$x_{k+1} = x_k + F(x_k, u_k) + G(x_k, u_k)\theta + \vartheta_k \tag{13.23}$$

where ϑ is a bounded disturbance that satisfies $\|\vartheta_k\| \leq M_\vartheta < \infty$. It is assumed that θ is uniquely identifiable and lies within an initially known compact set defined by the ball function $\Theta^0 = B(\theta_0, z_\theta)$ where θ_0 is an initial estimate of the unknown parameters and z_θ is the radius of the parameter uncertainty set.

Using the state predictor defined in (14.2) and the output filter defined in (14.3), the prediction error $e_k = x_k - \hat{x_k}$ is given by

$$\begin{aligned} e_{k+1} &= e_k + G(x_k, u_k)\tilde{\theta}_{k+1} - K_k e_k \\ &\quad + \omega_k(\hat{\theta}_k - \hat{\theta}_{k+1}) - K_k\omega_k(\hat{\theta}_k - \hat{\theta}_{k+1}) + \vartheta_k \\ e_0 &= x_0 - \hat{x}_0. \end{aligned} \tag{13.24}$$

The auxiliary variable η_k dynamics are as follows:

$$\begin{aligned} \eta_{k+1} &= e_{k+1} - \omega_{k+1}\tilde{\theta}_{k+1} + \vartheta_k \\ \eta_0 &= e_0. \end{aligned} \tag{13.25}$$

Since ϑ_k is unknown, it is necessary to use an estimate, $\hat{\eta}$, of η. The estimate is generated by the recursion:

$$\hat{\eta}_{k+1} = \hat{\eta}_k - K_k\hat{\eta}_k. \tag{13.26}$$

The resulting dynamics of the η estimation error are:

$$\tilde{\eta}_{k+1} = \tilde{\eta}_k - K_k\tilde{\eta}_k + \vartheta. \tag{13.27}$$

Let the identifier matrix Σ_k be defined as

$$\Sigma_{k+1} = \Sigma_k + \omega_k^T \omega_k, \quad \Sigma_0 = \alpha I \succ 0 \tag{13.28}$$

with an inverse generated by the recursion

$$\Sigma_{k+1}^{-1} = \Sigma_k^{-1} - \Sigma_k^{-1^T} \omega_k^T \left(I + w_k \Sigma_k^{-1} w_k^T\right)^{-1} \omega_k \Sigma_k^{-1}, \quad \Sigma_0^{-1} = \frac{1}{\alpha} I \succ 0. \tag{13.29}$$

From (14.2), (14.3), and (14.6) and based on the preferred parameter update law proposed in Reference 7, the parameter update law is

$$\hat{\theta}_{k+1} = \hat{\theta}_k + \Sigma_k^{-1} \omega_k^T \left(I + w_k \Sigma_k^{-1} w_k^T\right)^{-1} (e_k - \hat{\eta}_k). \tag{13.30}$$

To ensure that the parameter estimates remain within the constraint set Θ_k, we propose to use a projection operator of the form:

$$\bar{\hat{\theta}}_{k+1} = \text{Proj}\{\hat{\theta}_k + \Sigma_k^{-1} \omega_k^T \left(I + w_k \Sigma_k^{-1} w_k^T\right)^{-1} (e_k - \hat{\eta}_k), \Theta_k\}. \tag{13.31}$$

The operator Proj represents an orthogonal projection onto the surface of the uncertainty set applied to the parameter estimate. The parameter uncertainty set is defined by the ball function $B(\hat{\theta}_c, z_{\hat{\theta}c})$, where $\hat{\theta}c$ and $z_{\hat{\theta}c}$ are the parameter estimate and set radius found at the latest set update.

Following, Goodwin and Sin (1995), the projection operator is designed such that

- $\hat{\theta}_{k+1} \in \Theta_k$
- $\bar{\tilde{\theta}}_{k+1}^T \Sigma_{k+1} \bar{\tilde{\theta}}_{k+1} \leq \tilde{\theta}_{k+1}^T \Sigma_{k+1} \tilde{\theta}_{k+1}$

It can be shown that the parameter update law defined in (13.30) guarantees convergence of parameter estimates to the true values.

Lemma 13.5.1. [80] *Consider the system*

$$x_{k+1} = Ax_k + Bu_k \tag{13.32}$$

where A is a stable matrix with eigenvalues inside the unit circle and B is a matrix of appropriate dimension. Then it can be shown that

$$\sum_{k=0}^{K-1} x_{k+1}^T x_{k+1} \leq \delta^2 \sum_{k=0}^{K-1} u_k^T u_k \tag{13.33}$$

for some $\delta > 0$ *and* $K - 1 > 0$.

Let l_2 denote the space of square finitely summable signals and consider the following lemma.

Lemma 13.5.2. *The identifier* (13.29) *and parameter update law* (13.31) *are such that* $\tilde{\theta}_k = \theta_k - \hat{\theta}_k$ *is bounded. Furthermore, if*

$$\vartheta_k \in l_2 \;\; or \;\; \sum_{k=0}^{\infty} [\|\tilde{\eta}_k\|^2 - \gamma\|e_k - \hat{\eta}_k\|^2] < +\infty \tag{13.34}$$

and

$$\lim_{k\to\infty} \Sigma_k = \infty \tag{13.35}$$

are satisfied, then $\tilde{\theta}_k$ *converges to 0 asymptotically.*

Proof: Let $V_{\tilde{\theta}k} = \tilde{\theta}_k^T \Sigma_k \tilde{\theta}_k$. It follows from the properties of the projection operator that:

$$V_{\tilde{\theta}k+1} - V_{\tilde{\theta}k} = \bar{\tilde{\theta}}_{k+1}^T \Sigma_{k+1} \bar{\tilde{\theta}}_{k+1} - \tilde{\theta}_k^T \Sigma_k \tilde{\theta}_k \leq \tilde{\theta}_{k+1}^T \Sigma_{k+1} \tilde{\theta}_{k+1} - \tilde{\theta}_k^T \Sigma_k \tilde{\theta}_k.$$

Using the parameter update law, one can write $\tilde{\theta}_{k+1}$ as:

$$\begin{aligned}\tilde{\theta}_{k+1} &= \tilde{\theta}_k - \Sigma_k^{-1}\omega_k^T\left(I + w_k \Sigma_k^{-1} w_k^T\right)^{-1}(e_k - \hat{\eta}_k)\\ &= \tilde{\theta}_k - \Sigma_k^{-1}\omega_k^T\left(I + w_k \Sigma_k^{-1} w_k^T\right)^{-1}(w_k\tilde{\theta}_k + \tilde{\eta}_k)\end{aligned}$$

or,

$$\tilde{\theta}_{k+1} = \Sigma_{k+1}^{-1}\Sigma_k\tilde{\theta}_k - \Sigma_k^{-1}\omega_k^T\left(I + w_k \Sigma_k^{-1} w_k^T\right)^{-1}\tilde{\eta}_k. \tag{13.36}$$

Upon substitution of the parameter update law, the identifier matrix dynamics, the filter dynamics, and the auxiliary variable dynamics, the rate change of the $V_{\tilde{\theta}k}$ is given by:

$$V_{\tilde{\theta}k+1} - V_{\tilde{\theta}k} \leq -(e_k - \hat{\eta}_k)^T \left(I + w_k \Sigma_k^{-1} w_k^T\right)^{-1}(e_k - \hat{\eta}_k) + \tilde{\eta}_k^T\left(I + w_k \Sigma_k^{-1} w_k^T\right)^{-1}\tilde{\eta}_k \tag{13.37}$$

From the $\tilde{\eta}_k$ dynamics given in (13.27), it follows from Lemma 13.5.1 if $\vartheta_k \in l_2$ then $\tilde{\eta}_k \in l_2$. Taking the limit as $k \to \infty$, the inequality becomes

$$\lim_{k\to\infty} V_{\tilde{\theta}k} = V_{\tilde{\theta}0} + \sum_{k=0}^{\infty} V_{\tilde{\theta}k+1} - V_{\tilde{\theta}k} \tag{13.38}$$

$$\leq V_{\tilde{\theta}0} - \sum_{k=0}^{\infty}\left[(e_k - \hat{\eta}_k)^T\left(I + w_k \Sigma_k^{-1} w_k^T\right)^{-1}(e_k - \hat{\eta}_k)\right] \tag{13.39}$$

$$+ \sum_{k=0}^{\infty}\left[\tilde{\eta}_k^T\left(I + w_k \Sigma_k^{-1} w_k^T\right)^{-1}\tilde{\eta}_k\right] \tag{13.40}$$

By the boundedness of the trajectories of the system, it follows that there exists a number $\gamma > 0$ such that

$$1 \geq \|\left(I + w_k \Sigma_k^{-1} w_k^T\right)^{-1}\| \geq \gamma.$$

as a result, one obtains the following inequality:

$$\lim_{k\to\infty} V_{\tilde{\theta}k} \le V_{\tilde{\theta}0} - \gamma \sum_{k=0}^{\infty} \left[(e_k - \hat{\eta}_k)^T (e_k - \hat{\eta}_k)\right] + \sum_{k=0}^{\infty} \left[\tilde{\eta}_k^T \tilde{\eta}_k\right] \tag{13.41}$$

Therefore, if the conditions (13.34) are met then the right-hand side of (13.41) is finite. As a result, one concludes that

$$\lim_{k\to\infty} \tilde{\theta}_k = 0 \tag{13.42}$$

as required. ■

13.5.2 Set update

An update law that measures the worst-case progress of the parameter update law is adapted from the one proposed in Reference 7

$$z_{\hat{\theta}k} = \sqrt{\frac{V_{z\hat{\theta}k}}{4\lambda_{\min}(\Sigma_k)}} \tag{13.43}$$

$$V_{z\theta k+1} = V_{z\hat{\theta}k} - (e_k - \hat{\eta}_k)^T \left(I + w_k \Sigma_k^{-1} w_k^T\right)^{-1} (e_k - \hat{\eta}_k) + \left(\frac{M_\vartheta}{K_k}\right)^2 \tag{13.44}$$

$$V_{z\hat{\theta}0} = 4\lambda_{\max}(\Sigma_0)(z_{\hat{\theta}0})^2 \tag{13.45}$$

The parameter uncertainty set, defined by the ball function $B(\hat{\theta}_c, z_c)$, is updated using the parameter update law (4.10) and the error bound (13.43) according to the following algorithm.

Algorithm 13.5.3. *Beginning at time step $k = 0$, the set is adapted according to the following iterative process:*

1. ***Initialize*** $z_{\hat{\theta}c} = z_{\hat{\theta}0}, \hat{\theta}_c = \hat{\theta}_0$
2. *At time step k, using* (13.30) *and* (13.43) *perform the update*

$$(\hat{\theta}_c, z_{\hat{\theta}c}) = \begin{cases} (\hat{\theta}_k, z_{\hat{\theta}k}) & \text{if } z_{\hat{\theta}k} \le z_c - \|\hat{\theta}_k - \hat{\theta}_c\| \\ (\hat{\theta}_c, z_{\hat{\theta}c}) & \text{otherwise} \end{cases} \tag{13.46}$$

3. *Return to Step* 2 *and iterate, incrementing to time step $k + 1$.*

Lemma 13.5.4. *The algorithm ensures that*

1. *the set is only updated when updating will yield a contraction*
2. *the dynamics of the set error bound described in* (13.43) *are such that they ensure the non-exclusion of the true value $\theta \in \Theta_k$, $\forall k$ if $\theta_0 \in \Theta_0$.*

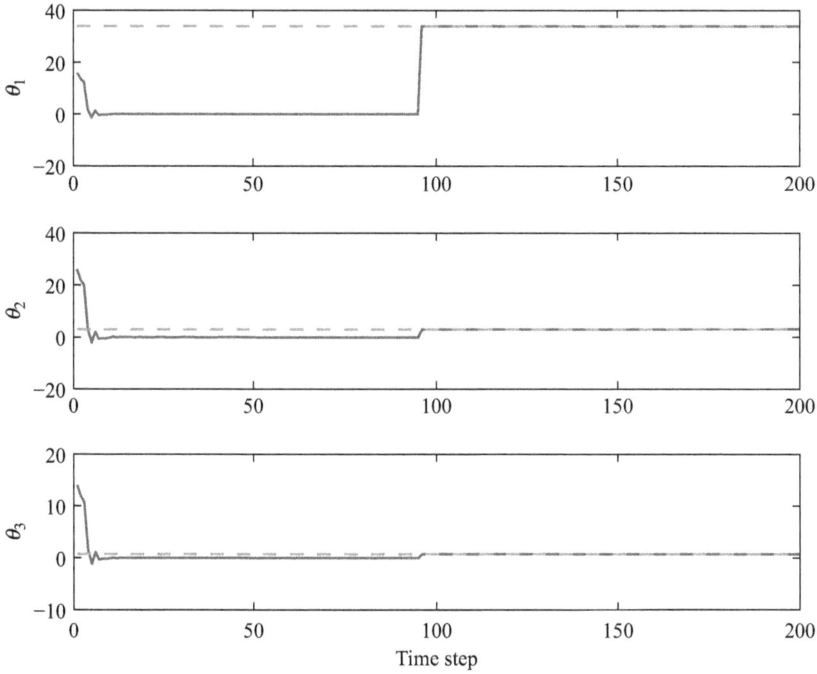

Figure 13.1 Time course plot of the parameter estimates and true values under the FT estimation algorithm: the dashed lines (- -) represent the true parameter values and the solid lines (–) represent the parameter estimates

Proof:

1. If $\Theta_{k+1} \not\subseteq \Theta_k$ then

$$\sup_{s\in\Theta_{k+1}} \|s - \hat{\theta}_k\| \geq z_{\hat{\theta}k}. \tag{13.47}$$

However, it is guaranteed by the set update algorithm presented, that Θ, at update times, obeys the following:

$$\sup_{s\in\Theta_{k+1}} \|s - \hat{\theta}_k\|$$

$$\leq \sup_{s\in\Theta_{k+1}} \|s - \hat{\theta}_{k+1}\| + \|\hat{\theta}_{k+1} - \hat{\theta}_k\| \tag{13.48}$$

$$\leq z_{\hat{\theta}k+1} + \|\hat{\theta}_{k+1} - \hat{\theta}_k\| \leq z_{\hat{\theta}k} \tag{13.49}$$

This contradicts (13.47). Therefore, $\Theta_{k+1} \subseteq \Theta_k$ at time steps where Θ is updated.

2. It is known, by definition, that

$$V_{\tilde{\theta}0} \leq V_{z\theta 0}, \quad \forall k \geq 0 \tag{13.50}$$

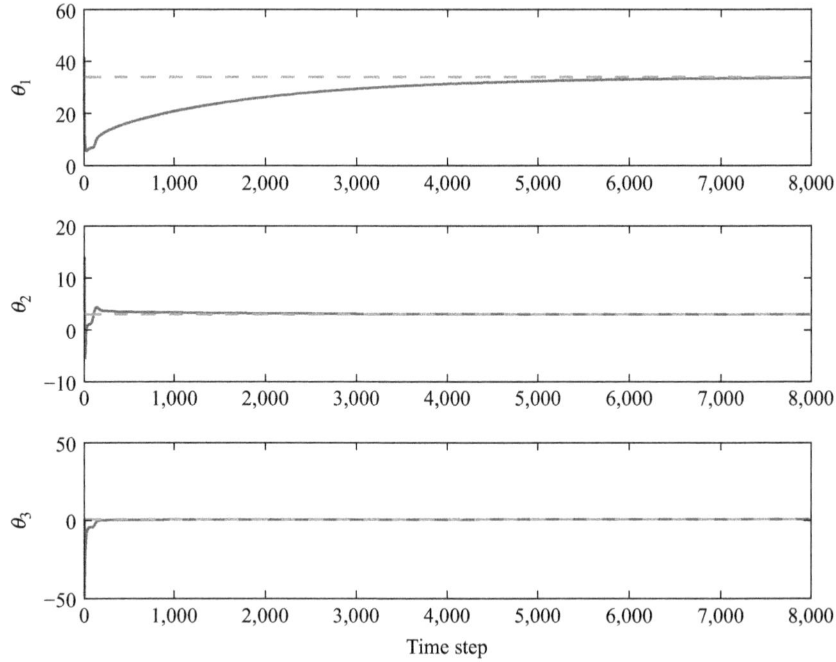

Figure 13.2 Time course plot of the parameter estimates and true values under the adaptive compensatory algorithm: the dashed lines (- -) represent the true parameter values and the solid lines (–) represent the parameter estimates

Since, $V_{\tilde{\theta}k} = \tilde{\theta}_k^T \Sigma_k \tilde{\theta}_k$,

$$\|\tilde{\theta}_k\| \leq \frac{V_{z\hat{\theta}k}}{\lambda_{\min}(\Sigma_k)} = 4z_{\hat{\theta}k}^2, \quad \forall k \geq 0 \tag{13.51}$$

Therefore, if $\theta \in \Theta_0$, then $\theta \in \Theta_k \ \forall k \geq 0$. ■

13.6 Simulation examples

Consider the following nonlinear system:

$$\begin{aligned} x_{1,k+1} &= 0.01\big(-x_{2,k} + u_{3,k} + x_{3,k}\theta_1 + \vartheta_{1,k}\big) \\ x_{2,k+1} &= 0.01\big((1 + x_{3,k})u_{1,k} - x_{1,k}\theta_2 + \vartheta_{2,k}\big) \\ x_{3,k+1} &= 0.01\big(-x_{1,k} + u_{2,k} + x_{2,k}\theta_3 + \vartheta_{3,k}\big) \end{aligned} \tag{13.52}$$

where $\theta^T = [\theta_1, \theta_2, \theta_3]$. The input is taken as constant, $u_k = [-0.1 \ \ 0.1 \ \ 0.2]^T$. The true parameter values are $\theta = [1.5 \ \ 3 \ \ 0.02]^T$.

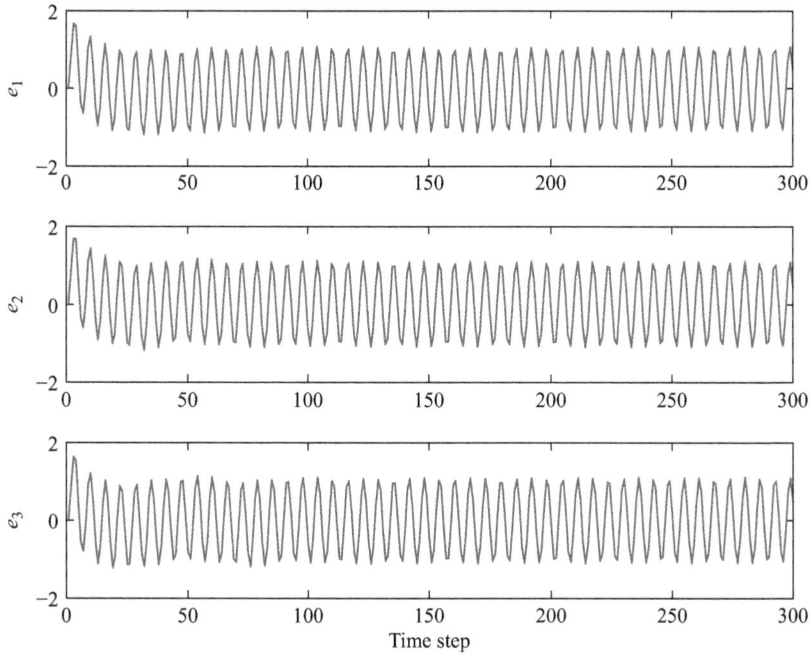

Figure 13.3 Time course plot of the state prediction error $e_k = x_k - \hat{x}_k$

In the first two examples (Figures 13.1 and 13.2), we consider system (13.52) with $\vartheta_{i,k} = 0,\ i = 1, 2, 3$. To demonstrate the parameter uncertainty set approach (Figure 13.4), a bounded noise term is added to the state equation, as shown in (13.23). The bounded noise term is

$$\vartheta_k = [\vartheta_{1,k}, \vartheta_{2,k}, \vartheta_{3,k}] = [\sin(k) \quad \sin(k) \quad \sin(k)]^T.$$

The input is taken as constant, $u_k = [-0.1 \ \ 0.1 \ \ 0.2]^T$. The true parameter values are $\theta = [34 \ \ 3 \ \ 0.02]^T$.

13.6.1 FT parameter identification

In the first simulation study, we consider the FT parameter identification. Figure 13.1 shows the parameter estimates converging to the true values almost immediately at the time step when the regressor matrix Q_k has full rank.

13.6.2 Adaptive compensation design

Consider the system described by (13.52). We consider the application of the adaptive compensator. The results are shown in Figure 13.2. Consistent with the result shown in (13.21), Figure 13.2 shows that after time step k_c the parameter estimate errors converge to their true values at an exponential rate.

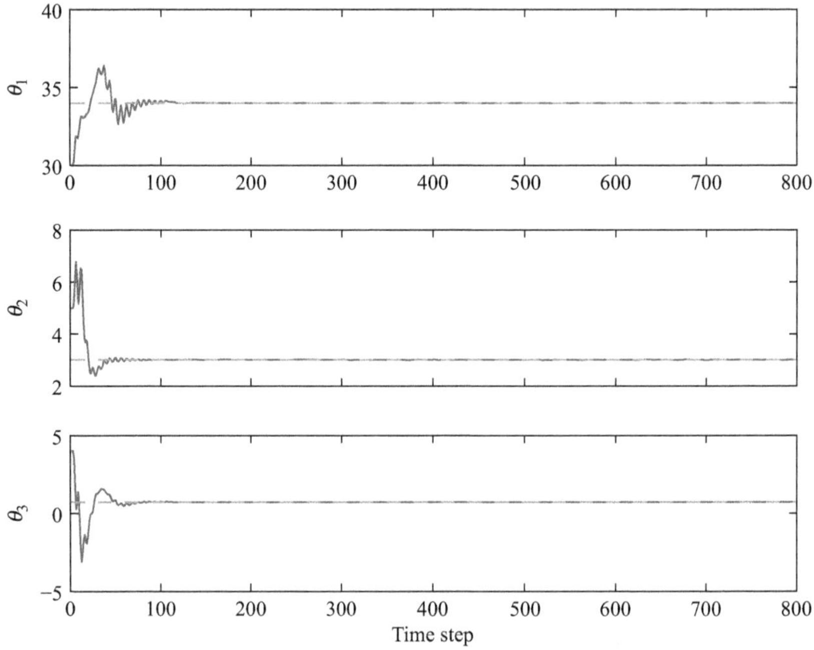

Figure 13.4 Time course plot of the parameter estimates and true values under the parameter uncertainty set algorithm: the dashed lines (- -) represent the true parameter values and the solid lines (–) represent the parameter estimates

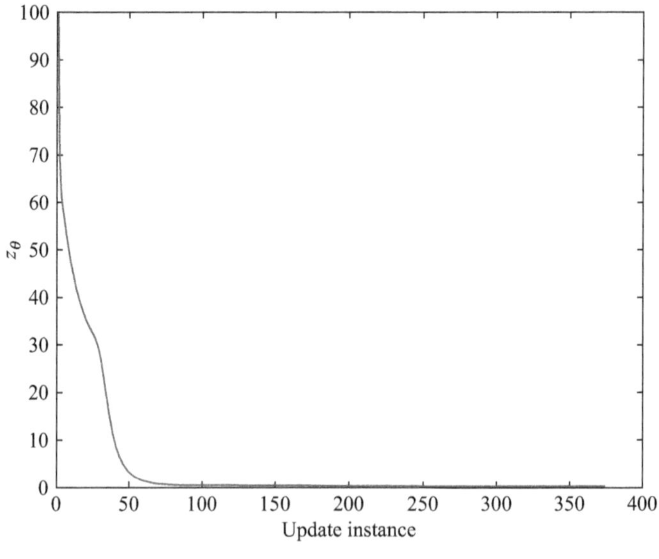

Figure 13.5 The progression of the radius of the parameter uncertainty set at time steps when the set is updated

13.6.3 Parameter uncertainty set estimation

In the third simulation, we consider the system (13.52) with the bounded noise term is ϑ. Figure 13.3 shows the trajectories of the state prediction error. As expected, the error is shown to converge to a neighborhood of zero. The parameter estimates are shown to converge to a neighborhood of their true values in Figure 13.4. The ability of the estimation routine to recover the true parameter value in the presence of exogenous disturbances is clearly demonstrated. The size of this neighborhood is limited by the magnitude of the injected noise. The size of the uncertainty set, z_θ is shown in Figure 13.5.

13.7 Summary

In this chapter, three methods for parameter identification for nonlinear systems were presented. Each method presented guarantees convergence of the parameter estimation error to zero, provided an appropriate PE condition is met. Each identification algorithm has been implemented to demonstrate its performance, each algorithm demonstrates convergence of the parameter estimation error.

Chapter 14

Robust adaptive MPC for discrete-time systems

14.1 Introduction

In this chapter, we establish a sound theoretical for the analysis of robust adaptive MPC control system subject to exogenous disturbances for a class of discrete-time nonlinear control systems. As in the previous chapters, no claims are made concerning the computational requirements of the proposed min–max approach to adaptive MPC technique. However, it is argued that a Lipschitz-based approach provides a conservative approximation of the min–max approach that retains all of the stability and robustness properties.

The uncertainties associated with the parameters is handled using the set-based estimation approach for a class of discrete-time nonlinear systems presented in this chapter. In this chapter, it is shown how this set-based approach can be formulated in the context of nonlinear adaptive MPC approach for discrete-time systems in the presence of parameter uncertainties and exogenous disturbances.

The chapter is structured as follows. The problem description is given in Section 14.2. The parameter estimation routine is presented in Section 14.3. Two approaches to robust adaptive MPC are detailed in Section 14.4. This is followed by a simulation example in Section 14.5 and brief conclusions in Section 14.6.

14.2 Problem description

Consider the uncertain discrete-time nonlinear system

$$x_{k+1} = x_k + F(x_k, u_k) + G(x_k, u_k)\theta + \vartheta_k \triangleq \mathcal{F}(x_k, u_k, \theta, \vartheta_k) \tag{14.1}$$

where the disturbance $\vartheta_k \in \mathcal{D} \subset \mathbb{R}^{n_d}$ is assumed to satisfy a known upper bound $\|\vartheta_k\| \leq M_\vartheta < \infty$. The objective of the study is to (robustly) stabilize the plant to some target set $\Xi \subset \mathbb{R}^{n_x}$ while satisfying the pointwise constraints $x_k \in \mathcal{X} \in \mathbb{R}^{n_x}$ and $u_k \in \mathcal{U} \in \mathbb{R}^{n_u}$, $\forall k \in \mathbb{Z}$. The target set is a compact set, contains the origin and is robustly invariant under no control. It is assumed that θ is uniquely identifiable and lie within an initially known compact set $\Theta^0 = B(\theta_0, z_\theta)$ where θ_0 is a nominal parameter value, z_θ is the radius of the parameter uncertainty set.

Remark 14.2.1. *In this study, the exogenous variable ϑ_k represents an unstructured bounded time-varying uncertainty. We do not provide any additional structure, such as a state-dependent disturbance matrix, since this is assumed to be expressed by the term $g(x_k, u_k)\theta$ in* (14.1).

14.3 Parameter and uncertainty set estimation

14.3.1 Parameter adaptation

In this section, we revisit the parameter update law proposed in Chapter 13.

Consider the following state predictor

$$\begin{aligned}\hat{x}_{k+1} &= \hat{x}_k + F(x_k, u_k) + G(x_k, u_k)\hat{\theta}_{k+1} \\ &\quad + K_k e_k - \omega_k(\hat{\theta}_k - \hat{\theta}_{k+1}) \\ &\quad + K_k \omega_k(\hat{\theta}_k - \hat{\theta}_{k+1})\end{aligned} \tag{14.2}$$

where $\hat{\theta}_k$ is the vector of parameter estimates at time step k given by any update law, K_k is a correction factor at time step k, and $e_k = x_k - \hat{x}_k$ is the state estimation error at time step k. The variable ω_k is the following output filter at time step k

$$\omega_{k+1} = \omega_k + G(x_k, u_k) - K_k \omega_k, \qquad \omega_0 = 0 \tag{14.3}$$

Using the state predictor defined in (14.2) and the output filter defined in (14.3), the prediction error $e_k = x_k - \hat{x}_k$ is given by

$$\begin{aligned}e_{k+1} &= e_k + G(x_k, u_k)\tilde{\theta}_{k+1} - K_k e_k \\ &\quad + \omega_k(\hat{\theta}_k - \hat{\theta}_{k+1}) - K_k \omega_k(\hat{\theta}_k - \hat{\theta}_{k+1}) + \vartheta_k \\ e_0 &= x_0 - \hat{x}_0.\end{aligned} \tag{14.4}$$

The auxiliary variable η_k dynamics are as follows:

$$\begin{aligned}\eta_{k+1} &= e_{k+1} - \omega_{k+1}\tilde{\theta}_{k+1} + \vartheta_k \\ \eta_0 &= e_0.\end{aligned} \tag{14.5}$$

Since ϑ_k is unknown, it is necessary to use an estimate, $\hat{\eta}$, of η. The estimate is generated by the recursion:

$$\hat{\eta}_{k+1} = \hat{\eta}_k - K_k \hat{\eta}_k \tag{14.6}$$

The resulting dynamics of the η estimation error are:

$$\tilde{\eta}_{k+1} = \tilde{\eta}_k - K_k \tilde{\eta}_k + \vartheta \tag{14.7}$$

Let the identifier matrix Σ_k be defined as

$$\Sigma_{k+1} = \Sigma_k + \omega_k^T \omega_k, \quad \Sigma_0 = \alpha I \succ 0 \tag{14.8}$$

with an inverse generated by the recursion

$$\Sigma_{k+1}^{-1} = \Sigma_k^{-1} - \Sigma_k^{-1^T} \omega_k^T \left(I + w_k \Sigma_k^{-1} w_k^T\right)^{-1} \omega_k \Sigma_k^{-1}, \quad \Sigma_0^{-1} = \frac{1}{\alpha} I \succ 0. \tag{14.9}$$

From (14.2), (14.3), and (14.6) and based on the the preferred parameter update law proposed in Reference 7, the parameter update law is

$$\hat{\theta}_{k+1} = \hat{\theta}_k + \Sigma_k^{-1}\omega_k^T\left(I + w_k\Sigma_k^{-1}w_k^T\right)^{-1}(e_k - \hat{\eta}_k). \tag{14.10}$$

To ensure that the parameter estimates remain within the constraint set Θ_k, we propose to use a projection operator of the form:

$$\bar{\hat{\theta}}_{k+1} = \text{Proj}\{\hat{\theta}_k + \Sigma_k^{-1}\omega_k^T\left(I + w_k\Sigma_k^{-1}w_k^T\right)^{-1}(e_k - \hat{\eta}_k), \Theta_k\}. \tag{14.11}$$

The operator Proj represents an orthogonal projection onto the surface of the uncertainty set applied to the parameter estimate. The parameter uncertainty set is defined by the ball function $B(\hat{\theta}_c, z_{\hat{\theta}c})$, where $\hat{\theta}_c$ and $z_{\hat{\theta}c}$ are the parameter estimate and set radius found at the latest set update.

Following, Goodwin and Sin (1995), the projection operator is designed such that

- $\hat{\theta}_{k+1} \in \Theta_k$
- $\bar{\tilde{\theta}}_{k+1}^T\Sigma_{k+1}\bar{\tilde{\theta}}_{k+1} \leq \tilde{\theta}_{k+1}^T\Sigma_{k+1}\tilde{\theta}_{k+1}$.

It can be shown that the parameter update law defined in (14.10) guarantees convergence of parameter estimates to the true values.

14.3.2 Set update

An update law that measures the worst-case progress of the parameter update law is adapted from the one proposed in Reference 7

$$z_{\hat{\theta}k} = \sqrt{\frac{V_{z\hat{\theta}k}}{4\lambda_{\min}(\Sigma_k)}} \tag{14.12a}$$

$$V_{z\theta k+1} = V_{z\hat{\theta}k} - (e_k - \hat{\eta}_k)^T\left(I + w_k\Sigma_k^{-1}w_k^T\right)^{-1}(e_k - \hat{\eta}_k) + \left(\frac{M_\vartheta}{K_k}\right)^2 \tag{14.12b}$$

$$V_{z\hat{\theta}0} = 4\lambda_{\max}(\Sigma_0)(z_{\hat{\theta}0})^2 \tag{14.12c}$$

The parameter uncertainty set, defined by the ball function $B(\hat{\theta}_c, z_c)$ is updated using the parameter update law (14.10) and the error bound (14.12) according to the following algorithm:

Algorithm 14.3.1. *Beginning at time step $k = 0$, the set is adapted according to the following iterative process.*

1. ***Initialize*** $z_{\hat{\theta}c} = z_{\hat{\theta}0}, \hat{\theta}_c = \hat{\theta}_0$

2. *At time step k, using* (14.10) *and* (14.12) *perform the update*

$$(\hat{\theta}_c, z_{\hat{\theta}c}) = \begin{cases} (\hat{\theta}_k, z_{\hat{\theta}k}) & \text{if } z_{\hat{\theta}k} \leq z_c - \|\hat{\theta}_k - \hat{\theta}_c\| \\ (\hat{\theta}_c, z_{\hat{\theta}c}) & \text{otherwise} \end{cases} \tag{14.13}$$

3. *Return to Step 2 and iterate, incrementing to time step* $k+1$

14.4 Robust adaptive MPC

14.4.1 A min–max approach

The formulation of the min–max MPC consists of maximizing a cost function with respect to $\theta \in \Theta$, $\vartheta \in \mathcal{D}$ and minimizing over feedback control policies κ. This formulation is a simple application of the min–max approach proposed in the continuous-time setting.

The proposed robust receding horizon control law is given by:

$$u = \kappa_{mpc}(x, \hat{\theta}, z_\theta) \triangleq \kappa^*(0, x, \hat{\theta}, z_\theta) \tag{14.14a}$$

$$\kappa^* \triangleq \arg\min_{\kappa(\cdot,\cdot,\cdot,\cdot)} J(x, \hat{\theta}, z_\theta, \kappa) \tag{14.14b}$$

where

$$J(x, \hat{\theta}, z_\theta, \kappa) \triangleq \max_{\theta \in \Theta,\ \vartheta \in \mathcal{D}} \sum_{k=0}^{T-1} L(x_k^p, u_k^p) d\tau + W(x_T^p, \tilde{\theta}_T^p) \tag{14.15a}$$

$$\text{s.t. } \forall k \in [0, T]$$

$$x_{k+1}^p = x_k + f(x_k^p, u_k^p) + g(x_k^p, u_k^p)\theta + \vartheta_k, \quad x_0^p = x \tag{14.15b}$$

$$\hat{x}_{k+1}^p = \hat{x}_k^p + F(x_k^p, u_k^p) + G(x_k^p, u_k^p)\hat{\theta}_{k+1}^p \tag{14.15c}$$

$$+ K_k e_k^p - \omega_k^p(\hat{\theta}_k^p - \hat{\theta}_{k+1}^p) + K_k \omega_k^p(\hat{\theta}_k^p - \hat{\theta}_{k+1}^p) \tag{14.15d}$$

$$w_{k+1}^p = w_k + G(x_k^p, u_k^p) - k_w w_k^p, \quad w_0^p = w \tag{14.15e}$$

$$(\Sigma_{k+1}^{-1})^p = (\Sigma_k^{-1})^p - (\Sigma_k^{-1^T})^p (\omega_k^p)^T \left(I + w_k^p (\Sigma_k^{-1})^p (w_k^p)^T\right)^{-1} \omega_k^p (\Sigma_k^{-1})^p \tag{14.15f}$$

$$(\Sigma_0^{-1})^p = \Sigma^{-1} \tag{14.15g}$$

$$\hat{\theta}_{k+1}^p = \text{Proj}\{\hat{\theta}_k^p + (\Sigma_k^{-1})^p (\omega_k^T)^p \left(I + w_k^p (\Sigma_k^{-1})^p (w_k^T)^p\right)^{-1} (e_k^p - \hat{\eta}_k^p), \Theta\} \tag{14.15h}$$

$$\tilde{\theta}^p = \theta - \hat{\theta}_k^p, \quad \hat{\theta}_0^p = \hat{\theta} \tag{14.15i}$$

$$u^p(\tau) \triangleq \kappa(\tau, x^p(\tau), \hat{\theta}^p(\tau)) \in \mathbb{U} \tag{14.15j}$$

$$x^p(\tau) \in \mathbb{X}, \quad x^p(T) \in \mathbb{X}_f(\tilde{\theta}^p(T)) \tag{14.15k}$$

As before, the effect of future parameter adaptation is also accounted for in this formulation but the proposed discrete-time parameter estimation and set-based approach is considered.

The conservativeness of the algorithm is reduced by parameterizing both W and $\mathbb{X}_f$ as functions of $\tilde{\theta}(T)$. While it is possible for the set Θ to contract upon θ over time, the robustness feature due to $\vartheta \in \mathcal{D}$ will still remain.

Algorithm 14.4.1. *The MPC algorithm performs as follows: At sampling instant k*

1. ***Measure*** *the current state of the plant* x_k *and obtain the current value of matrices* w *and* Σ^{-1} *from* (14.3) *and* (14.9), *respectively.*
2. ***Obtain*** *the current value of parameter estimates* $\hat{\theta}$ *and uncertainty bound* z_θ *from* (14.10) *and* (14.12), *respectively*

   If $z_{\theta_k} \leq z_{\theta_{k-1}} - \|\hat{\theta}_k - \hat{\theta}_{k-1}\|$

   $$\hat{\theta} = \hat{\theta}_k, \quad z_\theta = z_{\theta_k},$$

   Else

   $$\hat{\theta} = \hat{\theta}_{k-1}, \quad z_\theta = z_{\theta_{k-1}}$$

   End
3. ***Solve*** *the optimization problem* (14.14) *and apply the resulting feedback control law to the plant until the next sampling instant.*
4. ***Increment*** $k = k + 1$. *Repeat the procedure from Step* 1 *for the next sampling instant.*

The min–max approach guarantees robust stability but remains impractical in application. The next approach adopts a tube-based approach that retains robust stability.

14.4.2 Lipschitz-based approach

In this section, we present a Lipschitz-based method whereby the nominal model rather than the unknown bounded system state is controlled, subject to conditions that ensure that given constraints are satisfied for all possible uncertainties. State prediction error bound is determined based on the Lipschitz continuity of the model. A knowledge of appropriate Lipschitz bounds for the x-dependence of the dynamics $F(x, u)$ and $G(x, u)$ are assumed as follows:

Assumption 14.4.2. *A set of functions* $\mathcal{L}_j : \mathbb{X} \times \mathbb{U} \to \mathbb{R}^+$, $j \in \{F, G\}$ *are known which satisfy*

$$\mathcal{L}_j(\mathbb{X}, u) \geq \min\Big\{\mathcal{L}_j \;\Big|\; \sup_{x_1, x_2 \in \mathbb{X}} \big(\|j(x_1, u) - j(x_2, u)\| - \mathcal{L}_j\|x_1 - x_2\|\big) \leq 0\Big\},$$

where for $j \equiv G$ *is interpreted as an induced norm since* $G(x, u)$ *is a matrix.* ■

Assuming a knowledge of the Lipschitz bounds for the x-dependence of the dynamics $F(x, u)$ and $G(x, u)$ as given in Assumption 14.4.2 and let $\Pi = z_\theta + \|\hat{\theta}\|$, a worst-case deviation $z^p_{x,k} \geq \max_{\theta \in \Theta} \|x_k - x^p_k\|$ can be generated from

$$z^p_{x,k+1} = (\mathcal{L}_f + \mathcal{L}_g \Pi) z^p_{x,k} + \|G(x^p_k, u_k)\| z_\theta + M_\vartheta, \quad z^p_{x,0} = 0. \tag{14.17}$$

Using this error bound, the robust Lipschitz-based MPC is given by

$$u = \kappa_{mpc}(x, \hat{\theta}, z_\theta) = u^*(0) \tag{14.18a}$$

$$u^*(\cdot) \triangleq \arg\min_{u^p_{[0,T]}} J(x, \hat{\theta}, z_\theta, u^p) \tag{14.18b}$$

where

$$J(x, \hat{\theta}, z_\theta, u^p) = \sum_{k=0}^{T-1} L(x^p_k, u^p_k)d\tau + W(x^p_T, z^p_\theta) \tag{14.19a}$$

$$\text{s.t. } \forall k \in [0,\ T]$$

$$x^p_{k+1} = x^p_k + F(x^p_k, u^p_k) + G(x^p_k, u^p_k)\hat{\theta}, \quad x^p_0 = x \tag{14.19b}$$

$$z^p_{x,k+1} = (\mathcal{L}_f + \mathcal{L}_g \Pi) z^p_{x,k} + \|G^p(x^p_k, u^p_k)\| z_\theta + M_\vartheta, \quad z^p_{x,0} = 0 \tag{14.19c}$$

$$X^p(\tau) \triangleq B(x^p_k, z^p_{x,k}) \subseteq \mathbb{X}, \quad u^p_k \in \mathbb{U} \tag{14.19d}$$

$$X^p(T) \subseteq \mathbb{X}_f(z_\theta) \tag{14.19e}$$

The effect of the disturbance is built into the uncertainty cone $B(x^p_k, z^p_{x,k})$ via (14.19c). Since the uncertainty bound is no more monotonically decreasing in this case, the uncertainty radius z_θ which appears in (14.19c) and in the terminal expressions of (14.19a) and (14.19e) are held constant over the prediction horizon. However, the fact that they are updated at sampling instants when z_θ shrinks reduces the conservatism of the robust MPC and enlarges the terminal domain that would otherwise have been designed based on a large initial uncertainty z_{θ_0}.

Algorithm 14.4.3. *The Lipschitz-based MPC algorithm performs as follows: At sampling instant k*

1. ***Measure*** *the current state of the plant* $x = x_k$.
2. ***Obtain*** *the current value of the parameter estimates* $\hat{\theta}$ *and uncertainty bound* z_θ *from* (14.10) *and* (14.12) *respectively,*

   If $z_{\theta_k} \leq z_{\theta_{k-1}}$

   $$\hat{\theta} = \hat{\theta}_k, \quad z_\theta = z_{\theta_k}$$

   Else

   $$\hat{\theta} = \hat{\theta}_{k-1}, \quad z_\theta = z_{\theta_{k-1}}$$

   End
3. ***Solve*** *the optimization problem* (14.18a) *and apply the resulting feedback control law to the plant until the next sampling instant.*
4. ***Increment*** $k := k + 1$; *repeat the procedure from Step 1 for the next sampling instant.*

14.5 Closed-loop robust stability

Robust stabilization to the target set Ξ is guaranteed by appropriate selection of the design parameters W and X_f. The robust stability conditions require the satisfaction of the following criteria.

Criterion 14.5.1. *The terminal penalty function* $W : \mathbb{X}_f \times \tilde{\Theta}^0 \to [0, +\infty]$ *and the terminal constraint function* $\mathbb{X}_f : \tilde{\Theta}^0 \to \mathbb{X}$ *are such that for each* $(\theta, \hat{\theta}, \tilde{\theta}) \in (\Theta^0 \times \Theta^0 \times \tilde{\Theta}^0_\epsilon)$, *there exists a feedback* $k_f(\cdot, \hat{\theta}) : \mathbb{X}_f \to \mathbb{U}$ *satisfying*

1. $0 \in \Xi \subseteq \mathbb{X}_f(\tilde{\theta}) \subseteq \mathbb{X}, \;\; \mathbb{X}_f(\tilde{\theta})$ *closed*
2. $k_f(x, \hat{\theta}) \in \mathbb{U}, \forall x \in \mathbb{X}_f(\tilde{\theta})$
3. $W(x, \tilde{\theta})$ *is continuous with respect to* $x \in \mathbb{R}^{n_x}$
4. $\forall x \in \mathbb{X}_f(\tilde{\theta}) \backslash \Xi$, $\mathbb{X}_f(\tilde{\theta})$ *is strongly positively invariant under* $k_f(x, \hat{\theta})$ *with respect to* $x_+ \in x + F(x, k_f(x, \hat{\theta})) + G(x, k_f(x, \hat{\theta}))\Theta + \mathcal{D}$
5. $L(x, k_f(x, \hat{\theta})) + W(x_+, \hat{\theta}) - W(x, \hat{\theta}) \leq 0, \;\; \forall x \in \mathbb{X}_f(\tilde{\theta}) \backslash \Xi.$

The condition 5 from Criteria 14.5.1 require, W to be a local robust CLF for the uncertain system 14.1 with respect to $\theta \in \Theta$ and $\vartheta \in \mathcal{D}$.

Criterion 14.5.2. *For any* $\tilde{\theta}_1, \; \tilde{\theta}_2 \in \tilde{\Theta}^0$ *s.t.* $\|\tilde{\theta}_2\| \leq \|\tilde{\theta}_1\|$,

1. $W(x, \tilde{\theta}_2) \leq W(x, \tilde{\theta}_1), \;\; \forall x \in \mathbb{X}_f(\tilde{\theta}_1)$
2. $\mathbb{X}_f(\tilde{\theta}_2) \supseteq \mathbb{X}_f(\tilde{\theta}_1)$

14.5.1 Main results

Theorem 14.5.3. *Let* $X_{d0} \triangleq X_{d0}(\Theta^0) \subseteq \mathbb{X}$ *denote the set of initial states with uncertainty* Θ^0 *for which* (14.14) *has a solution. Assuming Criteria* 14.5.1 *and* 14.5.2 *are satisfied, then the closed-loop system state x, given by* (14.1–14.3, 14.6, 14.9, 14.10, 14.12, 14.14), *originating from any* $x_0 \in X_{d0}$ *feasibly approaches the target set* Ξ *as* $t \to +\infty$.

Proof:
Feasibility: The closed-loop stability is based upon the feasibility of the control action at each sample time. Assuming, at time t, that an optimal solution $u^p_{[0,T]}$ to the optimization problem (14.14) exist and is found. Let Θ^p denote the estimated uncertainty set at time t and Θ^v denote the set at time $t+1$ that would result with the feedback implementation of $u_t = u^p_0$. Also, let x^p represents the worst-case state trajectory originating from $x^p_0 = x_t$ and x^v represents the trajectory originating from $x^v_0 = x + v$ for $v \in \{F(x^a, u^p) + G(x^a, u^p)\Theta^b + \mathcal{D}\}$ under the same feasible control input $u^v_{[1,T]} = u^p_{[1,T]}$. Moreover, let $X^a_{\Theta^b} \triangleq \{x^a \mid x^a_+ \in x^a + F(x^a, u^p) + G(x^a, u^p)\Theta^b + \mathcal{D}\}$ which represents the set of all trajectories of the uncertain dynamics.

Since the $u^p_{[0,T]}$ is optimal with respect to the worst-case uncertainty scenario, suffice to say that $u^p_{[0,T]}$ drives any trajectory $x^p \in X^p_{\Theta^p}$ into the terminal region $\mathbb{X}^p_f$. Since

Θ is non-expanding over time, we have $\Theta^v \subseteq \Theta^p$ implying $x^v \in X^p_{\Theta^v} \subseteq X^p_{\Theta^p}$. The terminal region $\mathbb{X}^p_f$ is strongly positively invariant for the nonlinear system (14.1) under the feedback $k_f(\cdot,\cdot)$, the input constraint is satisfied in $\mathbb{X}^p_f$ and $\mathbb{X}^v_f \supseteq \mathbb{X}^p_f$ by Criteria 14.5.1(2), 14.5.1(4), and 14.5.2(2), respectively. Hence, the input $u = [u^p_{[1,T]}, k_{f[T,T+1]}]$ is a feasible solution of (14.14) at time $t+1$ and by induction, the optimization problem is feasible for all $t \geq 0$.

Stability: The stability of the closed-loop system is established by proving strict decrease of the optimal cost $J^*(x,\hat{\theta},z_\theta) \triangleq J(x,\hat{\theta},z_\theta,\kappa^*)$. Let the trajectories $(x^p,\ \hat{\theta}^p,\ \tilde{\theta}^p, z^p_\theta)$ and control u^p correspond to any worst-case minimizing solution of $J^*(x,\hat{\theta},z_\theta)$. If $x^p_{[0,T]}$ were extended to $k \in [0, T+1]$ by implementing the feedback $u^p_{T+1} = k_f(x^p_{T+1}, \hat{\theta}^p)$ then Criterion 14.5.1(5) guarantees the inequality:

$$L(x^p_T, k_f(x^p_T,\hat{\theta}^p_T)) + W(x^p_{T+1},\tilde{\theta}^p_T) - W(x^p_T,\tilde{\theta}^p_T) \leq 0. \tag{14.20}$$

The optimal cost

$$\begin{aligned} J^*(x_t,\hat{\theta}_t,z_{\theta_t}) &= \sum_{k=0}^{T-1} L(x^p_k,u^p_k) + W(x^p_T,\tilde{\theta}^p_T) \geq \sum_{k=0}^{T-1} L(x^p,u^p) + W(x^p_T,\tilde{\theta}^p_T) \\ &\quad + L(x^p_T, k_f(x^p_T,\hat{\theta}^p_T)) + W(x^p_{T+1},\tilde{\theta}^p_T) - W(x^p_T,\tilde{\theta}^p_T) && (14.21) \\ &\geq L(x^p_0,u^p_0) + \sum_{k=1}^{T} L(x^p_k,u^p_k) \\ &\quad + L(x^p_T, k_f(x^p_T,\hat{\theta}^p_T)) + W(x^p_{T+1},\tilde{\theta}^p_{T+1}) && (14.22) \\ &\geq L(x^p_0,u^p_0) + J^*(x_{t+1},\hat{\theta}_{t+1},z_{\theta_{t+1}}). && (14.23) \end{aligned}$$

Then, it follows from (14.23) that

$$J^*(x_{t+1},\hat{\theta}_{t+1},z_{\theta_{t+1}}) - J^*(x_t,\hat{\theta}_t,z_{\theta_t}) \leq -L(x_t,u_t) \leq -\mu_L(\|x\|). \tag{14.24}$$

where μ_L is a class $\mathcal{K}_\infty$ function. Hence $x(t)$ enters Ξ asymptotically.

Remark 14.5.4. *In the above proof,*

- (14.21) *is obtained using inequality* (14.20)
- (14.22) *follows from Criterion* 14.5.1.1 *and the fact that* $\|\tilde{\theta}\|$ *is non-increasing*
- (14.23) *follows by noting that the last three terms in* (14.22) *is a (potentially) suboptimal cost on the interval* $[\delta, T+\delta]$ *starting from the point* $(x^p(\delta),\hat{\theta}^p(\delta))$ *with associated uncertainty set* $B(\hat{\theta}^p(\delta), z^p_\theta(\delta))$.

The closed-loop stability is established by the feasibility of the control action at each sample time and the strict decrease of the optimal cost J^*. The proof follows from the fact that the control law is optimal with respect to the worst-case uncertainty $(\theta,\ \vartheta) \in (\Theta,\ \mathcal{D})$ scenario and the terminal region $\mathbb{X}^p_f$ is strongly positively invariant for (14.1) under the (local) feedback $k_f(\cdot,\cdot)$. ■

Theorem 14.5.5. *Let $X'_{d0} \triangleq X'_{d0}(\Theta^0) \subseteq \mathbb{X}$ denote the set of initial states for which* (14.18a) *has a solution. Assuming Assumption* 14.4.2 *and Criteria* 14.5.1 *and* 14.5.2 *are satisfied, then the origin of the closed-loop system given by* (14.1–14.3, 14.6, 14.9, 14.10, 14.12, 14.18a) *is feasibly asymptotically stabilized from any $x_0 \in X'_{d0}$ to the target set Ξ.*

The proof of the Lipschitz-based control law follows from that of Theorem 14.5.3.

Remark 14.5.6. *Note that the min–max approach can be prohibitively difficult to implement in practice to the computational complexity associated with the min–max optimization. However, the Lipschitz-based approach can be implemented using any standard RTO algorithm currently used for the solution of standard MPC problems. This latter technique will be employed in the simulation example presented in the next section.* ■

14.6 Simulation example

$$A \xrightarrow{k1} B \xrightarrow{k2} C$$
$$2A \xrightarrow{k3} D \tag{14.25}$$

In discrete-time, the system of equations that describe the non-isothermal dynamic response can be represented by (14.26). The system states are the concentrations of the components A (x_1) a B (x_2) and the reactor temperature (x_3). In addition, two manipulated variables were considered, the dilution rate (u_1) and the jacket temperature (u_2).

$$\begin{aligned}
x_1(k+1) &= x_1(k) + \Delta t u_1(k)[C_{ae} - x_1(k)] - \Delta t\theta_1 e^{-\alpha_1/x_3(k)} x_1(k) \\
&\quad - \Delta t\theta_3 e^{-\alpha_3/x_3(k)} x_1^2(k) \\
x_2(k+1) &= x_2(k) - \Delta t u_1(k) x_2(k) + \Delta t\theta_1 e^{-\alpha_1/x_3(k)} x_1(k) - \Delta t\theta_2 e^{-\alpha_2/x_3(k)} x_2(k) \\
x_3(k+1) &= x_3(k) + \frac{\Delta t}{\rho C_p}\Big(\theta_1 e^{-\alpha_1/x_3(k)} x_1(k)\Delta H_1 \\
&\quad + \theta_2 e^{-\alpha_2/x_3(k)} x_2(k)\Delta H_2 + \theta_3 e^{-\alpha_3/x_3(k)} x_1^2(k)\Delta H_3\Big) \\
&\quad + \Delta t u_1(k)[T_0 - x_3(k)] + \Delta t\frac{K_w A_R}{\rho c_p V}[u_2(k) - x_3(k)]
\end{aligned} \tag{14.26}$$

This system can be represented by:

$$\mathbf{x}(k+1) = \mathbf{x}(k) + \mathbf{F}(\mathbf{x}(k), \mathbf{u}(k)) + \mathbf{G}(\mathbf{x}(k), \mathbf{u}(k))\boldsymbol{\theta} + \mathbf{v}_k. \tag{14.27}$$

Table 14.1 Model parameters

Parameter	Value
α_1	9758.3 K
α_2	9758.3 K
α_3	8560.0 K
ΔH_1	$4.2\ \frac{\text{kJ}}{\text{molA}}$
ΔH_2	$-11\ \frac{\text{kJ}}{\text{molB}}$
ΔH_3	$-41.85\ \frac{\text{kJ}}{\text{molA}}$
ρ	0.9342 kg/l
C_p	$3.01\ \frac{\text{kJ}}{\text{kg} \cdot \text{K}}$
A_R	$0.215\ \text{m}^2$
k_w	$4032\ \frac{\text{kJ}}{\text{hm}^2\text{K}}$
T_0	403.15 K

For $\boldsymbol{\theta} = [\theta_1 \ \ \theta_2 \ \ \theta_3]^T$, these matrices can be defined as:

$$\mathbf{F}(\mathbf{x}(k), \mathbf{u}(k)) = \begin{pmatrix} u_1(k)[C_{ae} - x_1(k)] \\ -u_1(k)x_2(k) \\ u_1(k)[T_0 - x_3(k)] + \frac{K_w A_R}{\rho c_p V}[u_2(k) - x_3(k)] \end{pmatrix} \Delta t \tag{14.28}$$

$$\mathbf{G}(\mathbf{x}(k), \mathbf{u}(k)) = \begin{pmatrix} -e^{-\alpha_1/x_3(k)}x_1(k) & 0 & -e^{-\alpha_3/x_3(k)}x_1^2(k) \\ e^{-\alpha_1/x_3(k)}x_1(k) & -e^{-\alpha_2/x_3(k)}x_2(k) & 0 \\ \dfrac{e^{-\alpha_1/x_3(k)}x_1(k)\Delta H_1}{\rho C_p} & \dfrac{e^{-\alpha_2/x_3(k)}x_2(k)\Delta H_2}{\rho C_p} & \dfrac{e^{-\alpha_3/x_3(k)}x_1^2(k)\Delta H_3}{\rho C_p} \end{pmatrix} \Delta t \tag{14.29}$$

The parameters used in this work were obtained from Reference 93 and are reproduced in Table 14.1.

The true value of the parameter vector is $\boldsymbol{\theta}_r = [1.287 \ \ 1.287 \ \ 9.043]^T$.

The control objective is to regulate the desired product concentration (x_2) and the reactor temperature (x_3) to a setpoint and simultaneously estimating the frequency (or pre-exponential) factors of the Arrhenius equation that are assumed to lie inside of a ball of known radius. This is a real industrial problem that can be found in some reactors in which catalyst deactivation is present and, consequently, the kinetics parameters may change after the system start-up. An example of this kind of chemical system is the deactivation of hydrotreating catalysts by coke deposition [134].

The MPC cost function was quadratic and can be written in deviation variables as:

$$\ell(\tilde{\mathbf{x}}, \tilde{\mathbf{u}}) = \tilde{\mathbf{x}}^T \mathbf{Q} \tilde{\mathbf{x}} + \tilde{\mathbf{u}}^T \mathbf{R} \tilde{\mathbf{u}} \tag{14.30}$$

In which $\tilde{\mathbf{x}} = \mathbf{x} - \mathbf{x}_{eq}$ and $\tilde{\mathbf{u}} = \mathbf{u} - \mathbf{u}_{eq}$. The subscript *eq* denotes the equilibrium point. The terminal penalty function used was parameter dependent using quadratic stability:

$$W(\tilde{\mathbf{x}}, \boldsymbol{\theta}) = \tilde{\mathbf{x}}^T \mathbf{P}(\boldsymbol{\theta}) \tilde{\mathbf{x}} \tag{14.31}$$

It was obtained using the approach proposed by Reference 66, solving a finite set of LMIs. For this purpose, the MATLAB® LMI toolbox [122] was used to represent the system and to find the solution.

The terminal region was estimated by the algorithm presented in Reference 34, which the main step is to find the value α that satisfies $\tilde{\mathbf{x}}^T \mathbf{P}(\boldsymbol{\theta}) \tilde{\mathbf{x}} \leq \alpha$ by solving the optimization problem 14.32 decreasing the value of α until the optimal value is nonpositive.

$$\max_{\tilde{\mathbf{x}}} \{\tilde{\mathbf{x}}^T \mathbf{P}(\boldsymbol{\theta}) \boldsymbol{\phi}(\tilde{\mathbf{x}}) - \kappa \cdot \tilde{\mathbf{x}}^T \mathbf{P}(\boldsymbol{\theta}) \tilde{\mathbf{x}}\} \tag{14.32}$$

In the problem 14.32, the value κ must satisfy the inequality:

$$\kappa \leq -\lambda_{\max}(\mathbf{A}_k) \tag{14.33}$$

in which $\mathbf{A}_k$ is the closed-loop response of the linearized system under the local controller and $\boldsymbol{\phi}(\tilde{\mathbf{x}})$ is the difference between the nonlinear and linear response

$$\boldsymbol{\phi}(\tilde{\mathbf{x}}) = f(\tilde{\mathbf{x}}, \mathbf{K}\tilde{\mathbf{x}}) - \mathbf{A}_k \tilde{\mathbf{x}}. \tag{14.34}$$

For the initial nominal estimate, the matrix $\mathbf{P}(\boldsymbol{\theta})$ is given by

$$\mathbf{P}(\boldsymbol{\theta}) = \begin{bmatrix} 2.9776 & 0 & 0 \\ 0 & 2.9760 & 0 \\ 0 & 0 & 3.5916 \end{bmatrix}. \tag{14.35}$$

Using this matrix, the problem 14.32 was solved. The maximum point was $\tilde{\mathbf{x}}_{\max} = [2.9776 \;\; 2.9760 \;\; 3.5916]$. This solutions leads to the terminal region:

$$\tilde{\mathbf{x}}^T \mathbf{P}(\boldsymbol{\theta}) \tilde{\mathbf{x}} \leq 177. \tag{14.36}$$

14.6.1 Open-loop tests of the parameter estimation routine

The uncertainty-based estimation routine for discrete-time systems was tested for the frequency factors estimation in a open-loop test. Two scenarios were evaluated. In the first one, the disturbance added to the system is a pulse in the manipulated variables. In the second test, a persistent bounded periodical signal was added to the reactor temperature. In both cases it is showed that the true parameters values were recovered. For this simulations the initial parameter estimates were

$$\boldsymbol{\theta}_0 = [5 \;\; 6 \;\; 7]^T \tag{14.37}$$

In the first simulation, the disturbance inserted into the system is a 10% pulse in the jacket temperature as showed in Figure 14.1.

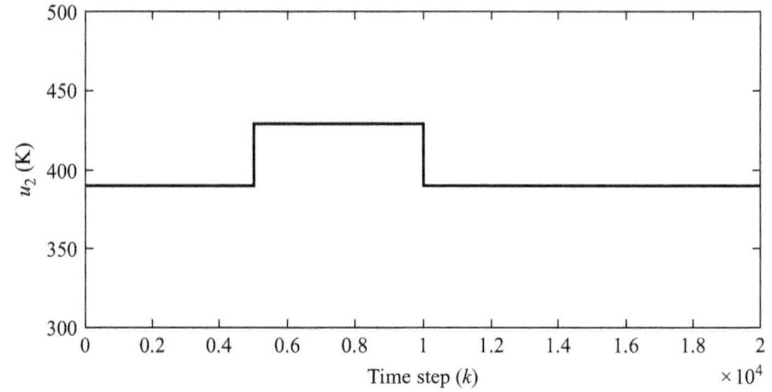

Figure 14.1 Disturbance inserted in the jacket temperature u_2

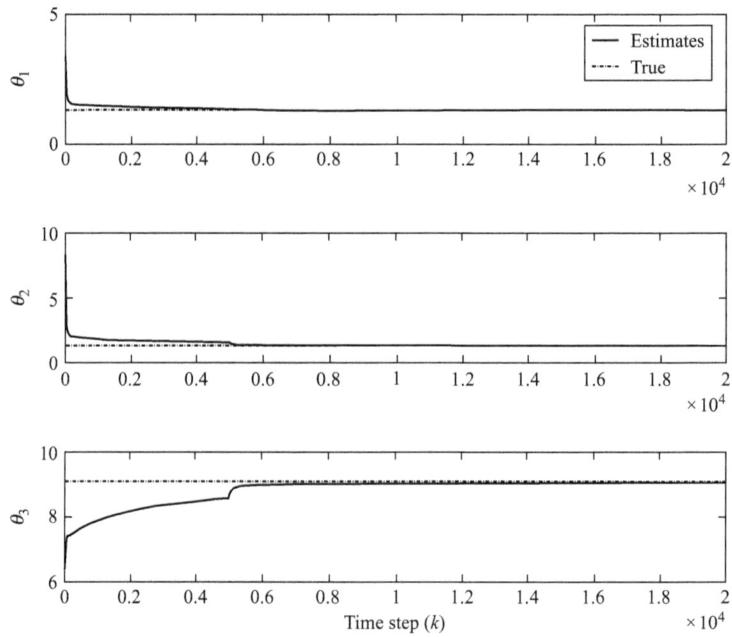

Figure 14.2 Time evolution of the parameter estimates and true values, using a pulse as disturbance

In Figure 14.2, the time evolution of the parameters during the simulation is showed. The excitation added by the pulse disturbance improves the convergence and accelerates the set contraction as showed in Figure 14.3. The true values of the parameters are recovered and the prediction error converges to zero (Figure 14.4).

In the next simulation, a persistent disturbance was added to the jacket temperature in the form of the function:

$$u_2(k) = u_{2,nom} + B \cdot \sin\left(\frac{k}{C}\right) \tag{14.38}$$

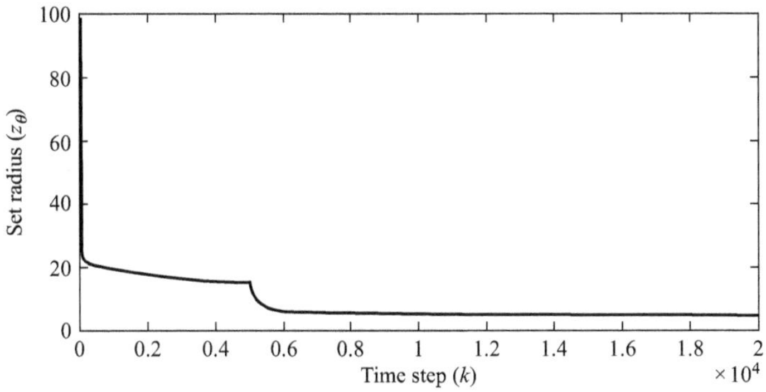

Figure 14.3 Progression of the set radius

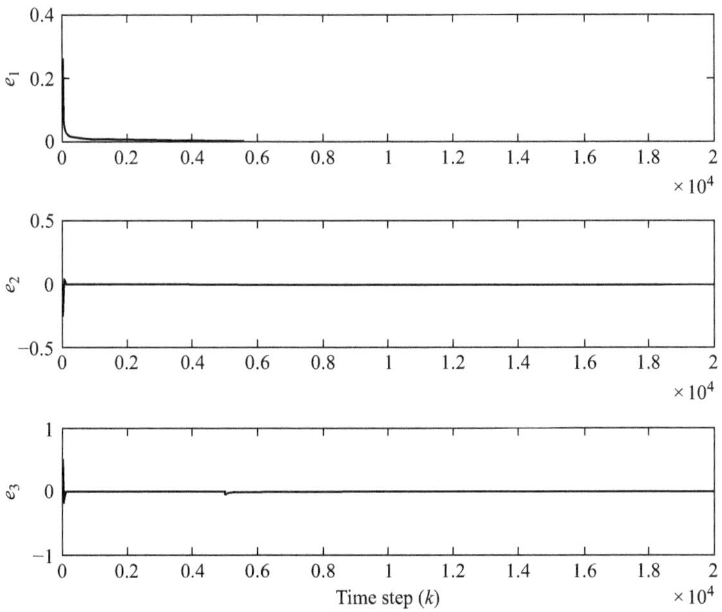

Figure 14.4 State prediction error $e_k = x_k - \hat{x}_k$ versus time step (k)

The parameters were $B = 1$ and $C = 50$, the time course of the jacket temperature is show in Figure 14.5.

The sine disturbance provides more excitation to the reactor system than the pulse signal, it leads to a faster convergence to the true values. The complete simulation is show in the Figure 14.6. Due to the difference in the rate of parameters convergence, the beginning of the simulation is show in Figure 14.7. As one can see, the kinetic constant of the side product reaction ($2A \xrightarrow{k3} D$), represented by the parameter θ_3, has

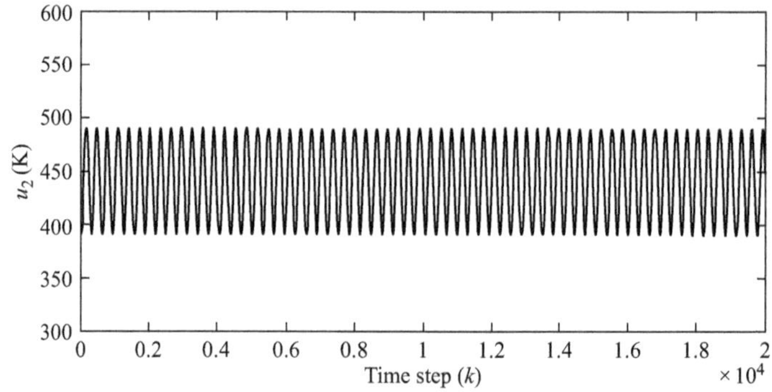

Figure 14.5 Sine disturbance inserted in the jacket temperature u_2

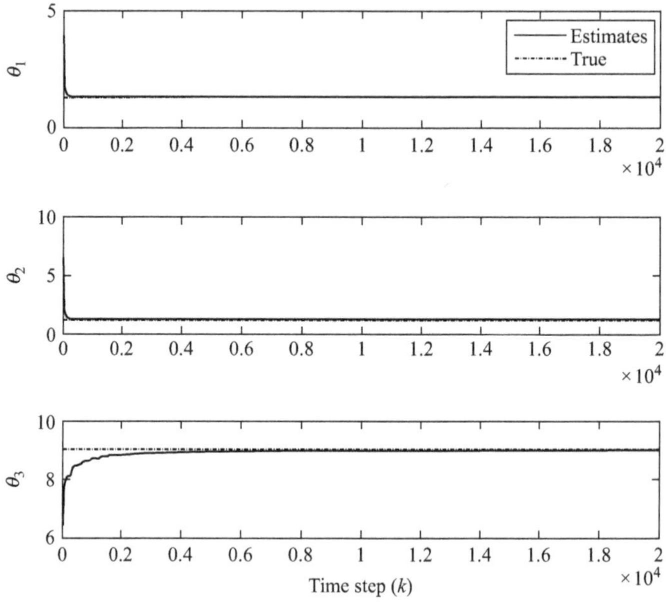

Figure 14.6 Time evolution of the parameter estimates and true values, using a sine function as disturbance

the slowest rate of convergence. The Figure 14.8 shows the estimator error and the Figure 14.9 displays the uncertainty set radius contraction.

14.6.2 Closed-loop simulations

For the closed-loop simulations, the Lipschitz constraints were used to replace the min–max problem and maintain the robustness of the controller. The initial values of the parameters are assumed to lie in a ball of radius $z_{\hat{\theta}0}$ and centered in the initial estimate $\boldsymbol{\theta}_0 = [5 \quad 6 \quad 7]^T$. The true value of the parameter vector is

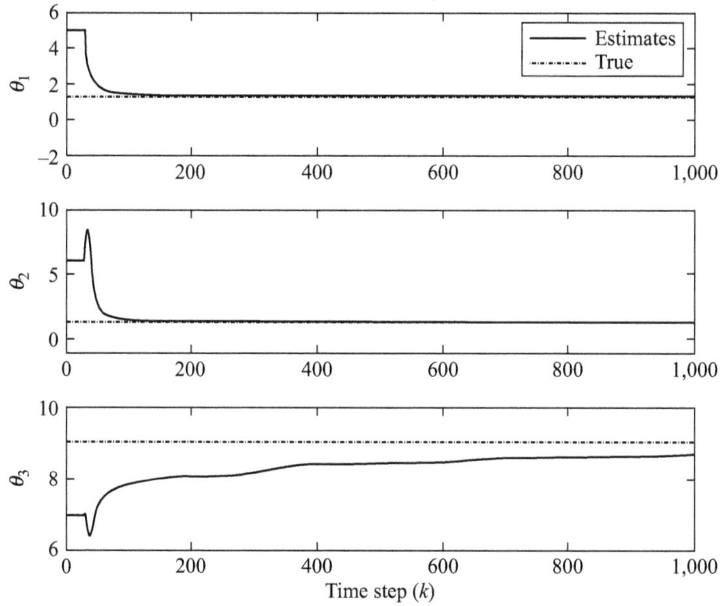

Figure 14.7 Time evolution of the parameter estimates and true values during the beginning of the simulation, using a sine as disturbance

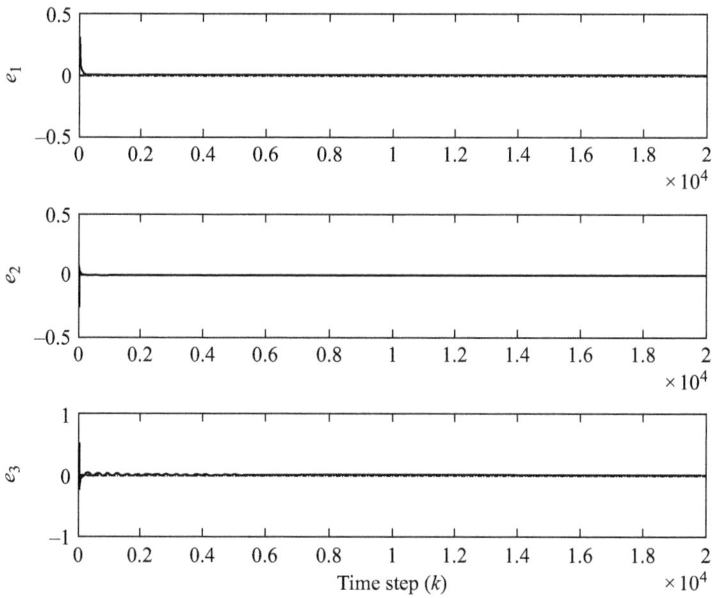

Figure 14.8 State prediction error $e_k = x_k - \hat{x}_k$ versus time step (k) for a sine disturbance

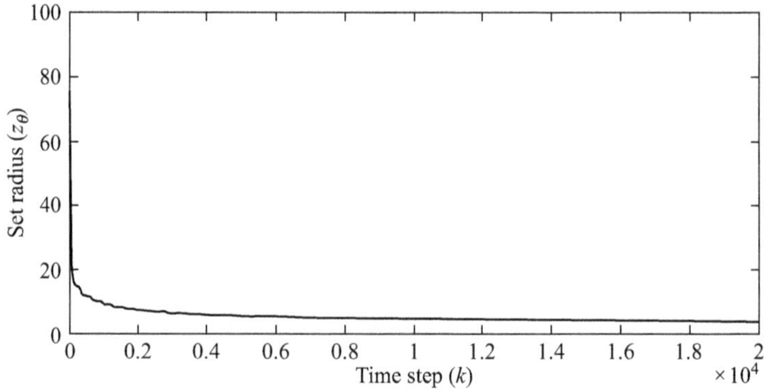

Figure 14.9 Progression of the set radius using a sine function as disturbance

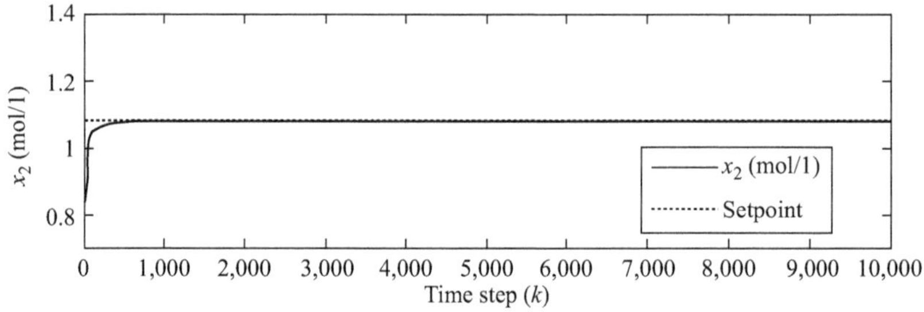

Figure 14.10 Concentration trajectory for the closed-loop system (x_2) versus time

$\boldsymbol{\theta}_r = [1.287 \;\; 1.287 \;\; 9.043]^T$. The matrices for the cost function 14.30 were chosen to be:

$$\mathbf{Q} = \begin{bmatrix} 35 & 0 \\ 0 & 7 \end{bmatrix} \quad \mathbf{R} = \begin{bmatrix} 0.1 & 0 \\ 0 & 1 \end{bmatrix} \tag{14.39}$$

The control objective is to regulate the system to the setpoint:

$$\mathbf{x}_{sp} = [x_{2,sp} \quad x_{3,sp}] = [1.079(\text{mol/l}) \quad 394.7(\text{K})] \tag{14.40}$$

while satisfying the constraints:

$$\begin{aligned}
&0 \le u_1 \le 500 \\
&0 \le u_2 \le 600 \\
&0 \le \Delta u_1 \le 10 \\
&0 \le \Delta u_2 \le 30 \\
&0 \le x_1 \le 10 \\
&0 \le x_2 \le 800
\end{aligned} \tag{14.41}$$

In Figures 14.10 and 14.11 the concentration and reactor temperature are showed, the states achieve the desired setpoint without offset. The manipulated variables are showed in Figures 14.12 (reactor flowrate) and 14.13 (jacket temperature). Moreover,

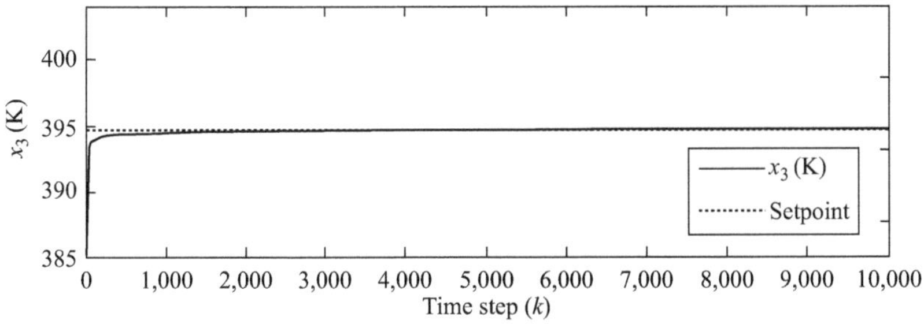

Figure 14.11 Reactor temperature trajectory for the closed-loop system (x_3) versus time

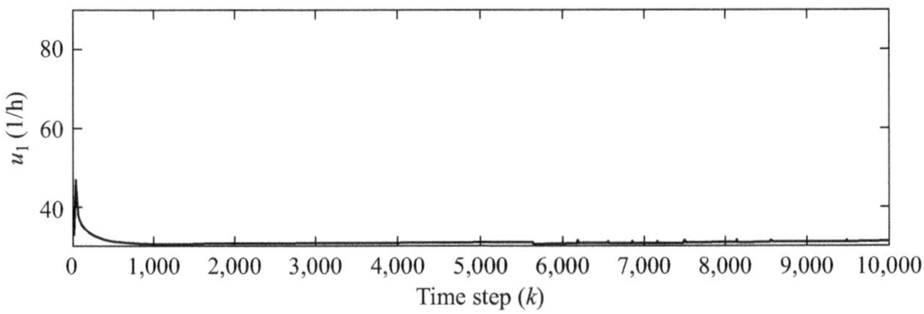

Figure 14.12 Manipulated flowrate (u_1) versus time

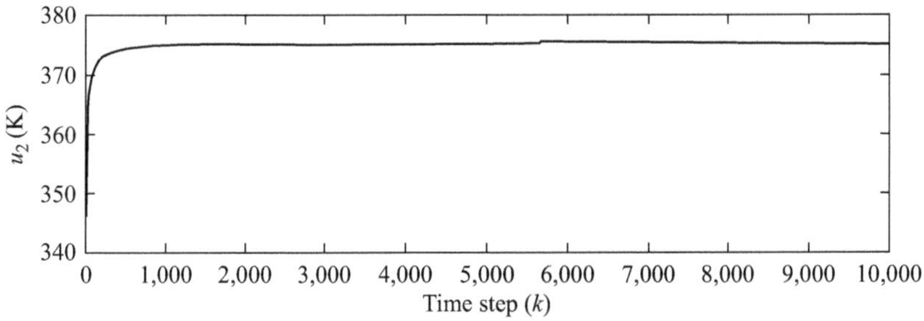

Figure 14.13 Manipulated jacket temperature (u_2) versus time

the parameter estimates converge to the true value, the parameter θ_3 has the slowest convergence rate (Figure 14.14). Figure 14.15 shows that the uncertainty set radius reduces over time.

14.6.3 Closed-loop simulations with disturbances

In order to simulate a disturbance in the closed-loop system, a fluctuation in the inlet temperature was inserted as a periodic function:

$$T_0(k) = 403.15 + \sin(k) \tag{14.42}$$

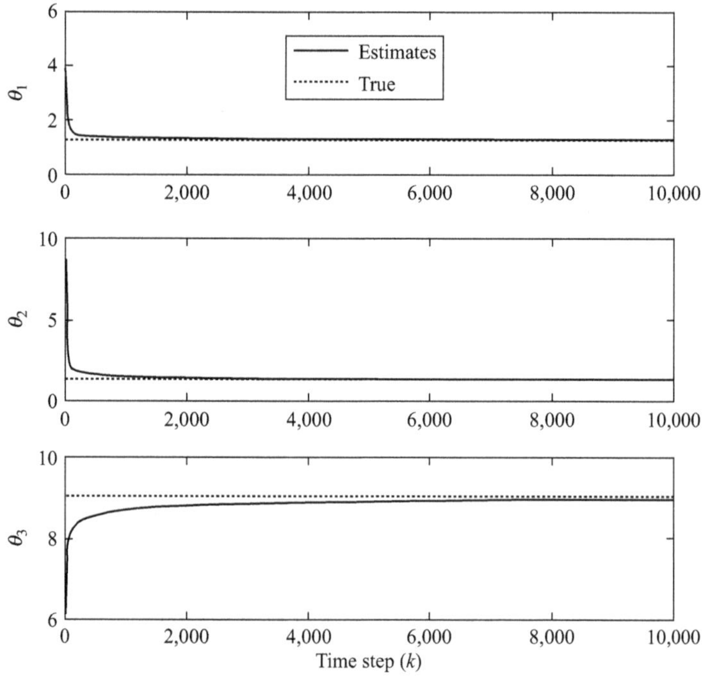

Figure 14.14 Time evolution of the parameter estimates and true values for the closed-loop system

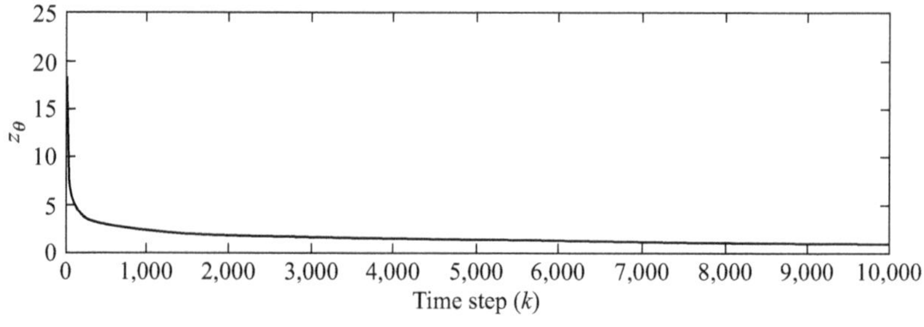

Figure 14.15 Closed-loop set radius versus time

The control objective is to regulate the system to the setpoint presented in Section 14.6.2. As depicted in Figures 14.16 and 14.17 the parameter convergence and the set radius reduction is more conservative in comparison with the disturbance-free case (Figures 14.14 and 14.15), however the true values of the parameters are recovered. The reactor temperature and concentration oscillate around the setpoint (Figures 14.18 and 14.19). In Figures 14.20 and 14.21 the control actions are showed. Finally, in Figure 14.22 a comparison between the two scenarios for the evolution of the uncertanity set radius is presented.

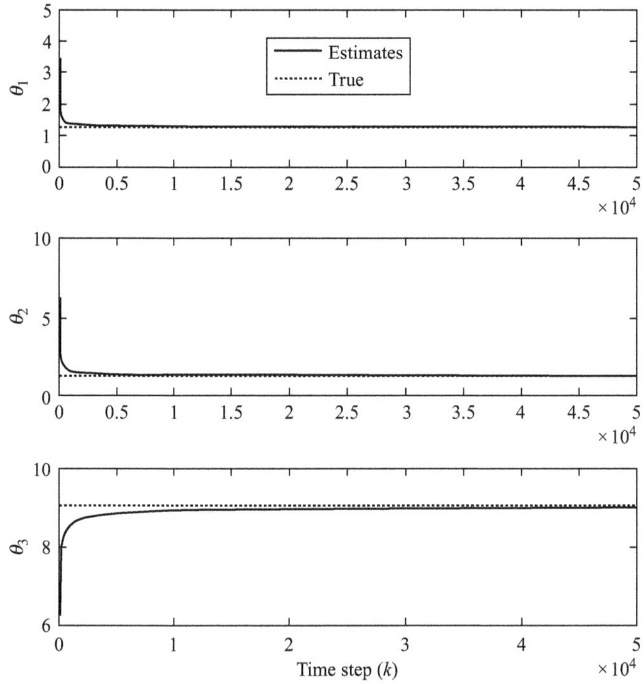

Figure 14.16 Time evolution of the parameter estimates and true values for the closed-loop system

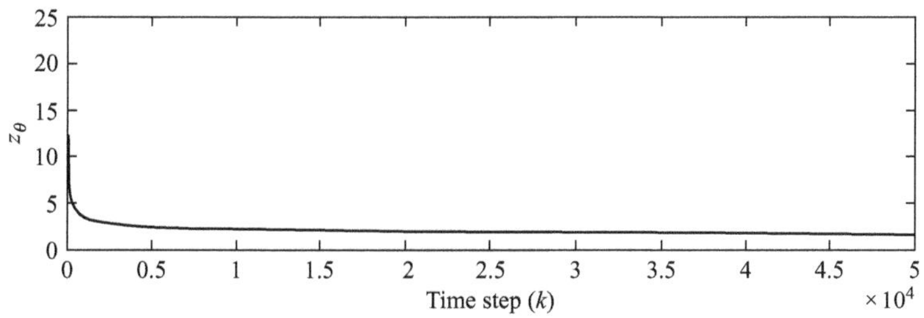

Figure 14.17 Closed-loop set radius versus time

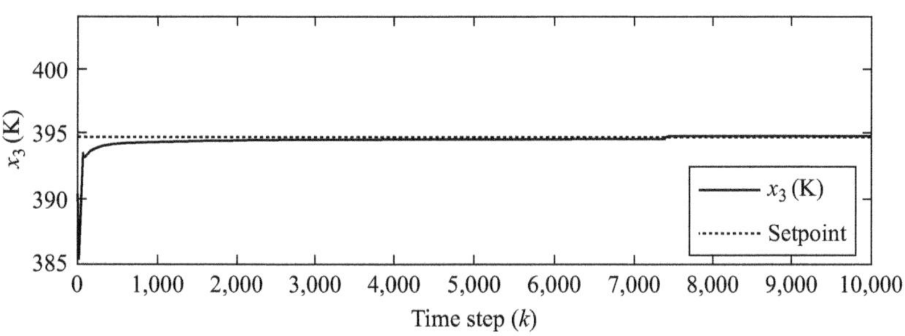

Figure 14.18 Reactor temperature trajectory for the closed-loop system (x_3) with disturbance versus time

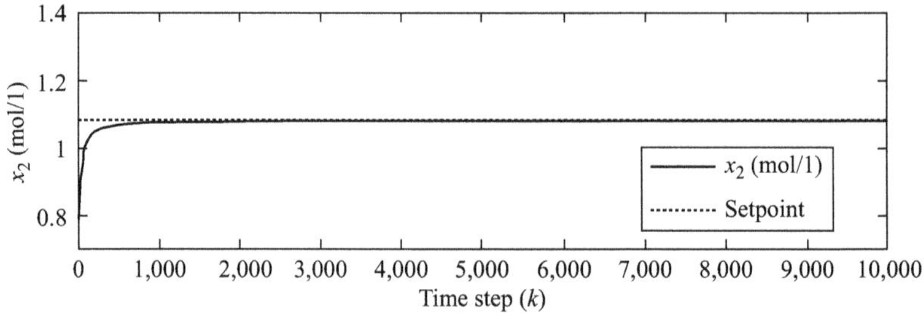

Figure 14.19 Concentration trajectory for the closed-loop system (x_2) with disturbance versus time

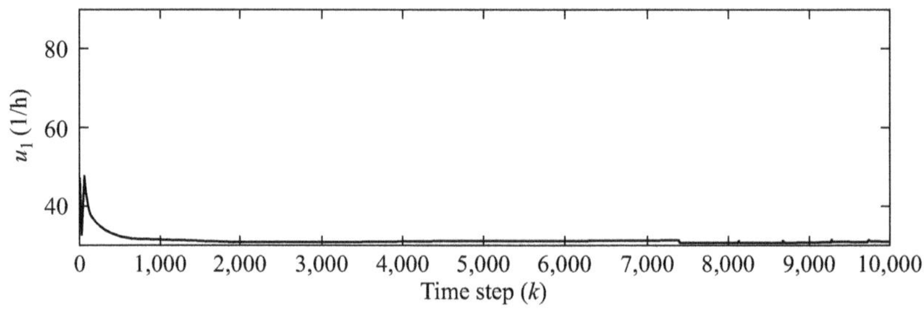

Figure 14.20 Manipulated flowrate (u_1) with disturbance versus time

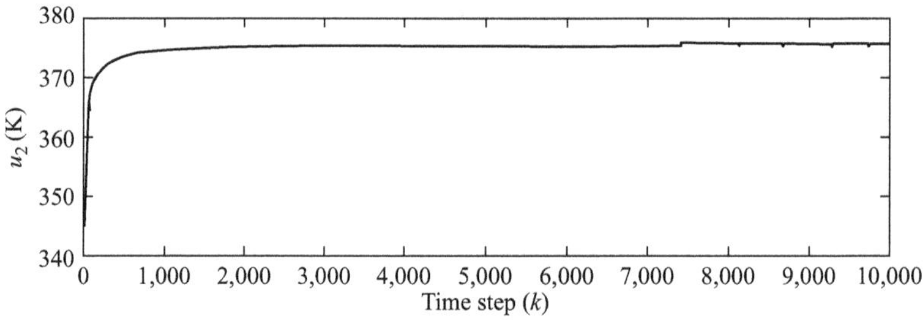

Figure 14.21 Manipulated jacket temperature (u_2) with disturbance versus time

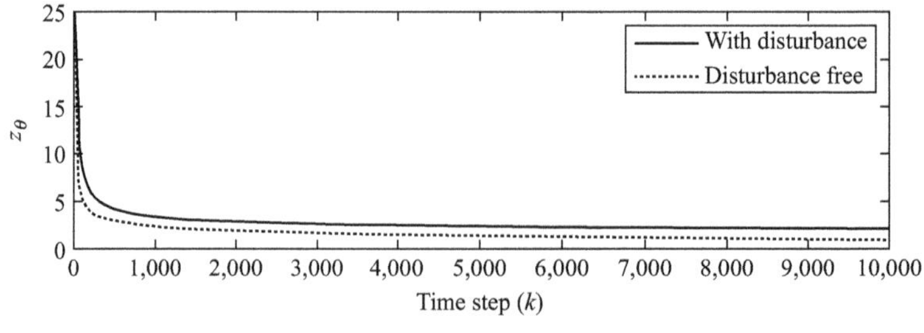

Figure 14.22 Comparison between the uncertainty set radius evolution for the inlet temperature disturbance and disturbance free cases

14.7 Summary

The adaptive MPC design technique is proposed for the control of constrained nonlinear systems subject to both parametric and time-varying disturbances. The proposed robust controller updates the plant model online when model improvement is guaranteed. The adaptation mechanism enables the construction of terminal design parameters based upon subsets of the original parametric uncertainty in a minimally conservative approach. The conservativeness and the complexity due to the parametric uncertainty is effectively reduced over time using a self-exciting mechanism arising from the adaptive MPC formulation. The portion due to the disturbance $\vartheta \in \mathcal{D}$ remains active for all time with guaranteed robust stability.

Bibliography

[1] V. Adetola and M. Guay. Adaptive receding horizon control of nonlinear systems. In *Proceedings of the IFAC Symposium on Nonlinear Control Systems*, pp. 1055–1060, Stuttgart, Germany, 2004.

[2] V. Adetola and M. Guay. Excitation signal design for parameter convergence in adaptive control. In *Proceedings of the IFAC ADCHEM*, pp. 447–452, San Diego, CA, 2006.

[3] V. Adetola and M. Guay. Finite-time parameter estimation in adaptive control of nonlinear systems. In *Proceedings of the American Control Conference*, New York, NY, 2007.

[4] V. Adetola and M. Guay. Parameter convergence in adaptive extremum seeking control. *Automatica*, 43(1):105–110, 2007.

[5] V. Adetola and M. Guay. Finite-time parameter estimation in adaptive control of nonlinear systems. *IEEE Transactions on Automatic Control*, 53(3): 807–811, 2008.

[6] V. Adetola and M. Guay. Performance improvement in adaptive control of nonlinear systems. In *Proceedings of the American Control Conference*, St. Louis, MO, 2009.

[7] V. Adetola and M. Guay. Robust adaptive mpc for systems with exogenous disturbances. In *Proceedings of the IFAC ADCHEM*, Istanbul, Turkey, 2009.

[8] V. Adetola and M. Guay. Integration of real-time optimization and model predictive control. *Journal of Process Control*, 20(2):125–133, 2010.

[9] V. Adetola and M. Guay. Robust adaptive mpc for constrained uncertain nonlinear systems. *International Journal of Adaptive Control and Signal Processing*, 25(2):155–167, 2011.

[10] V.A. Adetola and M. Guay. Excitation signal design for parameter convergence in adaptive control of linearizable systems. In *Proceedings of the 45th IEEE Conference on Decision and Control*, San Diego, CA, 2006.

[11] M. Alamir. Solutions of nonlinear optimal and robust control problems via a mixed collocation/dae's based algorithm. *Automatica*, 37:1109–1115, 2001.

[12] E.L. Allgower and K. George. *Numerical Continuation Methods*. Springer, Heidelberg, Germany, 1990.

[13] D. Angeli and J. Rawlings. Economic optimization using model predictive control with a terminal cost. *Annual Reviews in Control*, 35(2):178–186, 2011.

[14] D. Angeli and J.B. Rawlings. Receding horizon cost optimization and control for nonlinear plants. In *Proceedings of the 8th IFAC Symposium on Nonlinear Control Systems*, pp. 1217–1223, Bologna, Italy, 2010.

[15] Z. Artstein. Stabilization with relaxed controls. *Nonlinear Analysis, Theory, Methods and Applications*, 7(11):1163–1173, 1983.

[16] H. Attouch, X. Goudou, and P. Redont. The heavy ball with friction method, I. The continuous dynamical system: global exploration of the local minima of a real-valued function by asymptotic analysis of a dissipative dynamical system. *Communications in Contemporary Mathematics*, 2:1–34, 2000.

[17] J.P. Aubin. *Viability Theory. Systems & Control: Foundations & Applications*. Birkhäuser, Boston, MA, 1991.

[18] K. Basak, K.S. Abhilash, S. Ganguly, and D.N. Saraf. Online optimization of a crude distillation unit with constraints on product properties. *Industrial Engineering Chemistry Research*, 41:1557–1568, 2002.

[19] R. Bellman. *Dynamic Programming*. Princeton Press, Princeton, NJ, 1957.

[20] R. Bellman. On the theory of dynamic programming. *Proceedings of the National Academy of Sciences of the United States of America*, 38(8):716, 1952.

[21] H.Y. Benson, D.F. Shanno, and R.J. Vanderbei. Interior-point methods for nonconvex nonlinear programming: Jamming and numerical testing. *Mathematical Programming*, 99(1):35–48, 2004.

[22] H.Y. Benson, R.J. Vanderbei, and D.F. Shanno. Interior-point methods for nonconvex nonlinear programming: Filter methods and merit functions. *Computational Optimization and Applications*, 23(2):257–272, 2002.

[23] M. Bergounioux and K. Kunisch. Primal-dual strategy for state-constrained optimal control problems. *Computational Optimization and Applications*, 22:193–224, 2002.

[24] D. Bertsekas. *Dynamic Programming and Optimal Control*, vol. I. Athena Scientific, Belmont, MA, 1995.

[25] L.T. Biegler. Solution of dynamics optimization problems by successive quadratic programming and orthogonal collocation. *Computers and Chemical Engineering*, 8:243–248, 1984.

[26] L.T. Biegler. Efficient solution of dynamic optimization and nmpc problems. In F. Allgöwer and A. Zheng, editors, *Nonlinear Predictive Control*, volume 26 of *Progress in Systems Theory*, pp. 219–244. Birkhäuser, Berlin, Germany, 2000.

[27] B. Brogliato and A.T. Neto. Practical stabilization of a class of nonlinear systems with partially known uncertainties. *Automatica*, 31(1):145–150, 1995.

[28] A. Bryson and Y.C. Ho. *Applied Optimal Control*. Ginn and Co., Waltham, MA, 1969.

[29] M. Cannon and B. Kouvaritakis. Continuous-time predictive control of constrained non-linear systems. In F. Allgower and A. Zheng, editors, *Nonlinear Model Predictive Control: Assessment and Future Directions for Research*. Birkhäuser-Verlag, Berlin, Germany, 2000.

[30] M. Cannon and B. Kouvaritakis. Optimizing prediction dynamics for robust MPC. *IEEE Transaction on Automatic Control*, 50(11):1892–1897, 2005.

[31] C. Cao, J. Wang, and N. Hovakimyan. Adaptive control with unknown parameters in reference input. In *Proceedings of the 13th Mediterranean Conference on Control and Automation*, pp. 225–230, Limassol, Cyprus, 2005.

[32] C. Cao, J. Wang, and N. Hovakimyan. Adaptive control with unknown parameters in reference input. In *Proceedings of the 13th IEEE International Symposium on Mediterranean Conference on Control and Automation*, pp. 225–230, Limassol, Cyprus, June 2005.

[33] H. Chen and F. Allgöwer. A computationally attractive nonlinear predictive control scheme with guaranteed stability for stable systems. *Journal of Process Control*, 8(5–6):475–485, 1998.

[34] H. Chen and F. Allgöwer. A quasi-infinite horizon nonlinear model predictive control scheme with guaranteed stability. *Automatica*, 34(10):1205–1217, 1998.

[35] H. Chen, C. Scherer, and F. Allgöwer. A game theoretic approach to nonlinear robust receding horizon control of constrained systems. In *Proceedings of the American Control Conference*, 1997.

[36] W. Chen, D.J. Ballance, and J. O'Reilly. Model predictive control of nonlinear systems: Computational burden and stability. In *IEE Proceedings—Control Theory and Applications*, vol. 147, pp. 387–394, 2000.

[37] F. Clarke, Y. Ledyaev, E. Sontag, and A. Subbotin. Asymptotic controllability implies feedback stabilization. *IEEE Transaction on Automatic Control*, 42(10):1394–1407, 1997.

[38] F.H. Clarke, L. Rifford, and R.J. Stern. Feedback in state-constrained optimal control. *ESIAM: Control, Optimization and Calculus of Variations*, 7(1): 97–133, 2002.

[39] F.H. Clarke and R.J. Stern. State constrained feedback stabilization. *SIAM Journal on Control and Optimization*, 42(2):422–441, 2003.

[40] F.H. Clarke, Y.S. Ledyaev, R.J. Stern, and P.R. Wolenski. *Nonsmooth Analysis and Control Theory*. Grad. Texts in Math. 178, Springer-Verlag, New York, NY, 1998.

[41] M.J. Corless and G. Leitmann. Continuous state feedback guaranteeing uniform ultimate boundedness for uncertain dynamic systems. *IEEE Transaction on Automatic Control*, AC-26(5):1139–1144, 1981.

[42] J.M. Coron and L. Rosier. A relation between continuous time-varying and discontinuous feedback stabilization. *Journal of Mathematical Systems, Estimation, and Control*, 4(1):67–84, 1994.

[43] C.R. Cutler and B.L. Ramaker. Dynamic matrix control: A computer control algorithm. In *Proceedings of the Joint Automatic Control Conference*, San Francisco, CA, 1980.

[44] G. De Nicolao, L. Magni, and R. Scattolini. On the robustness of receding horizon control with terminal constraints. *IEEE Transaction on Automatic Control*, 41:454–453, 1996.

[45] N. de Oliveira and L.T. Biegler. An extension of Newton-type algorithms for nonlinear process control. *Automatica*, 31(2):281–286, 1995.

[46] N. de Oliveira and L.T. Biegler. Newton-type algorithms for nonlinear process control: Algorithm and stability restults. *Automatica*, 3(2):281–286, 1995.

[47] D. DeHaan, V. Adetola, and M. Guay. Adaptive robust mpc: An eye towards computational simplicity. In *Proceedings of the IFAC Symposium on Nonlinear Control Systems*, South Africa, 2007.

[48] D. DeHaan and M. Guay. Extremum seeking control of state constrained nonlinear systems. *Automatica*, 41(9):1567–1574, 2005.

[49] D. DeHaan and M. Guay. Adaptive robust mpc: A minimally-conservative approach. In *Proceedings of the American Control Conference*, July 2007.

[50] D. DeHaan and M. Guay. A real-time framework for model-predictive control of continuous-time nonlinear systems. *IEEE Transactions on Automatic Control*, 52(11):2047–2057, 2007.

[51] C.A. Desoer and M. Vidyasagar. *Feedback Systems: Input–Output Properties*. Academic Press, New York, NY, 1975.

[52] M. Diehl. *Real-Time Optimization for Large Scale Processes*. PhD thesis, University of Heidelberg, Germany, 2001.

[53] M. Diehl, R. Amrit, and J.B. Rawlings. A lyapunov function for economic optimizing model predictive control. *IEEE Transactions on Automatic Control*, 56(3):703–707, 2011.

[54] M. Diehl, G. Bock, J. Schloder, R. Findeisen, Z. Nagy, and F. Allgöwer. Real-time optimization and nonlinear model predictive control of processes governed by differential–algebraic equations. *Journal of Process Control*, 12(4):577–588, 2002.

[55] M. Diehl, H. Bock, and J.P. Schloder. A real-time iteration scheme for nonlinear optimization in optimal feedback control. *SIAM Journal on Control and Optimization*, 43(5):1714–1736, 2005.

[56] M. Diehl, R. Findeisen, F. Allgöwer, H.G. Bock, and J.P. Schloder. Nominal stability of real-time iteration scheme for nonlinear model predictive control. *IEE Proceedings: Control Theory and Applications*, 152(3):296–308, 2005.

[57] R. Findeisen and F. Allgöwer. Computational delay in nonlinear model predictive control. In *Proceedings of the IFAC Symposium on Advanced Control of Chemical Processes*, Hong Kong, 2003.

[58] R. Findeisen and F. Allgöwer. Stabilization using sampled-data open-loop feedback: A nonlinear model predictive control perspective. In *Proceedings of the IFAC Symposium on Nonlinear Control Systems*, pp. 735–740, Stuttgart, Germany, 2004.

[59] R. Findeisen, L. Imsland, F. Allgöwer, and B.A. Foss. Towards a sampled-data theory for nonlinear model predictive control. In C. Kang, M. Xiao, and W. Borges, editors, *New Trends in Nonlinear Dynamics and Control, and Their Applications*, vol. 295, pp. 295–313. Springer-Verlag, New York, NY, 2003.

[60] W.H. Fleming and R.W. Rishel. *Deterministic and Stochastic Optimal Control. Applications of Mathematics*. Springer-Verlag, Berlin, Germany, 1975.

[61] M. Fliess, S. Fuchshumer, M. Schuberl, K. Schlacher, and H. Sira-Ramirez. An introduction to algebraic discrete-time linear parametric identification with a concrete application. *Journal Européen des Systèmes Automatisés*, 42:210–232, 2008.

[62] F. Floret-Pontet and F. Lamnabhi-Lagarrigue. Parameter identification and state estimation for continuous-time nonlinear systems. In *Proceedings of the American Control Conference*, vol. 1, pp. 394–399, 8–10 May 2002.

[63] F. Fontes. A general framework to design stabilizing nonlinear model predictive controllers. *Systems and Control Letters*, 42(2):127–143, 2001.

[64] R. Freeman and P. Kokotović. Inverse optimality in robust stabilization. *SIAM Journal on Control and Optimization*, 34:1365–1391, 1996.

[65] R. Freeman and P. Kokotović. *Robust Nonlinear Control Design*. Birkhäuser, Berlin, Germany, 1996.

[66] P. Gahinet, P. Apkarian, and M. Chilali. Affine parameter-dependent lyapunov functions and real parametric uncertainty. *IEEE Transactions on Automatic Control*, 41(3):436–442, 1996.

[67] R. Goebel and A.R. Teel. Solutions to hybrid inclusions via set and graphical convergence with stability theory applications. *Automatica*, 42(4):573–587, 2006.

[68] M.A. Golberg. The derivative of a determinant. *The American Mathematical Monthly*, 79(10):1124–1126, 1972.

[69] G. Grimm, M. Messina, S. Tuna, and A. Teel. Nominally robust model predictive control with state constraints. In *Proceedings of the IEEE Conference on Decision and Control*, pp. 1413–1418, 2003.

[70] G. Grimm, M. Messina, S. Tuna, and A. Teel. Examples when model predictive control is non-robust. *Automatica*, 40(10):1729–1738, 2004.

[71] G. Grimm, M. Messina, S. Tuna, and A. Teel. Model predictive control: For want of a local control lyapunov function, all is not lost. *IEEE Transaction on Automatic Control*, 50(5):617–628, 2005.

[72] L. Grune and D. Nešić. Optimization-based stabilization of sampled-data nonlinear systems via their approximate discrete-time models. *SIAM Journal on Control and Optimization*, 42(1):98–122, 2003.

[73] M. Guay, D. Dochain, and M. Perrier. Adaptive extremum seeking control of continuous stirred tank bioreactors with unknown growth kinetics. *Automatica*, 40:881–888, 2004.

[74] M. Guay, D. Dochain, and M. Perrier. Adaptive extremum seeking control of nonisothermal continuous stirred tank reactors. *Chemical Engineering Science*, 60:3671–3681, 2005.

[75] M. Guay and T. Zhang. Adaptive extremum seeking control of nonlinear dynamic systems with parametric uncertainties. *Automatica*, 39:1283–1293, 2003.

[76] W.M. Haddad. *Nonlinear Dynamical Systems and Control*. Princeton University Press, Princeton, NJ, 2008.

[77] H. Hermes. Discontinuous vector fields and feedback control. In J.K. Hale and J.P. LaSalle, editors, *Differential Equations and Dynamical Systems*, pp. 155–166. Academic Press, New York, NY, 1967.

[78] M. Hestenes. *Calculus of Variations and Optimal Control*. John Wiley & Sons, New York, NY, 1966.

[79] J.T. Huang. Sufficient conditions for parameter convergence in linearizable systems. *IEEE Transactions on Automatic Control*, 48:878–880, 2003.

[80] P.A. Ioannou and J. Sun. *Robust Adaptive Control*. Prentice Hall, Upper Saddle River, NJ, 1996.

[81] A. Jadbabaie, J. Yu, and J. Hauser. Unconstrained receding-horizon control of nonlinear systems. *IEEE Transaction on Automatic Control*, 46(5):776–783, 2001.

[82] H.T. Jongen and O. Stein. Constrained global optimization: Adaptive gradient flows. In C.A. Floudas and P.M. Pardalos, editors, *Frontiers in Global Optimization*, pp. 223–236. Kluwer Academic, Boston, MA, 2003.

[83] R.E. Kalman. Contributions to the theory of optimal control. *Boletin de la Sociedad Matemática Mexicana*, 5:102–119, 1960.

[84] R.E. Kalman. Mathematical description of linear dynamical systems. *SIAM Journal on Control*, 1:152–192, 1963.

[85] S.S. Keerthi and E.G. Gilbert. Optimal, infinite horizon feedback laws for a general class of constrained discrete time systems: Stability and moving-horizon approximations. *Journal of Optimization Theory and Applications*, 57:265–293, 1988.

[86] C.M. Kellet, H. Shim, and A.R. Teel. Further results on robustness of (possibly discontinuous) sample and hold feedback. *IEEE Transactions on Automatic Control*, 49(7):1081–1089, 2004.

[87] C.T. Kelley. *Iterative Methods for Optimization*, vol. 18. Siam, Philadelphia, PA, 1999.

[88] H. Khalil. Adaptive output feedback control of nonlinear systems represented by input–output models. *IEEE Transactions on Automatic Control*, 41(2):177–188, 1996.

[89] H.K. Khalil. *Nonlinear Systems*. Macmillan, New York, NY, 1992.

[90] H.K. Khalil. *Nonlinear Systems*. Prentice Hall, Englewood Cliffs, NJ, 3rd edition, 2002.

[91] J.-K. Kim and M.-C. Han. Adaptive robust optimal predictive control of robot manipulators. *IECON Proceedings*, 3:2819–2824, 2004.

[92] T.H. Kim, H. Fukushima, and T. Sugie. Robust adaptive model predictive control based on comparison model. In *Proceedings of the IEEE Conference on Decision and Control*, pp. 2041–2046, 2004.

[93] K. Klatt and S. Engell. Gain-scheduling trajectory control of a continuous stirred tank reactor. *Computers and Chemical Engineering*, 22(4–5):491–502, 1998.

[94] M. Kothare, V. Balakrishnan, and M. Morari. Robust constrained model predictive control using linear matrix inequalities. *Automatica*, 32(10):1361–1379, 1996.

[95] B. Kouvaritakis, J.A. Rossiter, and J. Schuurmans. Efficient robust predictive control. *IEEE Transactions on Automatic Control*, 45(8):1545–1549, 2000.

[96] G. Kreisselmeier. Adaptive observers with exponential rate of convergence. *IEEE Transactions on Automatic Control*, 22:2–8, 1977.

[97] T. Kronseder, O. von Stryk, and R. Bulirsch. Towards nonlinear model based predictive optimal control of large-scale process models with application to separation plants. In J. Rambau, M. Groetschel, and S.O. Krumke, editors, *Optimization of Large-Scale Systems: State of the Art*. Springer, Berlin, 2001.

[98] M. Krstic, I. Kanellakopoulos, and P. Kokotovic. *Nonlinear and Adaptive Control Design*. John Wiley & Sons Inc., Toronto, Canada, 1995.

[99] M. Krstic, I. Kanellakopoulos, and P.V. Kokotovic. *Nonlinear and Adaptive Control Design*. John Wiley & Sons, New York, NY, 1995.

[100] I.D. Landau, B.D.O. Anderson, and F. De Bruyne. Recursive identification algorithms for continuous-time nonlinear plants operating in closed loop. *Automatica*, 37(3):469–475, March 2001.

[101] W. Langson, I. Chryssochoos, S. Raković, and D.Q. Mayne. Robust model predictive control using tubes. *Automatica*, 40(1):125–133, 2004.

[102] U.E. Lauks, R.J. Vasbinder, P.J. Valkenburg, and C. van Leeuwen. On-line optimization of an ethylene plant. *Computers and Chemical Engineering*, 16:S213–S220, 1992.

[103] E.B. Lee and L. Markus. *Foundations of Optimal Control Theory*. Wiley, New York, NY, 1967.

[104] J.H. Lee and Z. Yu. Worst-case formulations of model predictive control for systems with bounded parameters. *Automatica*, 33(5):763–781, 1997.

[105] J.R. Leis and M.A. Kramer. The simultaneous solution and sensitivity analysis of systems described by ordinary differential equations. *ACM Transactions on Mathematical Software*, 14(1):45–60, 1988.

[106] S. Li and L. Petzold. Software and algorithms for sensitivity analysis of large-scale differential algebraic systems. *Journal of Computational Applied Mathematics*, 125:131–145, 2000.

[107] W.C. Li and L.T. Biegler. Multistep, Newton-type control strategies for constrained nonlinear processes. *Chemical Engineering Research and Design*, 67:562–577, 1989.

[108] D. Limon, J.M. Bravo, T. Alamo, and E.F. Camacho. Robust MPC of constrained nonlinear systems based on interval arithmetic. In *IEE Proceedings on Control Theory and Applications*, vol. 152, pp. 325–332, 2005.

[109] J.-S. Lin and I. Kanellakopoulos. Nonlinearities enhance parameter convergence in output-feedback systems. *IEEE Transactions on Automatic Control*, 43(2):204–222, 1998.

[110] J.-S. Lin and I. Kanellakopoulos. Nonlinearities enhance parameter convergence in strict feedback systems. *IEEE Transactions on Automatic Control*, 44:89–94, 1999.

[111] J. Lygeros, K.H. Johansson, S.N. Simić, J. Zhang, and S.S. Sastry. Dynamical properties of hybrid automata. *IEEE Transactions on Automatic Control*, 48(1):2–17, 2003.

[112] L. Magni, G. De Nicolao, L. Magnani, and R. Scattolini. A stabilizing model-based predictive control for nonlinear systems. *Automatica*, 37(9):1351–1362, 2001.

[113] L. Magni, G. De Nicolao, R. Scattolini, and F. Allgöwer. Robust model predictive control for nonlinear discrete-time systems. *International Journal of Robust and Nonlinear Control*, 13(3–4):229–246, 2003.

[114] L. Magni, H. Nijmeijer, and A.J. van der Schaft. Receding-horizon approach to the nonlinear h_∞ control problem. *Automatica*, 37(3):429–435, 2001.

[115] L. Magni and R. Scattolini. Model predictive control of continuous-time nonlinear systems with piecewise constant control. *IEEE Transactions on Automatic Control*, 49(6):900–906, 2004.

[116] L. Magni and R. Sepulchre. Stability margins of nonlinear receding-horizon control via inverse optimality. *Systems and Control Letters*, 32:241–245, 1997.

[117] R. Marino and P. Tomei. *Nonlinear Control Design*. Prentice Hall, Hertfordshire, UK, 1995.

[118] R. Marino and P. Tomei. Adaptive observers with arbitrary exponential rate of convergence for nonlinear systems. *IEEE Transactions on Automatic Control*, 40:1300–1304, 1995.

[119] D. Marruedo, T. Alamo, and E. Camacho. Input-to-state stable MPC for constrained discrete-time nonlinear systems with bounded additive uncertainties. In *Proceedings of the IEEE Conference on Decision and Control*, pp. 4619–4624, 2002.

[120] D.L. Marruedo, T. Alamo, and E.F. Camacho. Input-to-state stable MPC for constrained discrete-time nonlinear systems with bounded additive uncertainties. In *Proceedings of the IEEE Conference on Decision and Control*, pp. 4619–4624, 2002.

[121] F. Martinsen, L. T. Biegler, and B. A. Foss. A new optimization algorithm with application to nonlinear MPC. *Journal of Process Control*, 14(8):853–865, 2004.

[122] MATLAB. *version 7.8.0 (R2008a)*. The MathWorks Inc., Natick, MA, 2008.

[123] D. Mayne. Optimization in model based control. In *Proceedings of the IFAC Symposium on Dynamics and Control, Chemical Reactors and Batch Processes*, pp. 229–242. Oxford: Elsevier Science, 1995. Plenary address.

[124] D.Q. Mayne and H. Michalska. Receding horizon control of non-linear systems. *IEEE Transactions on Automatic Control*, 35(5):814–824, 1990.

[125] D.Q. Mayne and H. Michalska. Adaptive receding horizon control for constrained nonlinear systems. In *Proceedings of the IEEE Conference on Decision and Control*, pp. 1286–1291, San Antonio, TX, 1993.

[126] D.Q. Mayne, J. B. Rawlings, C. V. Rao, and P. O. M. Scokaert. Constrained model predictive control: Stability and optimality. *Automatica*, 36:789–814, 2000.

[127] A. M'hamdi, A. Helbig, O. Abel, and W. Marquardt. Newton-type receding horizon control and state estimation. In *Proceedings of the IFAC Triennial World Congress*, 1996.

[128] H. Michalska and D.Q. Mayne. Robust receding horizon control of constrained nonlinear systems. *IEEE Transactions on Automatic Control*, 38(11):1623–1633, 1993.

[129] K.S. Narendra and A.M. Annaswamy. *Stable and Adaptive Systems*. Prentice Hall, NJ, 1989.

[130] D. Nešić and A.R. Teel. A framework for stabilization of nonlinear sampled-data systems based on their approximate discrete-time models. *IEEE Transactions on Automatic Control*, 49(7):1103–1122, 2004.

[131] M. Niethammer, P. Menold, and F. Allgöwer. Parameter and derivative estimation for nonlinear continuous-time system identification. In *5th IFAC Symposium Nonlinear Control Systems*, Russia, 2001.

[132] T. Ohtsuka. A continuation/GMRES method for fast computation of nonlinear receding horizon control. *Automatica*, 4(40):563–574, 2004.

[133] T. Ohtsuka. Quasi-Newton-type continuation method for nonlinear receding horizon control. *Journal of Guidance, Control, and Dynamics*, 25(4):685–692, 2002.

[134] M.E. Pacheco, V.M.M. Salim, and J.C. Pinto. Accelerated deactivation of hydrotreating catalysts by coke deposition. *Industrial & Engineering Chemistry Research*, 50(10):5975–5981, 2011.

[135] L.S. Pontryagin. Optimal regulation processes. *American Mathematical Society Transactions, Series 2*, 18:321–339, 1961.

[136] J. Primbs. *Nonlinear Optimal Control: A Receding Horizon Approach*. PhD thesis, California Institute of Technology, Pasadena, CA, 1999.

[137] J. Primbs, V. Nevistic, and J. Doyle. A receding horizon generalization of pointwise min-norm controllers. *IEEE Transactions on Automatic Control*, 45(5):898–909, 2000.

[138] S. Raković and D.Q. Mayne. Robust time optimal obstacle avoidance problem for constrained discrete time systems. In *Proceedings of the IEEE Conference on Decision and Control*, 2005.

[139] D. Ramirez, T. Alamo, and E. Camacho. Efficient implementation of constrained min–max model predictive control with bounded uncertainties. In *Proceedings of the IEEE Conference on Decision and Control*, pp. 3168–3173, 2002.

[140] C.V. Rao, S.J. Wright, and J.B. Rawlings. Application of interior point methods to model predictive control. *Journal of Optimization Theory and Applications*, 99(3):723–757, 1998.

[141] J. Richalet, A. Rault, J. Testud, and J. Papon. Algorithmic control of industrial processes. In *Proceedings of the IFAC Symposium on Identification and System Parameter Estimation*, pp. 1119–1167, 1976.

[142] J. Richalet, A. Rault, J. Testud, and J. Papon. Model predictive heuristic control: Applications to industrial processes. *Automatica*, 14:413–428, 1978.

[143] A.P. Sage and C.C. White. *Optimum Systems Control*. Prentice Hall, NJ, 2nd edition, 1977.

[144] R. Sanfelice, R. Goebel, and A. Teel. Results on convergence in hybrid systems via detectability and an invariance principle. In *Proceedings of the American Control Conference*, pp. 551–556, 2005.

[145] L. Santos, P. Afonso, J. Castro, N. Oliveira, and L. Biegler. Online implementation of nonlinear MPC: An experimental case study. *Control Engineering Practice*, 9:847–857, 2001.

[146] L.O. Santos. *Multivariable Predictive Control of Nonlinear Chemical Processes*. PhD thesis, Universidade do Coimbra, 2000.

[147] S. Sastry and M. Bodson. *Adaptive Control Stability, Convergence, and Robustness*. Prentice Hall, NJ, 1989.

[148] P.O.M. Scokaert, D.Q. Mayne, and J.B. Rawlings. Suboptimal model predictive control (feasibility implies stability). *IEEE Transactions on Automatic Control*, 44(3):648–654, 1999.

[149] P.O.M. Scokaert and D.Q. Mayne. Min–max feedback model predictive control for constrained linear systems. *IEEE Transactions on Automatic Control*, 43(8):1136–1142, 1998.

[150] R. Sepulchre, J. Jankovic, and P. Kokotovic. *Constructive Nonlinear Control*. Springer, NY, 1997.

[151] M. Shouche, H. Genceli, P. Vuthandam, and M. Nikolaou. Simultaneous constrained model predictive control and identification of darx processes. *Automatica*, 34(12):1521–1530, 1998.

[152] M.S. Shouche, H. Genceli, and M. Nikolaou. Effect of on-line optimization techniques on model predictive control and identification (MPCI). *Computers and Chemical Engineering*, 26(9):1241–1252, 2002.

[153] E. Sontag. A "universal" construction of Artstein's theorem on nonlinear stabilization. *Systems and Control Letters*, 13:117–123, 1989.

[154] E.D. Sontag. *Mathematical Control Theory: Deterministic Finite Dimensional Systems*. Texts in Applied Mathematics. Springer-Verlag, Berlin, 1990.

[155] E.D. Sontag. Lyapunov-like characterization of asymptotic controllability. *SIAM Journal on Control and Optimization*, 21(3):462–471, 1983.

[156] Y. Tang. Simple robust adaptive control for a class of non-linear systems: An adaptive signal synthesis approach. *International Journal of Adaptive Control and Signal Processing*, 10(4–5):481–488, 1996.

[157] S. Tuna, R. Sanfelice, M. Messina, and A. Teel. Hybrid MPC: Open-minded but not easily swayed. In *International Workshop on Assessment and Future Directions of Nonlinear Model Predictive Control*, pp. 169–180, Freudenstadt-Lauterbad, Germany, August 2005.

[158] R.J. VanderBei and D.F. Shanno. An interior-point algorithm for nonconvex nonlinear programming. *Computational Optimization and Applications*, 13:231–252, 1999.

[159] R. Vinter. *Optimal Control*. Springer Science + Business Media, 2010.

[160] H.-H. Wang, M. Krstic, and G. Bastin. Optimizing bioreactors by extremum seeking. *International Journal of Adaptive Control and Signal Processing*, 13:651–669, 1999.

[161] A.G. Wills and W.P. Heath. Barrier function based model predictive control. *Automatica*, 40(8):1415–1422, 2004.

[162] J.-X. Xu and H. Hashimoto. Parameter identification methodologies based on variable structure control. *International Journal of Control*, 57(5):1207–1220, 1993.

[163] J.-X. Xu and H. Hashimoto. VSS theory-based parameter identification scheme for MIMO systems. *Automatica*, 32(2):279–284, 1996.

[164] W. Yu. Nonlinear system identification using discrete-time recurrent neural networks with stable learning algorithms. *Information Sciences*, 158:131–147, 2004.

[165] J. Zhao and I. Kanellakopoulos. Active identification for discrete time nonlinear control—Part I: Output-feedback systems. *IEEE Transactions on Automatic Control*, 47:210–224, 2002.

[166] J. Zhao and I. Kanellakopoulos. Active identification for discrete time nonlinear control—Part II: Strict-feedback systems. *IEEE Transactions on Automatic Control*, 47:225–240, 2002.

Index

www.ingramcontent.com/pod-product-compliance
Ingram Content Group UK Ltd.
Pitfield, Milton Keynes, MK11 3LW, UK
UKHW020801290926
15470UKWH00003B/887

9 781849 195522